T0396561

PHENOMENOLOGY/ONTOPOIESIS RETRIEVING GEO-COSMIC
HORIZONS OF ANTIQUITY

ANALECTA HUSSERLIANA

THE YEARBOOK OF PHENOMENOLOGICAL RESEARCH

VOLUME CX

Founder and Editor-in-Chief:

ANNA-TERESA TYMIENIECKA

The World Institute for Advanced Phenomenological Research and Learning
Hanover, New Hampshire, USA

For further volumes:
http://www.springer.com/series/5621

PHENOMENOLOGY/ONTOPOIESIS RETRIEVING GEO-COSMIC HORIZONS OF ANTIQUITY

LOGOS AND LIFE

Volume CX / Part I

Edited by

ANNA-TERESA TYMIENIECKA

The World Institute for Advanced Phenomenological Research and Learning,
Hanover, New Hampshire, USA

Published under the auspices of
The World Institute for Advanced Phenomenological Research and Learning
A-T. Tymieniecka, President

 Springer

Editor
Prof. Anna-Teresa Tymieniecka
The World Institute for Advanced
    Phenomenological Research
    and Learning
Ivy Pointe Way 1
03755 NH Hanover
USA
wphenomenology@aol.com

Printed in 2 volumes
ISBN 978-94-007-1690-2   e-ISBN 978-94-007-1691-9
DOI 10.1007/978-94-007-1691-9
Springer Dordrecht Heidelberg London New York

Library of Congress Control Number: 2011934269

Printed on acid-free paper

Springer is part of Springer Science+Business Media (www.springer.com)

TABLE OF CONTENTS

v

SECTION III   LOGOS AND EDUCATION

SECTION IV   HUSSERL IN THE CONTEXT OF TRADITION

SECTION V   COGNITION, CREATIVITY, EMBODIMENT

ACKNOWLEDGEMENTS

We present this volume to the public with considerable pride. The title: *Phenomenology/Ontopoiesis Retrieving Geo-Cosmic Horizons of Antiquity: Logos and Life* plans to bring to light the long chain of fragmentized issues along which we have retrieved the full Greek intuition of man, earth, cosmos which has been in its entire horizon, forgotten since. The horizon of the cosmos called to be retrieved. It is in the ontopoietic foundation of the logos that "the soul is in the cosmos, and the cosmos is in the soul".

Papers collected here were read at the 60th International Congress at the University of Bergen, *Logos and Life: Phenomenology/Ontopoiesis Reviving Antiquity*, held August 10–13, 2010. We owe sincere thanks to all the authors.

The local organizers, chaired by Professor Konrad Rokstad, representing *The Research Group of Phenomenology and Existential Philosophy* at the Department of Philosophy at the University of Bergen, Ane Faugstad Aaro, Anne Granberg, Egil H. Olsvik, Johannes Servan have offered us a warm hospitality for which we thank them wholeheartedly.

Jeff Hurlburt deserves our usual appreciation for his preparation of the volume.

A-T.T.

INAUGURAL LECTURE

ANNA-TERESA TYMIENIECKA

INSPIRATIONS OF HERACLITUS FROM EPHESUS FULFILLED IN OUR NEW ENLIGHTENMENT

Prologue

Nihil sub sole novum.

ABSTRACT

Reviewing with a keen eye the history of philosophy, we would be struck by the continued filiations of contradictions between the adverse perspectives on issues that repeat themselves, albeit in novel formulations. In these latter they are amplified by fresh insights, approaches, and refinements that bring about fuller and clearer visions of the real. These new soundings of old themes do not proceed from comparisons of concepts but, rather, from genuinely new pursuits that contribute the benefits of the progress of knowledge.

The great question striking the human mind from the beginning of reflection, one which in its numerous interpretations still remains open, is the question of flux and stasis as it concerns the deepest nature of reality. As raised over some six centuries by Greek thinkers, the question has been expressed in three essential differentiations: in considerations of the media of becoming, of the first generative elements, and of composition amid everlasting transformation. From these root understandings in the classical Greek thinkers there has been transmitted a fascinating puzzle pondered throughout the entire history of Occidental philosophy down to the most recent times. It is through interpretation of the striking teaching of Heraclitus of Ephesus that it has found expression in most of history's great philosophical systems.

Heraclitus' penetrating and prophetic style, having informed and fascinated innumerable minds, penetrates even now the metaphysical imagination. Though interpreted variously in the advancing avenues of Occidental thought, today the great advances in contemporary science, our penetrating probing of reality, is answering the mental quest inspired by Heraclitus and so variously expressed.

Already at the initial phase of formulating the main lines of our New Enlightenment limning the web of discoveries, insights, dynamisms at work in forming the new spirit of humanity, I was struck by the points of contact Heraclitean inspirations, insights, and wisdom have with our new reality.

In the present study I will in turn attempt to show succinctly how my *phenomenology/ontopoiesis of life* is reformulating the questions emerging from this ancient

3

A.-T. Tymieniecka (ed.), Analecta Husserliana CX, 3–12.
DOI 10.1007/978-94-007-1691-9_1, © Springer Science+Business Media B.V. 2011

inspiration and offering an ultimate answer to perennial questions.[1] In this brief study, I will concentrate on unraveling the stream of my reflection in its innermost affinity with the main Heraclitean insights, retrieving them through the millennia of progressing thought in the New Enlightenment, as stated above.

If we may say that what is sought ever anew is to reach at its deepest level the all-underlying unity of life, man, and the cosmos, we can attain this only in our unique enlightenment by tortuous paths, step by step, advancing by jumps in one or the other direction and thus retrieving the hidden key. The three doors spoken of above will open forthwith.

PART ONE: FLUX VERSUS STASIS

HERACLITUS. THE PRIMOGENITAL PHILOSOPHICAL ISSUES OF FLUX VERSUS STASIS

We may say that the first Greek philosophers, in arriving at the basic insights into nature, reality, knowledge, arrived basically at insights revealing flux versus stasis to be the ground issue of reality. Philosophers such as Thales, Anaximander, Anaximenes, etc. pondered three main questions. First of all, there was the question of what might be the first generative elements of reality, with the stress here on flux followed by stability. The second great question concerns the composition or arrangement of elements in ceaseless transformation. Third, there is the question, given that flux is the principle of becoming, by what media is the flux brought to a stability?

Heraclitus, as we know, in flourishing at the end of the Fifth Century, was naturally introduced to these three enigmas at the origin and heart of reality by his contemporaries. But unlike his contemporaries Anaximander and Anaximenes, his was not simply the scientific attitude of the School of Miletus but also a flair for poetic artistry and a seer's wisdom.

Strongly influenced by Anaximander, who pioneered in viewing the cosmos in terms of the play of natural powers, forces, and qualities, with these being involved in the constant interactions of the "aggression" and "counter-aggression" of opposites, Heraclitus, while following this intuition, apprehended it in symbolic terms. And unlike Parmenides, who emphasized the "being" that the cosmos manifests, Heraclitus emphasized the everlasting change in which the cosmos is caught up. In contrast to his contemporaries, who attempted to grasp the order in the indisputable flux, change, and transformation by basing the cosmos in more fundamental elements such as earth, air, water, and fire, he symbolically singled out fire as the fundamental element—in contrast to Thales, who had chosen water, and Anaximenes, who had chosen air, with both of these seeing these as physical elements.[2]

As a matter of fact, with the first sentence opening the scant collection of the fragments preserved in his only book, Heraclitus comes out as a seer issuing a call to all human beings, "**Listen to the Logos!**"

(a) Now, Heraclitus' understanding of the "logos" was strikingly different from that common among his contemporaries. "Logos" was seen by him as the rational, the "true *account* of the nature of things," but this account in his understanding calls for the *discovery* of what things are, because "nature likes to hide itself." That is to say, discovery of the *logos* means the revelation of an *independent, objective state of affairs*. This account/report is a language, or a speech, and the author has to formulate it for himself, *according to his enlightenment.* What Heraclitus seeks is, in fact, an inherent *state of affairs.* What is meant in the linguistic garb is that which is *independent* of *any* account. Only when in an *enlightenment* do these two understandings come together, do we reach the complete sense of the logos.

(b) It is this view on "true nature" that strikes the stringent chord in the harmony in the disharmony of All. Flux remains an everlasting state of All, but this harmony perdures in this transformation. The most striking expression in Heraclitus' philosophical milieu, one characterizing his teaching universally is *panta rei*: all things.

"One cannot step into the same river, nor can one grasp any mortal substance in a stable condition, other and still other waters flow upon them"; "nor can one grasp any mortal substance in a stable condition. But it scatters and again gathers; it forms and dissolves, and approaches and departs."(So quoted in Plutarch.) In brief, "It rests by changing."

In these fragments is stated the crucial insight into the nature of everything. Countering the fleeting nature of everything is a universal order that captures it (whether this order be derivable from the physical forces of the "opposites"—as it was for earlier Greek philosophers of nature—or by a symbolic permanence in the changing fire). It is the logos which sustains the order of change and repose.

(c) *Heraclitus conceives of the logos,* and of the illumination that it yields for the recognition of the deepest level of things and nature, *as the underlying unity of the life of the cosmos and human life.* Deeply influenced by the Miletian philosophers involved in astronomical investigations, he apprehends the question of the nature of the logos as a question concerning man and the cosmos interchangeably.

Traversing the entire Heraclitean quest, the axis passing through the entire enterprise of this vision is Heraclitus' teaching of the correlation of the individual human soul (psyche) with the wider realm of the entire cosmos. Beginning with the disclosure "I found in myself the universal logos, the cosmic law," we see him emphasize that the search for oneself, for "self-knowledge" when extrapolated brings understanding of the universal logos.

In these insights we find, indeed, that the human soul which grows "without limits" in its logos is a microcosm interchangeable with the all-engulfing macrocosm.

The human soul, understood by Heraclitus as the center of personality and as caught in elemental transformation is essentially the measureless logos. In seeking one's own self one finds one's identity with the universe, for the logos of the soul goes so deep that it coincides with the logos that structures "everything" (cosmos). It is cosmos.

(d) This vision of Heraclitus takes in human conduct—moral, psychological, social—and ascends to final tie of the universal picture by referring to the cosmic "wisdom" who orders the continuance of the entire edifice: god.

HERACLITUS' TEACHING BEQUEATHS US THE OUTLINE OF A UNIVERSAL APPROACH FOR SEARCHING OUT THE INNERMOST DEPTHS OF REALITY IN THE LOGOS

(1) In its dominion the logos embraces: the human being, earth, the cosmos;
(2) The logos is present in the innermost bearing of all and so explicates irresistible change, constructive and transformatory becoming, and stasis in flux;
(3) Logos is the transmitter in the interchangeable communication of nature, man, and the cosmos.

Heraclitus' insights, these striking metaphysical as well as prophetic claims, had, as we know, a profound impact upon his contemporaries and successors in Greek philosophy. But even more widely, they have had influence through history through innumerable channels of reflection down to the most recent times. Their profundity and vigor have arrived at the crux of the human quest after truth and are in one or another way inherent to Western and World-Wide Philosophy.

It is my intent in this paper to show the pervasive inheritance of Heraclitean insights into the nature of things embedded in the conception of the logos as they come to light in our *mathesis universalis* of the phenomenology/ontopoiesis of life. They are exfoliated in my ontopoiesis of logos along these major lines:

(i) *The quest after the true nature of things.* This quest has to be reported/expressed in accord with an appropriate experience/illumination and only in the accord of both is reached the discovery of the logos as the innermost depth of the All.
(ii) Logos manifests itself through the transformative measures of the ever flowing flux, through the operations of all beingness as becoming, in differentiation and coalescence, in a strife of opposites that issues in an irresistible change of All, manifested in the subjacent oneness of all things that are one.
(iii) Logos sustains the underlying unity of human life and the cosmos.

PART TWO: THROUGH LIFE TO TRUTH

THE NEW ENLIGHTENMENT IN THE PRESENT PHASE OF PHILOSOPHICAL REFLECTION

The development of culture and the sciences through the centuries has not only corroborated the unfolding of human wisdom in its successive foci, but has also unfolded the understanding of the great questions we reviewed above. We could say that it is the enigma of reason, of rationalizing, of discerning cause, of measuring, extrapolating, simply cognizing that has been and remains at the center of scientific and practical inquiry. As I have discussed in my various studies, scientific inquiry has at various periods differentiated various approaches to the real, which variety raises the question of reason to an enigma and brings that question to a culminating point in inquiry into the real, *to a new critique of reason* in the new post-Kantian and post-Husserlian period, one in which the quintessential faculty of creativity leads the way, the creative faculty being indispensible for disentangling the knots by which human cognition, scientific experimentation, rationalizing, etc. have tied reality.[3] Our new critique of reason enters the vortex and the context of the Human Condition opening the vast dimensions and realms into which contemporary science has expanded its reach and so allows for grasping together their otherwise dispersed results.

In fact, the expansion of scientific rationalities and in particular their corroboration has imminently extended into a sphere of wonder and troublesomely dispersed queries carried on throughout the centuries, a sphere that in our age has been recognized as an existential counterpart of human reality, namely, the skies, that is, the heavens. The advances in the different branches of astronomical research have made them a central focus of our attention.

We enter into a new phase of the understanding of the world of life, of our earth, and of the cosmic completing counterpart of life. This apprehension is the dawn of *a New Enlightenment of Reason*—an Enlightenment allowing reason to emerge as an *all-illuminating logos*.

LIFE. THE PHILOSOPHICAL QUEST AND ITS RADICAL BEGINNING

Descartes' original starting point for philosophy was the apprehension "cogito ergo sum." Three centuries after him, Ortega y Gasset reversed that and declared, "I live, therefore I think." My new starting point declares: "I live, therefore I am." From the beginning of human times men caught in crisis, thrust to the very edge by various predicaments, seeing that life itself is at stake, have pondered about life. But, in fact, human life is at stake in each and every instant, in each and every concern of our existence. "To be" is to be living, to be alive. No speculative metaphysical category like existence or being can substitute for the unique experience of life's inward/onward orientation, the streaming rays of life's emotions from its the incipient "existence" to the last exhausted release of breath. Our physiological, psychic, intellective propulsions, function to subtend the stream of the most complex self of the living agent that carries and individualizes living beingness, organizing and directing it, while being involved in its functioning in the processes of circumambient beingnesses that struggle to initiate or maintain life. Growing into a creative potency as a human mind, the life experience embarks upon a fantastic circuit of imaginative undertakings in constituting the circumambient world in what becomes its intimately "own" embodiment in life, oneself.

Our feeling alive—feeling we are ourselves—elicits "everything there is alive," our life-experience and its concomitant spheres. I live, therefore, I am!

The living being at all levels of development is ceaselessly concentrated and constantly attentive. This is the crucial all-penetrating experience of being oneself, of living one's identity. One is constantly aware of and fixated upon one's vital, psychic, spiritual needs and on the evaluation of their validity, concreteness, nature, reasons. One scrutinizes the situation to find out whether one's subjective experience corresponds to the facts as observed by others—that is, to their objective "truth": *the truth of facts*. The truth of facts, which may appear differently from varied valid perspectives, points of view, and which might play a decisive role in pragmatic and utilitarian matters of life.[4]

Our first and foremost, urgent and immediate commitment to advance our living at every instant—to act—is bound precisely to the eventful circumstances of the truth of facts. Only rarely, and in special attunements of our mind, do we query beyond the truth of facts, after the *truth of things*—the truth of life.

The philosophic, metaphysically inclined mind will ask just how the matter in question is, how apart from the circumstantial evidences, it is "in itself." What is the truth of things, of life? Leaving to the side personal, pragmatic, objectively valid, circumambient demands, prospects, and expectations, we seek the mere facts that account for the naked truth of things, what is life per se? Why do we seek the "true" validity of things in life in depths to be uncovered, if not to find the ultimate reason of everything, the *logos* of each and all.

This "truth" we seek in life's proceedings, in its networks and avenues as such is, in fact, unwittingly one our perception is making, or creating. In its interconnections it is molding their sense, their intrinsic reason, their logos: the logos of life. The truth of life is the logos of life.

The logos of life carries the continuity of life's incessant flow, with transformation in becoming onwards pouring all its forces into the sense of becoming, its underlying truth.

ILLUMINATING TRUTH ALONG THE WAY: THE INTRODUCTION OF THE CREATIVE SPIRIT IN THE NEW CRITIQUE OF REASON

To prepare our way, we propose a fundamental critique of reason. Through cognition of the vital order of our existence we acquire/constitute the common knowledge of the world and of things—but their true nature is hidden—and only through further recognition can we gain truth, only through the fullness of human experience in creative insights that illuminate reflection.

The access to the ultimate truth of things cannot proceed through the singular channel of sentient and intentional consciousness with its intellective operations and its corresponding vital, empirical, and constructively reduced lines. It is the specific creative condition of humans that opens a wider horizon allowing communication with realities beyond those that our narrowly structured intentional schemata reach. The human creative act emerging *sua sponte* within the setup of the human mind enters into the fulgurating flux of individualizing becoming. We pursue backwards the trajectory of the genesis of that flux's objective aim step by step down to its incipient level in the mind. There, at the origins of our experience as such, where the genesis of being and becoming is initiated, we find the platform from which the ontological-metaphysical-poetic level of true beingness emerges. This level emerges in the *self-individualizing* of beingness. *Self-individualization* moves along with both force and order, step by step, as prompted by the generative *logos of life,* emerging thereby in the flux of its formative becoming and appearing throughout the broadening stream that ties together the cognitive links through which the mind proceeds in structuring its circumambient milieu. Immersing ourselves in this stream, we are open to all the horizons of the human creative condition, and it is by plunging into its intricacies that we see emerge revealing rays of light: the Logos of Life which surges from the gyres of the Human Condition through the ontopoietic process of life.

THE INDIVIDUALIZING-ONTOPOIETIC PROCESS OF LIFE: SENSE AND ORDERING

It is the ontopoiesis of life that is *life's individualization, accomplished through the intrinsic ordering of all that is and by the processing of sense* that carries on the relative stabilizing of spheres into becoming from the anonymous flux. It is the spine of progress' individualization, establishing a simultaneously perduring as well as fluctuating condition by which the ontopoietic advance proceeds *measuring out* life's flux into a constructive becoming.

The underlying stream of the ontopoietic logos carrying the constructive becoming in an individualizing schema by its going beyond the anonymous flux on the one side answers the perennial issue of flux versus stasis that provokes all the previously mentioned questions about beingness and reveals on the other side the spine of the primordial truth of everything that is otherwise hidden to the ordinary sight of humans. The continuity of the first order of things is given in this discovery of the ontopoietic origination of reality and the processes of its genesis. This unveiling of the ontopoietic logos, which we have discussed above while speaking of the phenomenology of the critique of reason, brings together as well as sunders all the generic elements that flow together or coincide in a formative progress in which the logos acts simultaneously as the *prompting force*, energy that continues the coherent progress as well as that progress' formative differentiation. That is to say, the coalescing cooperation of the elements arrived at by a step by step selection and directed formatively by the ontopoietic sequence is revealed. The ontopoietic sequence acts as the principle of the logos of life prompting, carrying, and directing the self-individualization of the living beingness.

It is, indeed, by constructively ordering *becoming* in its primal force that the logos of life conducts the otherwise anonymous flux into significant fragments of sense, that is, into *an ontopoietic sequence* simultaneously differentiating and coalescing previously anonymous elements toward a progressively shaping telos of individualized becoming; flux and stasis are grasped as aspects of an everlasting stream of transformation that draws all available resources into becoming.

Panta rei (nothing lasts—all flies), *and yet all passes away in measured transformations in which the constructive/creative flux is reconciled with stability as the ontopoietic logos expands, it being the crux of life, a pipeline which binds flux and stasis throughout reality—life, earth, and the cosmic spheres.*

THE HUMAN-CONDITION-IN-THE-UNITY-OF-EVERYTHING-THERE-IS-ALIVE AND ITS COROLLARY

This flux operates within logos' field of forces and not *eo ipso* but in close cogeneration amid the circumambient situations that participate in the human condition-in-the-unity-of-everything-there-is-alive and its corollary, the cosmos. Thereby, an infinitely extensive network is projected by the ever self-transforming logos of life in its ever changeable senses. Thus, the constructive, converting transformative becoming of the life process energized and instrumentalized by the logos brings in the ordering and the sense of individualizing becoming. Flux and stasis appear then as abstract notions that hide the coalescing and dissipating game of the generative/regressing process of life.

(a) *The logoic impulse of the forces converting life in its singular ordering and individualizing sequence* draws into its constructive/generative network of logoic energies a circumambient array of elements that are then transformed into coalescing proficiencies that as they are drawn into the constructive network extend into the circumambient generative potentialities of the individualizing beingness. Thereby, the individualizing life is immersed in its generative course in the entire network of the unifying logoic links progressively unveiled, within which the living individual unfolds and passes away.

It is the logoic spread of the Human Condition that comprises all of reality— the existential reality in all its perspectives, but also the realm of the vital conundrum of generative cycles, and the psychic, social, and communal cycles of coexistence among all living beings, and which lift the logos to it intellective and spiritual heights.

(b) *The Fullness of the Human Soul Fashioned by the Human Condition*

Having discerned the life factor's encirclement of the human agent / living agent, constituting the existential circumference of the human condition, we turn to focus now on that inner, specifically human expansion, in which the logoic rays penetrate into the vital conditions of individualizing life in one direction and into the spiritual outgrowth of the human soul in the other. These rays center the expanding powers of the mind as the soul's crucial constitutive faculty. It is, indeed, in this imaginative, creative, and governing faculty that the living individual centralizes its entire existential spread in this condition.

The human soul, having the mind, consciousness as its instrument, projects and negotiates with the life horizons. And so the living agent, with the influx into individualizing life of the specifically human inventive/creative/imaginative system through which the human mind operates, projects not only necessary links for the vitally expanded functioning of ontopoietic becoming, but also and foremost unfolds in tandem innumerable morally, aesthetically, emotionally, imaginatively evaluative threads as well as other lines of sense through which the living agent progressively acquires a human mind. The human mind continues the work of the vegetatively-vitally subservient agent now unfolded in a self that imaginatively projects, in a determined, self-selective, and self-decisive human individual. The intentional system of consciousness directing this entire apparatus of life and compassing its full extent—from primitive sensing, feeling, desiring, evaluating to constituting the world, to esoteric longings to escape all that existentially binds and to transcend it—that is, the living agent who incorporates the prerogatives of the human mind amounts to what we call the "soul," in whose fulgurating symphony of life's becoming, the entire course of life resounds into infinite realms.

(c) *The Soul in the Cosmos and the Cosmos in the Soul*

Through its innumerably ramified rays the human soul reflects her entire horizon: her originary ties are reflected in the "passions of the earth," the earth that is her ground and from which her subtending forces come, but reflected as well, it is presumed from time immemorial, in the "passions of the skies," are her psychic and spiritual forces and propulsions.[5] We can say that in this way the self-prompted and self-oriented human soul reflects the universal ordering of the All: from her originary ties to the earth's soil, to the congenital influences exerted on life on earth by the firmament.[6]

Proceeding from our generic roots in the earth's soil, the logos of life upon which the entirety of ordering and vital sense is suspended traverses intentional-creative-ontopoietic becoming *within* and expands into an unconfined horizon of becoming *without,* and so the ontopoietic becoming of life finds a completion. The generic system of the ontopoietic design of life finds correlative logoic support from above—in the skies. Earth, our womb of life, is itself linked in its existential conditions to the dynamic architectonics of an orderly cosmos.

The great developments of the New Enlightenment have led to an essential transformation of the positioning of life and human constitution.

Transcendental intentional consciousness has lost its dominant role, which has now been accorded the dynamic architectonic of the cosmos.

This Copernician turn of the present day philosophical orientation due to our ontopoiesis of life marks the cosmic integration of a too long neglected path of metaphysics which we now pursue (Note 6).

The World Institute for Advanced Phenomenological Research Learning, Hanover, NH, USA
e-mail: wphenomenology@aol.com

NOTES

[1] In my quintessential work, *Logos and Life, Impetus and Equipoise in the Life-Strategies of Reason,* Analecta Husserliana LXX (Dordrecht: Kluwer, 2000), pp. 291–322, I have presented some traits of my phenomenology of life as a new ontopoietic answer to the perennial questions by its unfolding of the logos in a manner having affinity with the Heraclitean intuitions.

[2] Quotations and direct references to Heraclitus' utterances are taken from *The Art and Thought of Heraclitus, an Edition of the Fragments with Translation and Commentary,* Charles H. Kahn (Cambridge: Cambridge University Press, 1979).

[3] See by the present author, *Logos and Life,* Book 2: *Creative Experience and the Critique of Reason,* Analecta Husserliana XXIV (Dordrecht: Kluwer, 1987).

[4] Analecta Husserliana, 2006, pp. 109–123.

[5] See A-T.T. Editor, Analecta Husserliana, Volume 107, *The Passions of the Skies: Astronomy and Civilization in the New Enlightenment*; and Analecta Husserliana, Volume 108, *Transcendentalism Overturned: From Absolute Power of Consciousness Until the Forces of Cosmic Architectonics* (Dordrecht: Springer).

[6] Ibid., pp. 7–8.

SECTION I
PHENOMENOLOGY OF LIFE IN THE CRITIQUE OF REASON

KONRAD ROKSTAD

WAS PLATO A PLATONIST?

ABSTRACT

My paper will pose the seemingly rather trivial question – was Plato a Platonist? and it will demonstrate that it is possibly not as trivial as it first appears to be. Thus, the definition of "Platonism" becomes an issue – particularly as regards the understanding of the concept of "idea" and its constitution: is it to be understood in a realistic manner giving the idea (or quite generally, the object) total independence of the subject conceiving it? Or is some other conception possible? And then we have Plato's metaphor of re-membering getting into the core of what philosophizing means. As memory and all the functions of consciousness related to it thus come into the field of interrogation, this might give rise to a phenomenological interpretation of what happens in the constitution of the idea-object. This might, of course, seem problematic in regard to the traditional conception of Plato, but the traditional conception might also be questioned, and my paper will do exactly that. In this way the phenomenological field of problems and possibilities – the constitutional together with history and the understanding of history – will come into focus so that the problem of Platonism and Plato becomes "originally" reflected and interrogated upon this grounding.

The question "Was Plato a Platonist?" might perhaps cause surprise; it might seem puzzling and even provocative to pose this apparently trivial question in front of a highly scholarly audience such as you are – probably all of you having spent much time in intellectual company with Plato. But perhaps we should not consider the surprise and puzzling only in a negative manner not even as regards the so called trivial and obvious, particularly not in a context in which the philosophy of Plato is an issue. Some would even view the genuine beginning of philosophizing as generated by amazement, wondering and surprise.[1] The answer to the question – Was Plato a Platonist or a Platonic, is perhaps not that obvious after all.

The fact that we here have two different words or concepts at our disposal might indicate this, but I will not be considering this in our context here. I will only consider the concept of "Platonist" and even then find different meanings and interpretations that might be provided thus grounding different answers to the question. If for example the concept of "Platonism" is understood as "the major characteristics of the philosophy that the ancient Greek philosopher Plato developed 3–400 years before Christ, which afterwards became a major movement of thinking that deeply influenced our scientific-philosophical tradition", then fairly trivially – if this is the definition grounding the answer – Plato was a Platonist. But if we choose another definition, for example this: "Platonism is the manner of philosophical thinking developed after the death of the historical Plato (but) inspired by and based upon his

15

A.-T. Tymieniecka (ed.), Analecta Husserliana CX, 15–22.
DOI 10.1007/978-94-007-1691-9_2, © Springer Science+Business Media B.V. 2011

achievements", then the answer would not be that trivial – or rather, it would be trivial the other way around; if Platonism regarded as an "-ism" came into history after the death of Plato, he could scarcely have been a Platonist – at least not if you are not plainly unhistorical. And it seems we can say this even without entering the core substance of Platonism and what it means to be a Platonist. Maybe therefore it is not only the answer that has more depth to it than it first appears – maybe also the question has!

Let us look a bit closer at what is commonly regarded the core or the philosophical substance of Platonism, particularly as it is associated with and also is regarded as some prototype for philosophical realism. One definition might then be the following one: "attempts to supply concepts, mathematical and other abstract entities with an autonomous status and existence independent of our knowledge of and our interaction with them". This definition is then delimited to what is ontological-epistemic, and it tells us nothing about what might be the ethical-moral, aesthetical or political aspects in the philosophy of Plato. Nor does it say anything about the dialogical-communicative or pedagogical-educational aspects it might entail. And this might very well represent a delimitation that causes problems.

Of course, other interesting definitions might be provided, but let us now stick to this one exposing some philosophical realism – after all it hits one core of what is most often labeled Platonism. And it is not only by incident that it is so – Plato himself seems to be nourishing such a conception as he in many of his Dialogues uses metaphors and speaks of the forms or ideas which timelessly or all-timely exist in a world for themselves. That means that they are not here or there in any concrete space-sensing meaning, but they are "really" present in some pre-existence and then given to the soul's or the inner eye's "pure" intuition of them – so that when the soul has become incarnated in a sensing and sensible body within the existence of time and space, it is able to conceive of these forms only by remembering or by "recalling" them. Thus remembering becomes one major epistemic function that dwells in our mind's depth. Everything depends on the ability for memory so to speak, and Plato even seems to extend the ability for memory back into the time before we – each one of us – are born! This might, of course, appear pretty fantastic – but we have to remember that Plato expressed himself in a mythic-metaphoric, or even poetic, language and that an explanation given in such a language is not necessarily inferior to one given in a more objectivistic language. It is also by this manner of reasoning that we are able to recognize (and really understand) the possible power the forms possess for constituting actuality. As cognition and dialectical thinking thus becomes remembering the principles and the forms for the constitution of actuality, it will at the same time have genuine educational, ethical and even political implications. And again by this we also get a grip on what is the broader field for Plato's philosophical interests and actions; it entails fairly detailed thoughts about the good life, justice, the organizing and governing of society etc. But let us once more remind ourselves that to Plato the stories about pre-existence and re-membering are metaphors and pictures that scarcely can be taken literally. Due to the quite ingenious manner Plato uses them these metaphors have, however, obtained respect and admiration within a great and leading European tradition, which, motivated by this theme, has continued doing critique for more than 2000 years, dedicatedly following Plato and further developing the field.

Before getting more specifically into this, let us first return for a while to the definition of Platonism we already have been discussing – and particularly look at its internal consistency. It entails something which might give rise to both wondering and critique; as it is an attempt to supply concepts and abstract entities with an autonomous status and existence which are to be independent of our knowledge and interaction with them, then at least I find that I face a problem here – because I cannot really understand how to conceive of the expression "attempts to supply …" as it – the attempt – is done by humans themselves thinking and attempting by such attempts to seek knowledge – not either in this regard always with success. If the concept-mark "attempt to supply" is an essential part of the concept "Platonism" so that the tentative, or even the "experimental" – thus the process-character of it all that constitutes the autonomous status becomes explicit, then I will have problems in regard to *understanding the independence* of such general, abstract entities and forms from our knowledge of and interaction with them. It seems to me an inconsistency, at least in tendency, embedded in this position which would have me think (at least) twice before I became a "Platonist".

But perhaps the "genuine" Platonist would say – talking of "attempt" is only a manner of speech that represents a view from the outside; and that may in some respect prove correct. If, however, you view it as the genuine Platonist does – from inside, then it is something more like some revelation, pure intuition – in which what is intuited (by itself) marks itself onto the intuiting consciousness and thus becomes independent from my own act of cognition and other things I might draw from it and make use of. Consciousness is in this regard a pure and passive subordinating receptor and as regards the constitution of actuality (including the forms), it is nothing further. It receives the mark from the reality of forms, and whatever else might interfere such as preconditions provided by the actual, historical situation – also embodying the depth of a tradition – do not matter at all as regards truth and universal validity.

Thus might, of course, the Platonist attempt to defend his position, taking action to convince others to accept the forms with their presumed or stated independence from our knowledge of them and otherwise how we act and relate towards them. But exactly as he is attempting it in this manner, he also demonstrates – at least in the eyes of us who are able to view what happens in the space between humans – the dependence upon an activity of arguing and understanding; and this he does in his attempt to prove the opposite, namely the presumed essential independence and autonomy of the forms. What the forms could appear to be without such activity (and Platonism happens to be correct with the only passively receptive re-membering function) is known only to the souls not yet born or to the dead, or finally, perhaps, "only the gods know".

What we attempt to demonstrate by this exposition is that either (a) we are confronting a contradiction – at least in tendency, or (b) we are confronting a (presumed) universal objectivity which only the gods are able to know and we as humans living and acting, sensing, being sensed, thinking, communicating, seeking and always again attempting . . . ., cannot reach and learn to know that objectivity concretely in our capacity as humans. The question I now would like to pose by putting more stress upon our question in the beginning, is plainly to ask if Plato ever could have held such a position. Because of his enormously broad orientation, his intelligence and logical genius he could not have overlooked such a contradiction in tendency and could not have located the grounding of knowledge so far beyond the living human field for activity. It is more likely he knew what he was doing – even though he did not know it all and he knew this: Plato never pretended to be God – not even in the *Timaeus*, cf. the line: This is "only a likely story".[2] And this is likely because Plato knew very well what knowledge can mean in the space of human existence – in which each and every one of us is struggling for the right and for the good and is seeking knowledge that enables us to reach it as far as our life permits. I believe this is more what his major concern was all about. Therefore I think of Platonism such as it is here exposed in its realistic form or in its more naïve idealistic form, as a construction which might appear natural if you only take half of Plato and otherwise built on preconditions that are not really his.[3]

But would not this in an extreme manner misjudge a large, in a way dominating and utterly well documented tradition; in this regard I am now perhaps, as we say in Norwegian, about to "banne i kjerka" which means that I am "swearing in the Church"? And how could I provide documentation for my thesis here? Of course, I can not – and will not even try to provide substantial documentation in this short paper, but I will only suggest a possible strategy for an argumentation and interrogation of the issue. Roughly I will (a) point at what might be viewed as one core of the philosophy of Plato, namely the importance of the dialogical, interrogating and seeking character – the knowing of the not-knowing – as it at the same time provide commitment in regard to the human life and its social context, and (b) point at the tradition of mediating which encompasses problems and elements that might make opportune the half-cutting that makes Plato into a Platonist; this is also part of the picture that requires a critique from the ground – whatever it may be.[4]

Let's start in this context, then, by reminding ourselves about the fact that at least Plato himself in using metaphors speaks of re-membering (or re-calling) as a conscious process in which time and some time-consciousness are preconditions. And he also uses Socrates in a manner that embodies persons situated in dialogs seriously carried out by logical communication, but then entailing concrete ethical problems and a whole world which commits the participants to their own concrete lives. Or should we rather say are (teleologically) meant to commit – because there always seems to be an open end to it all? All this which has now only been indicated makes me doubt that Plato would have committed himself to the Platonism we have been speaking about. He was not that naïve, dreaming or logically blind. On the other hand, I would say we may find crystallizing in his philosophy two opposite directed tendencies. The one of them which in a way sustains the Platonism and partly makes it correct, is the tendency toward the pure intuiting, receptive "theoria" in which the subject as precondition in a way forgets itself as founding functioning field for cognition and knowledge. The other is the one in which the subject not only learns to know itself but also is able to view itself in a community with others – particularly as regards ethical and practical affairs pertaining to human life. And finally, it is this functioning and living "totality" which might provide and sustain the genuine community and universal truth in their interweaving.

Given this, the question which more particularly comes to the fore is if these two tendencies are able to intertwine in an interpretation and understanding of Plato that does not make him a Platonist in the sense we have been criticizing. And starting from and leaning on the second tendency I would say this is possible. For even if we lower the significance of "the one" and of pure intuition and, conversely, increases the significance of plurality (and even relativity), it is possible to think – and maybe also to live – the common, the universal objective and true with some form of independence. I am of course not referring to my private inclination for meaning, my private doxa – even as it may be an expression for some community (some ideology) which, often based on narrow self-centered interests, excludes others and seeks domination on that basis. When confronting such particulars you may sometimes have to make the independence "absolutely" independent in a mythical or dogmatic manner – and maybe this was Plato's point and situation (in the ancient Greece where he lived). But as regards another kind of subject, the subject I believe both Socrates and Plato were seeking, then the independence has to obtain an other character – which is quite the opposite from both the mythical and dogmatism. What it is all about is a subject who is reflectively able to know and to commit itself within an open community which is headed for the universal life exactly as this may both motivate and define the subject in its particularity. And is it not this subject Plato with his Socrates is leading us into community with – those of us who seriously wish to seek in order to realize community "in itself"[5]?

But now we have to stop for a little while – has not this become an all too modern way of expression? To speak of "subject", "universal community in tendency", "communication" etc., are not these modern concepts and constructions which would be strangers in the ancient context? Maybe that is the case. But would we at all be able to understand Plato if we did not understand him within our own horizon for understanding? And is not this actually also the only manner of finding and understanding what he "really" meant? I will say so – and ground that upon the premises laid by Plato such as they are provided by that tradition in which he meaningfully may be present in our situation – as a genuine partner for communication. I therefore stick to the Socratic Plato, the one so eloquently practicing dialogue between live persons who both wonder and are motivated by the problems provided by the situation they are living – asking questions and seriously seeking answers which morally embody commitment in regard to action and personal life. In other words, I will focus on this aspect in order both to understand the "ascendance" up to that universal objectivity so that they appear with their original appearance as "patterns" or ideals which are *morally motivating* – and thereby the kind of "independence" it thus takes on is a historical teleological independence.

But how is it possible to demonstrate this more concretely? Let's provide some suggestions, firstly by looking at Socrates who had been an active practicing philosopher and was a person Plato got acquainted with before he had formulated his own philosophy. What Plato does in his dialogues is to display what Socrates in the eyes of Plato had seen and represented – so to speak lift it up in a new and (more) permanent form. Socrates had first and foremost performed his work by his speech and, as is well known, he terminated his life by drinking the cup of poison. By transforming what Socrates had represented – such as Plato had experienced it and then remembered – maybe even "recollects" it – into a more solid form (the written Dialogues), Socrates is still alive, perhaps even more so than in his actual life. The story and history of Socrates does thus not end with the cup of poison, it begins there – by showing the spiritual power provided by the consequent faithful-ness until death with regard to the always ongoing search to know justice and truth; this yields the permanent form (of this history) provided by the written dialogues. It radiates a carrying power and the distinction between appearance and essence ap-pears with historical ideal leading power which people in the generations afterwards may relate to and participate in for themselves.

Over and above the admiration we may have for Plato's logical and limitless intel-lect – it is probably here his genius is located: no matter how it may in fact have been with regard to the historical Socrates – was he a seducer, a silly martyr or was he the genuine seeker of truth and justice – in either case Plato let us realize that life has one primary foundation for value and meaning within the serious commitment in the search for correlating knowledge and action as an always living, struggling unity of theory and praxis that never becomes finally completed. The death of Socrates was of course not a universal ideal for others, but others can, grounded upon the example of his, realize the need for a historical, thus permanent foundation, which, appears in the form of community. This community is genuinely and universally valid by grounding my own rightful freedom to think and act so that it does not limit and obstruct the freedom of the others in doing the same.[6] Thus the road opens for un-derstanding how ideality becomes constituted within the dialogical interrogations presented in Plato's many dialogues. The destiny of Socrates was of course more than the story of the termination of one person. It was an example which also mir-rors an entire society's situation of crisis entailing depression, inner disintegration, fighting against both inner and external enemies, plague and all together a situation in which brutal power and egoism defines what is right. Plato relates to and engages with all this as he is also engaged in the tradition that was present at his time, both what was philosophy and what was the historical situation more generally. In my perspective of historicity which I have presented in this paper, all of these elements will be of substantial interest. And this again will certainly constitute a veritable break with the understanding provided by the realistic – or the mythic-religious Platonism. If Plato is still to be living among us, we have to leave those behind.[7]

What we have been doing is really not more than to indicate and to some extent to explicate a horizon for understanding and a perspective for interpretation that could be helpful to an interrogation into the sense in which the Platonic forms and ideals might obtain objective character and status. And we have thus taken the historical aspect into this in a manner much deeper than what is commonly done: ideality does not transcend this historical grounding, but obtains its universal objectivity within it.[8] Then also such elementary phenomena as for example the writing which, of course, might be regarded as something without philosophical significance and "only" obvious, becomes essentially philosophical particularly the way Plato fills it with his powerful content which extends beyond and between generations. With a modern expression we can now say it is the historicity of ideality that has become our field for understanding. My point is that this provides a better and more adequate grounding than to understand the ideal objectivity in some sort of mythical or religious analogy to physical objects provided by nature. And this is the case not only because it is in better accordance with modern thinking; my bold thesis is that it will also be better in accordance with what we find in the dialogues of Plato. But, of course, then we have to relate to a very long and powerful tradition of mediation and interpretation in which strong elements of both mythical-religious and objectivistic transcendence-thinking have been at work.

This also has consequences in regard to the prevalent understanding of tradition and history. In regard to what I have been arguing, someone certainly will object that this cannot pass simply because it does not fit in with the facts of history and the ancient spirit as it "really" was. It is far more likely that Plato "really" thought in an ahistorical and quite naïve objectivistic manner which clearly proves that it is the realistic or the mythic-religious interpretation which is valid. And this is of course something that might be discussed on the preconditions which we actually want to base our interrogative argumentation upon. But if one wants to argue this way, one had better also realize that one has entered the *historical field of thinking* – exactly thus making what might be *historical facts the grounding field for the definition and judgment of the Platonic ideality*. I would not protest against this – it is what my paper actually has been dealing with! But I will in that case hope we do it in a manner which understands both what philosophical and historical facts are – as they at the same time are reflected upon this grounding and seeks to level with that which constitutes the core of Plato's philosophy.

University of Bergen, Bergen, Norway
e-mail: Konrad.Rokstad@fof.uib.no

NOTES

[1] This is, of course, true for both Plato and Aristotle; in the *Theaetetus* Soctrates says the following: "This sense of wonder is the mark of the philosopher. Philosophy indeed has no other origin, and he was a good genealogist who made Iris the daughter of Thaumas" (155D). But it is also true for modern phenomenological philosophers such as E. Fink who questions the obvious and makes the wondering something starting philosophizing but also a measure for the quality of it as he says: "The degree of the wondering's creative power does finally decide about the rank and the result of a philosophy [. . . .] The draft of the problem, the essential fundamental action of a philosophy, is not, however, the posing of the question – rather it is the actual living out the wondering question. The 'radical character' of a philosophy is entailed in the radicalization of its Problem." My translation from E. Fink, *Studien zur Phänomenologie 1930–1939*, Martinus Niehoff, Den Haag, 1966, p. 184.

[2] Speaking of the creation of the physical world Timaeus says (to Socrates): "Don't therefore be surprised, Socrates, if on many matters concerning the gods and the whole world of change we are unable in every respect and on every occasion to render a consistent and accurate account. You must be satisfied if our account is as likely as any, remembering that both I and you who are sitting in the judgment on

it are merely human, and should not look for anything more than a likely story in such matters." And
Socrates agrees on this. *Timaeus* p. 41/29, now quoted from the Penguin Books, translated by Desmond
Lee, 1976.

[3] I am here thinking about the influence from the Neo-platonic and Christianity and upon that the
dominating objectivistic manner of thinking having invaded the spiritual climate in Europe as the modern
research on Plato's philosophy developed during the 1800's.

[4] I am perhaps expressing myself a bit cryptic here, but the "ground" of which I am speaking will
finally be that historicity (of our existence) which the later Husserl speaks of. The concept of historicity
per se will, however, not be explicitly developed in this context – it will only be developed implicitly or
indirectly by the discussion of major lines in the philosophy of Plato, thus making it an "example" for
phenomenological analysis.

[5] This is in a way literary meant and even though it is ambiguous: "in itself" means, on the one hand,
in the individual, integrated in the person – something with which the person may identify him/her-self;
on the other, it is the philosophical "in itself" – the essential, the "real thing", the plurality of individuals
etc. As regards community in itself, the point then will be that these two aspects have to be present and
functioning together so that there is harmony both within and between the persons.

[6] This is what also Kant says about Plato as he in his *Critique of Pure Reason* discusses the philosophy
of Plato, especially the *Republic* but also meaning this is the core of his whole philosophy of which
Kant seems to be in full agreement. Kant says: "A constitution of the greatest possible human freedom
according to laws, by which the liberty of every individual can consist with the liberty of every other
(not the greatest possible happiness for this follows necessarily from the former), is, to say the least, a
necessary idea, which must be placed at the foundation not only of the first plan of the constitution of a
state, but of all its laws." *Critique of Pure Reason,* p. 220, translated by J. M. D. Meiklejohn, J.M. Dent
& Sons LTD, London, 1974.

[7] In her book *Postmoderen Platos* Cathrine H. Zuckert develops the perspectives of five (post)modern
philosophers onto Plato's philosophy. Those are Nietzche, Heidegger, Gadamer, Strauss and Derrida
each one holding different and more or less radical views on this philosophy. Neither Husserl nor Fink
is among these and my modest contribution in this context is to provide a "supplement" which in at
least some major lines exposes a way of looking which is inspired by these two. Analogically, as the
philosophy of Plato – viewed in a phenomenological perspective – will be about the constitution of
ideality, it is perhaps "The Origin of Geometry" which comes closest and now have been used as a
"model" because in it Husserl (in collaboration with Fink) develops the constitution of the ideality of
geometry actually, then, exposing a strategy for the constitution of ideality quite generally.

[8] This actually is the essence of "The Origin of Geometry". In it Husserl says – as he is speaking of a
"ruling dogma", the following: "The ruling dogma of the separation in principle between epistemological elucidation and historical, even humanistic-psychological explanation, between epistemological and
genetic origin, is fundamentally mistaken, unless one inadmissibly limits, in the usual way, the concept
of 'history,' 'historical explanation,' and 'genesis.' Or rather, what is fundamentally mistaken is the limitation through which precisely the deepest and the most genuine problems of history are concealed." *The
Crisis of European Sciences and the Transcendental Phenomenology*, Northwestern University Press,
Evanston, IL, 1970, p. 370.

D A N I E L A   V E R D U C C I

THE LIFE OF BEING REFOUND WITH THE PHENOMENOLOGY OF LIFE OF ANNA-TERESA TYMIENIECKA

A B S T R A C T

In the heart of the more objectivistic line of Rationalism, Leibniz planted an ontological seed of vital spontaneity that would bear fruit three centuries later in the reflective conversion of Husserlian phenomenology into the one subjective/objective field of research of the *Erlebnisse* and of the *Sachen selbst*. Anna-Teresa Tymieniecka carries out and gives structure to these ideas of philosophical solidarity between spirit and life, both pursuing the subjective road in empathizing with the profound intentionality of her masters, Leibniz and Ingarden *in primis*, and applying herself to the objective level to give rise to a phenomenology of phenomenology, through which she intends to realize an intuitive re-seeding of phenomenology itself. The surprising result of this phenomenological work has been the discover of the ontopoietic logos of life, which runs through and pervades every sphere of being, from the physical to the metaphysical level, with its expansive and evolutive dynamic of *impetus* and *equipoise*. Thus Tymieniecka threw open the ancient Parmenidean concept of being as a "mass of well rounded sphere" to the spectacle of being that gushes and runs in history, as if surging from an inexhaustible spring.

THE LIFE OF BEING MARGINALIZED

As Hans Jonas teaches us, Modernity was inaugurated with the intention of unchaining itself from the limitations that the recognition of the teleological order of life imposes on the analytical dominion of scientific and mechanistic reason.[1] According to Jonas, this happened in connection with the XVII century rise of astronomical physics, the science "of inanimate masses and forces," to the dominant and leading epistemological position, and because of the concomitant affirmation of a mentality that held that in order to guarantee the best and most correct scientific observation, it was necessary to bring the uncontrollable dynamism of the living being to the state of masterable immutability of the dead. In the passage to Modernity, therefore, "from the physical sciences there spread over the conception of all existence an ontology whose model entity is pure matter, stripped of all features of life." Thus it was that "also in terms of ontological genuineness, non-life [was] the rule, life the puzzling exceptions in physical existence."[2]

The same inquiry, undertaken by Descartes, for a method "of rightly conducting the reason" moves from the dissatisfaction about the cognitive results of his living

23

A.-T. Tymieniecka (ed.), Analecta Husserliana CX, 23–37.
DOI 10.1007/978-94-007-1691-9_3, © Springer Science+Business Media B.V. 2011

experience during the years of formation. His youthful was spent in deference to tradition, first of all studying under the guidance of the tutors of the La Fléche college[3]
and later thumbing through "the great book of the world," gaining bit by bit greater
reflexiveness and critical aptitude but without ever managing to respond adequately
to the "excessive desire to learn to distinguish the true from the false, in order to
see clearly in [his] actions and to walk with confidence in this life."[4] From his disappointment with the meager opportunity for rational self-determination implied by
the spontaneous teleology of immediate lived experience of tradition and sociality,
Descartes decided to apply himself to establishing his own method of theoretical research that, in imitation of "those long chains of reasoning, simple and easy as they
are, of which geometricians make use in order to arrive at the most difficult demon-
strations," would enable him to intercept and lay in founded logical sequence, and
therefore rationally controllable, the succession of "all those things which fall under
the cognizance of man," provided only that "we abstain from receiving anything
as true which is not so and always retain the order which is necessary in order to
deduce the one conclusion from the other."[5]

  In his enthusiasm at the possibility of establishing on the basis of his own reason a
mathesis universalis[6] that would take the place of worn-out scholastic metaphysics,
more adequately accomplishing the task of giving a rational foundation to the empirical sciences and the experience of all of life, Descartes relaxed the theoretical
vigilance that up to this point had pervaded his work and, almost without realizing
it, made ontologically permanent the condition artificially produced in the existent
by methodological suspension (*epochè*) of its validity through questioning. Passing
through successive reductions of his concrete subjective experience, Descartes came
to the point of exhibiting *cogito* as adequate principle of being, starting from which
to rebuild with a geometric method the entire ontological field. Once having reached
the indubitable *ego cogito ergo sum* ("I think, therefore I am"),[7] Descartes went on
to protect and consolidate it by examining attentively the essence and existence of
his "I", first of all observing "that [he] could conceive that [he] had no body and that
there was no world nor place where [he] might be;" later showing "that [he] was a
substance the whole essence or nature of which is to think, and that for its existence
there is no need of any place, nor does it depend on any material thing;"[8] and finally
reaching the conclusion that his "I" was *res cogitans*,[9] inasmuch as: "this 'me,' that
is to say, the soul by which I am what I am, is entirely distinct from body, and is
even more easy to know than is the latter; and even if body were not, the soul would
not cease to be what it is."[10]

  Through this proto-phenomenological road of inquiry into the meaning of lived
experiences through doubt and suspension of their ontic validity, Descartes found
himself inaugurating a sort of metaphysical reform in which the multiplicity of substances of Aristotle and Scholasticism was substituted by the fundamental duality of
the thinking substance and the extended substance, which in these terms translated
the dual dislocation of existence, which belongs, on the one hand, to consciousness and on the other, to a world external to consciousness, even if now entirely
submitted to mechanistic-causal rationality of consciousness' science, founded
in God.[11]

In doing so, however, Descartes, working "to drain the spiritual elements off the physical realm"[12] and to reinforce the supremacy of the *res cogitans* by strengthening the separation from the *res extensa*, also reduced the chances of success of the longed-for *mathesis universalis*, by which he intended to restore metaphysical unity and therefore overcome the just emerged polarity of the *res cogitans* and *res extensa*, making it "absorb into a higher unity of existence from which the opposites issue as faces of its being or phases of its becoming."[13] In fact, as H. Jonas notes, the living being itself was crushed between the two reigns of "consciousness" and the "extended world," in which the ontological whole as immediately lived was split by Descartes in order to reconstitute it on a scientific-rationalistic basis: since then the organism and the living body, as places of encounter of animate/thinking being and inanimate/extended being, represented "a problematical specialty in the configurations of extended substance" and, because of their exceptionality, were reduced to the inorganic "general being of the world" and stripped of their peculiar characteristics. "Precisely this," continues Jonas "is the task set to modern biological science by the goal of 'science' as such."[14]

Even if interpretable as a transition phase in the pursuit of a scientifically rigorous ontological reunification, the dualistic form of metaphysics therefore produced the paradoxical effect of marginalizing precisely that form of being that was instead crucial for Descartes' objective of the *mathesis universalis*. In fact, the living being, inasmuch as carrier at various levels of the actuality of coexistence and the synergy of consciousness and world, is the unavoidable ontic place of investigation of the adequate reasons for the hoped-for reconstitution of the ontological whole; and this even more so if, like Descartes, one wants it to be modeled on the new geometrical analysis, by which, unlike the practice in classic mathematics and Eleatic philosophy, the properties of the figures are shown in their generating according to the rational law of construction deposited in consciousness.[15] More in particular, our living body is the only form of being that documents to us the spontaneous convergence of the two spheres of the *res cogitans* and of the *res extensa* and we constantly experience that "our living body constitutes that very self-transcendence in either direction:" it "must be described as extended and inert, but equally as feeling and willing – and neither of the two descriptions can be carried to its end without trespass into the sphere of the other and without prejudging it." Thereby – Jonas clarifies – it is the experience itself of our living body that "makes methodological *epochè* founder on its rock," every time we wrongfully attribute ontological consistence to the reductions produced by it. "The fact of life, as the psycho-physical unity which the organism exhibits renders the reduction illusory" and "the actual coincidence of inwardness and outwardness in the body compels the two ways of knowledge [knowledge of consciousness and knowledge of world] to define their relation otherwise than by separate subjects."[16]

Perhaps for this reason, the reference to a naturalistic background of living forces, even if one claims to outdistance it, represents a constant in Rationalism. Christian Wolff, for example, is not satisfied with showing the rational self-evidence of the principle of non-contradiction, by which it is impossible for the same thing contemporaneously to be and not to be (*impossibile est, idem simul esse et non esse*).

Regarding such a source of every certainty, inasmuch as setting it one places certainty in human knowledge, removing it one takes away all certainty (*fontem omnis certitudinis, quo posito ponitur certitudo in cognitione humana; quo sublato tollitur omnis certitudo*),[17] Wolff wants to add as further foundational factor the datum of psychological experience according to which "we experience such a nature of our mind, that, while it judges that something exists, it cannot at the same time judge that it does not exist" (*eam experimur mentis nostrae naturam, ut dum ea iudicat aliquid esse, simul iudicare nequeat, idem non esse*).[18] In the same way Wolff proceeds with the principle of sufficient reason, according to which "nothing is without sufficient reason because it exists rather than does not exist" (*nihil est sine ratione sufficiente, cur potius sit quam non sit*);[19] in fact, "we experience such a nature of our mind, that in the individual case not easily someone admitted that something is without sufficient reason" (*eam experimur mentis nostrae naturam, ut in casu singulari non facile quis admiserit aliquid esse sine ratione sufficiente*).[20]

And what is to be said about I. Kant? The critical conclusion about the impossibility of a metaphysics as science and the consequent use in the merely regulative sense of the ideas of reason is drawn from the basis of the preliminary acknowledgment of metaphysics as natural disposition/ tendency of reason (*Naturanlage des Menschen, seiner Vernunft hinsichtlich der Metaphysik*).[21]

Truly, as Jonas observed, "the organic body signifies the latent crisis of every known ontology and the criterion of 'any future one which will be able to come forward as a science';" "this body is the *memento* of the still unsolved question of ontology, 'What is being?' and must move beyond the partial abstractions ('body and soul', 'extension and thought', and the like) toward the hidden ground of their unity and thus strive for an integral monism on a plane above the solidified alternative".[22]

But there is more: the contraction of life from the whole of nature into its distinct singularity that was promoted by modern rationalistic idealism against the ancient primordial monism which made life coextensive with being, was the vehicle for both the becoming of the lifeless coextensive with the objective being, and the isolation of pure consciousness, which now has no share in the objectified world, nor acts there, having become bodyless, merely contemplative, beholding consciousness.[23] In effect, without the body by which we are ourselves an actual part of the world and experience the nature of force and action in self-performance of them, on the one hand our knowledge-of-the-world is reduced to "a merely beholding knowledge," and on the other hand the world becomes "a strictly 'external world' with no real transition from myself to it;" our knowledge would thus be reduced to Hume's model in which causality has become "a fiction" that stands on a psychological basis, which in turn is left groundless itself.[24] In any case, the rationalistic paradigm has made pure consciousness as little alive as the pure matter confronting it. Accordingly, the one can as little generate the aliveness of active connection in its understanding as the other can present it to perception. "Both are fission products of the ontology of death to which the dualistic anatomy of being had led !"[25]

THE VITAL ASPIRATION SURFACES ANEW ON THE HORIZON OF BEING

Geometrizing Rationalism had thus backed itself into a blind alley, and still in the XX century it was at a loss for how to come out; in 1913, M. Scheler viewed as pioneeristic and incomplete the attempts at transformation of the European *Weltanschauung* and thus also of the idea of the world, undertaken by Nietzsche, Dilthey and Bergson with the intent to establish a philosophy that flows from the fullness of the experience of life.[26] Heidegger himself, before facing in 1929 the arduous topic of the relationship between the being of ontology and the time of life, had to work to take leave of the so-called theory of two worlds, psychological-subjective and logical-objective, and to root the predicative in the ante-predicative, since the world of ideas and of logical meanings must be able to manifest themselves in the empirical lived experiences, in order to enter into the life of man.[27]

Actually, already in the heart of rationalistic Modernity, G. W. Leibniz had cultivated the "proto-generic and proto-genetic" ontological seed of vital spontaneity that alone could "open the integral field of the real"[28] and bring to flower the *mathesis universalis* that did not germinate in the unfolded Cartesian system; and this notwithstanding that it was precisely Descartes, dealing with the general problem of tangents, who introduced to classic geometry the new genetic/generative logic by which knowledge no longer had to lose itself in the multiplicity of spacial forms, having now discovered the access to the *logos* that presides over their generation and that is reproduced by the original unitary activity with which the figures are set (*Setzung*) in consciousness.[29] For this reason, Leibniz had defended Descartes from the accusation of the thinkers of Cambridge that he was affected by the *morbus mathematicus*, pointing out that the great principle of mathematical explanation of nature conquered by Descartes for science must not be touched or limited in any way, because the doctrine of life not only did not contradict the principles of knowledge in physics and mathematics,[30] but rather, was supported by them. Instead, it was necessary to go more deeply with the "mathesic" intuition to the point of leading mechanical intelligence of phenomena to the "sufficient reason" or in other words the "ground" for their being and becoming, implicit in their mode of origination. Extension, form and movement, explicative of the phenomena of nature, in fact are not enough to explain the mechanism itself, as global phenomenon, expressive not only of cause-effect connections but also of a background of continuity and harmony that cause one to intuit deeper "inner workings of nature."[31] In this way Leibniz developed the generative bud of being left implicit in the scientific work of Descartes, and moves the *mathesis universalis* forward, showing the organic connection of the derived mechanic forces to the living metaphysical forces, primitive and original, that for him are the monads. In fact, every event can be traced back to these simple, individual substances, of infinite number and endowed with spontaneous auto-movement: they are entelechies, or in other words living forces, inasmuch as they are bearers of the principle of their own activity and of their own progressive evolution, in the course of which their essence unfolds, rising from one degree of formation to another, more perfect one. Mechanical becoming, in this "pluralistic universe,"[32] therefore, is nothing other than the exterior side, the manifestation of that becoming that takes place in the substantial units, in the intimately and spontaneously active energies for a self-given purpose. As Leibniz himself communicates by letter to Wolff, the extension in which Descartes believed to have recognized the substance of the body is based on what is non-extended: what is extended is founded on what is intense, the mechanical level of being on the vital one.[33] Through the latter then we are introduced to the "primitive force," in an ontological-metaphysical sense that is pregnant with possibilities, because the individual subject or substance, foundation of the extension, contains "the principles of all that which can be attributed to it, and the principle of its changes and its actions."[34]

Jonas warns that Leibniz' ingenious attempt to correct the Cartesian position of psycho-physical dualism is nonetheless couched in the problematic terms of Descartes' approach, drawing upon the motives and general determinations of his dichotomy.[35] Cassirer also underlines that "the concepts and basic tendencies of the Leibnizian system are transmitted [...] with certain limitations," "by way of the transformation they underwent in the system of Wolff."[36] Nevertheless, one cannot help but acknowledge that the Leibnizian idea of submitting all the mechanical conditions to the needs of self-deployment of the individual metaphysical subject's existential content, that is preformed even in its organic seeds, is influential even today in the scientific field.[37]

LIFE REFOUND IN A.-T. TYMIENIECKA'S PHENOMENOLOGY OF LIFE

E. Husserl also finds himself contextualizing "the egological Cartesian structure within a monadologic universe close to Leibnizian thought," when he comes in Cartesian French homeland, with the dual intent of honoring Descartes, taking up again the theoretical form of meditation, and at the same time of radicalizing his subjectivistic turn in order to verify whether from the sphere of the thinking substance one can with phenomenology "reach the transcendental connection with intersubjectivity and extended substance."[38] In fact, traces of Cartesian dualism continue to accompany Husserlian phenomenology itself, which still appears both held back by the "impossible situation of the subject's constituting the world and being simultaneously an objective element of it," and incapable of advancing "to unearth the universal logos and solve the quandry that puzzled Husserl."[39]

REVITALIZATION OF THE PHENOMENOLOGICAL METHOD

Anna-Teresa Tymieniecka feels strongly the reflective unease of this situation: for this reason she undertakes to subject the phenomenological enterprise to an inner "critique" that however will be far from the one proposed in E. Fink's

Sixth Cartesian Meditation, as "last" transcendental reduction of transcendentality, or in other words, of transcendental constitution as such. In fact, Anna-Teresa Tymieniecka intends to verify whether the phenomenological pursuit has not ultimately been hiding an ampler conception of rationality than was acknowledged by its founder Husserl and his followers. Consequently, rather than proceeding with one more effort to interpret phenomenology through its own method, in conformity with the Husserlian proposal of a self-critique of phenomenology upon its very own trascendental/subjective assumptions, she sets out to achieve an enlarged inquiry that will advance in virtue of rationalities that are not identical with constitutive/cognitive/intentional transcendentality.[40]

Pushing beyond the confines of essential givenness, assured by the constitutive genesis of objectivity, and establishing a phenomenology of phenomenology, A.-T. Tymieniecka manages to establish contact with the vital and creative "inner workings" that she intuited subtended on the level of constitution and hosted in the profundity of human living experience *(Erlebnis)*, in "the locus whence *eidos* and fact simultaneously spring," in the conviction that "not constitutive intentionality but the constructive advance of life which carries it may alone reveal to us the first principles of all things."[41]

What Tymieniecka set into motion with the whole movement of thought derived from Husserl was a true "intuitive re-sowing"[42] through intentional empathy. She approached it as an organic phenomenon in vital expansion, as one living and expressive body that had reached and touched her with its generative/propulsive energy, involving her empathetically in its productive logos. In accepting to use this "twist" of thought on experience and "to take into consideration insights from any of them that fall within our purview," A.-T. Tymieniecka, guided by the radical need "to follow the progress of the method in order to inquire into its very logos and its yielding,"[43] concentrated her attention on the "late breakthrough to the plane of nature-life," opened by the final phase of Husserlian *Phänomenologisieren*, introjecting it, however, according to "the seminal virtualities engendered by [Husserlian] thought;"[44] in this way she made a philosophically organic connection, through phenomenological dissemination rather than by mere speculation,[45] between "the historical body of phenomenological learning and the horizons for future programs".[46] In doing so, she succeeded, especially because of the previous work she had done to recontextualize conscious reflection in the sphere of life and to discover a further and more original talent/disposition of consciousness *(Uranlage des Bewusstsein)*[47] with consequent updating of philosophical discourse,[48] now directed to take on, beyond the "sequential 'therefore' order of writing" and "the stereotypical language of so-called 'scholarly' discourse that would ape science but be merely pseudo-scientific," an adequate approach to living life: it "streams in all directions and will at any point refract its modalities and their apparatus into innumerable rays that flow concurrently onward" and therefore requires the installation of "all modes of human functioning, all human involvement in the orbit of life".[49]

In any case, according to the "philosophical testament" of Husserl,[50] did not precisely that establishment of a living empathetic relationship[51] in the sphere of the "community of monads",[52] represent that source of "reproduction" *(Fortpflanzung)* of philosophising through the succession of generations?[53] Isn't empathetic relationship the only one that leaves hope for the passing beyond of "historically degenerated metaphysics"[54] of the twentieth century?

"Probing from within the phenomenological horizon of accomplishments",[55] A.-T. Tymieniecka realizes that not even Husserl, in his complex and fruitful reflective proceeding, kept to the logic of the "speculative thinker who seeks to unify his various insights"; rather, he, too, followed the simple logic of human experience, that "follows an analysis to an obvious end and then takes up deeper questions". In the same overall "developmental sequence" of Husserlian thought, therefore, still often considered "without [...] apparent links between its phases" and therefore strongly disorienting for students and followers, Tymieniecka instead discovers that

the planes of human reality are intrinsically legitimated in that sequence, for Husserl adjusted his assumptions as he went without dismissing any set of them.[56]

In other words, presiding at the succession of phases of the "integral Husserl"[57] is the same *logos* that is at work in the formation of "the planes of human reality" and that, in the temporal continuity of experience, builds each individual human being and opens him to ever-new cognitive and practical conquests. It is with exactly this living and temporally constructive logos that "carries on the great streaming edifice of life"[58] that Anna-Teresa Tymieniecka syntonizes herself, grasping the "thread of the iron necessity of the logos" of self-individualizing life that runs through the various phases of Husserlian thought and determines the reciprocal congruence of it in such a way that each level of it acts as a "springboard" for inquiry in a more profound direction. Responding to the many who see in this way of doing philosophy a vice of self-founding, Tymieniecka points out that the logos of life engaged in the Husserlian investigation is the same that is daily at work in every effective execution of descriptive inquiry, which phenomenology also is; it means that once an area has been cognitively traveled, one finds oneself at its borders and from there one can lean forward to grasp new dimensions, now within our reach. For that matter, it was precisely the marked heuristic-constructive value of this spontaneous cognitive human behavior that moved the progress of scientific knowledge in the twentieth century.[59]

A.-T. Tymieniecka is profoundly struck by the "rational framework" that sustains the advancement from time to time of the stages of Husserlian phenomenology "that ever expands its horizon". In fact, she realizes that in "this inquiry into reality, the human being, and the world, it is not only the validity of each phase of phenomenology that is preserved but also the promise each offers": the vital *logos*, that animates it, makes possible that phenomenology "effectively retains its assumptions as it proceeds even as it stepwise supercedes them", since "it rejects earlier work only in the sense that it digs deeper furrows into reality as successive layers of that reality become intuitively visible".[60]

The phenomenological *logos* that guides the evolutive sequence of the integral Husserl, "at deeper and deeper levels, establishing novel frameworks of legitimation as he went: eidetic, transcendental, the lifeworld, intersubjectivity, bodily participation in the constitutive process etc.",[61] is therefore rooted in the constructivism of life itself, that is, on that organic dynamic that, according to "the interrogative mode of the *logos* of life", "proceeds by throwing itself from the already achieved to the presumed". In this way, "each step posited throws up a 'question' for the next, that is, establishes an order for the dynamic" and "the logos of life [...] transforms the stream of its forces from a chaos into an organized becoming, the becoming of life".[62] This natural poiesis, or autopoiesis, according to U. Maturana and F. Varela,[63] however, observes A.-T. Tymieniecka, gained voice only when life reached the level of the human condition; only and exclusively at this level can it also mature its flowering in the ontopoiesis of life, operated by the living "enaction" of the human subjectivity that "expands life into possible world of life",[64] beyond the limits of natural determinism. Tymieniecka comments:

Thus, man's elementary condition – the same one which Husserl and Ingarden have attempted in vain to break through to, by stretching the expanse of his intentional bonds as well as by having recourse to prereduced scientific data – appears to be one of blind nature's elements, and yet at the same time, this element shows itself to have virtualities for individualization at the vital level and, what is more, for a specifically human individualization. These latter virtualities we could label the *subliminal spontaneity*.[65]

THE LIVING METAPHYSICS OF THE LOGOS OF LIFE

Indeed, A.-T. Tymieniecka has attained the pre-ontological position of being, that in which being generates itself and regenerates. From this point of view, she has been able to untangle the *logos*, which presides over the evolution of the life of being, indicating it, with a term of her own coinage, as "ontopoiesis", that is, "production/creation of being."[66]

Therefore, while in the past we traced the tracks of being, now we can follow the traces that beings, living and non, leave in their becoming: they pursue a road of progressive and growing individualization in existence, that is, in the environmental context of resources, strengths, and intergenerative energies; life itself, inasmuch as *vis vitale*, pushes them along this road, promoting their unfolding and controlling their course. Also from within the human condition, in fact, there radiates, grafted on the natural self-individualizing flow of life itself, a dynamic of creative vital expansion, upon which every intellectual dimension is based. For this, the cognitive act, which points to the structures of beings and things, in order to give rise to static ontologies, must give way to the creative act, during which man manifests the same *vis vitale* at work in the becoming of beings: establishing ourselves on the level of creativity, it is possible to follow the *poiein* of those same essential structures that knowledge identifies, isolating them.[67]

Establishing a bridgehead on the ontopoietic plane of life, Anna-Teresa Tymieniecka finds herself in the condition, which had seemed lost, of setting up anew that *mathesis universalis* to which Descartes, Leibniz, and Husserl had equally aspired.

The "ontopoietic plane of life" is, in fact, "a plane of inquiry that combines the dynamic ontology of beingness in becoming with metaphysical insight and conjectural reaching beyond toward the great enigmas of the Universal Logos". Now, within this proto-ontological field, it is a matter of showing "how the timing of life and temporality as such belong to the essential ways in which the vital spheres of life emerge and unfold, and the specifically human moral and intellective spheres also", to the point of "the sphere of the sacred that lay beyond and toward the Fullness of the All".[68]

But will the driving force of the logos that sustains and pushes life in the complete deployment of its self-individualizing dynamic be able to conduct it from "the incipient instance of originating life in its self-individualizing process" all the way to "the subsequent striving toward the abyss of the spirit"?[69]

The logos that is intrinsic to life has manifested itself as "a primogenital force striving without end, surging in its impetus and seeking equipoise": it promotes the constructive prompting that determines the progress of life and it prepares its own means/organs for its own advance. This advance means the fulfillment of constructive steps toward transformations, that is, "step by step unfolding projects of progressive conversion of constructive forces into new knots of sense". Therefore, "the crucial factum of life" has not appeared without reason, brought [...] out of "nowhere"; on the contrary, the logoic force of life has its purpose[70] – just like Schelling's living nature, that embodies the "scheme of freedom"[71] – and that purpose reveals itself as ontopoietic inasmuch as it expresses itself "in preparing scrupulously in a long progression the constructive route of individualizing life so that *Imaginatio Creatrix* emerges as an autonomous modality of force with its own motor, the human will". Crowning its development, the force of the logos of life, with the will as new modality of force, finds itself able to advance from the vital/ontopoietic round of significance into two new dominions of sense: that of the creative/spiritual and that of the sacral. In the terms of traditional ontology, this means that " 'substances' undergo a 'transubstantial' change" and also that "the inner modality of the logoic force undergoes an essential transmutation". Therefore, "Life, [...] as a manifestation of the ontopoietic process" "is far from a wild Heraclitean flux, for it articulates itself"; in addition and first of all "[life] 'times' itself",[72] because time reveals itself as "the main artery through which life's pulsating propensities flow, articulating themselves, intergenerating".[73]

In the metamorphic capacity that intrinsically qualifies the ontopoietic logos of life, there is the possibility for "the new metaphysical panorama"[74] that delineates itself to transcend "the timeless pattern of surrender to nature" and go beyond "the equipoise established through millennia of life between nature and human beings and between the gifts of nature and their use by living beings",[75] also establishing new nexuses between time as *chronos* and *kairos*.[76] The fulcrum of this metamorphosis is that "unique phase of evolutive transmutation", in which the "mature" phase of the platform of life manifests an extraordinary character and gives rise to the Human-Condition-within-the-unity-of-everything-there-is-alive. Paradoxically, the human being appears to be integrally part and parcel of nature yet to reach levels "beyond nature", levels of life that endow the human being with special unique significance that is no longer simply vital but is also spiritual.[77]

The appearance of the living human being sets off in natural life "a watershed event, essentially a transformation of the significance of life": the "enigmatic" surging of *Imaginatio Creatrix* in the middle of ontopoietic sequence, surging freely as it floats above the inner working of nature. Here we reach – observes Tymieniecka – the most surprising and enigmatic turn of logos of life, because this great shift was being prepared by the logos' constructive steps, starting at the very beginning of self-individualizing of life, but it produces a "countervailing move" that "brings about a complete conversion of its hold on life's individualization and opens the entire horizon of freedom".[78] *Imaginatio Creatrix*, rooted within the functioning of Nature-life and yet an autonomous sense giver, introduces three new sense giving factors: the intellective sense, the aesthetic sense, and the moral sense. With them life is endowed with meaning beyond what is geared to and strictly limited to survival; there comes about an inner transformation of the vitally oriented and single-minded functional system of reference into the *novum* of specifically human creativity. Within the creative modus of human functioning in its specifically creative orchestration there occurs a metamorphosis of the vital system of ontopoiesis.

The moral sense lies at the core of the metamorphosis of the life situation from vital existence into the advent of Human Condition:[79] here we have the entrance into the game of life of a specific thread of logos of life, that involves human communion and also the sacral quest.[80] The quest prompted by the moral sense is a mode of becoming but of an absolutely "spontaneous" becoming, one that does not follow a pre-programmed sequence to be accomplished but is "freely" projected becoming, building on the accomplishments of each actor. While the human creative condition and moral sense both develop in ontopoietic time, the quest for ultimate understanding goes in a direction reverse to that of the ontopoietic unfolding of life and work to undo its own accomplishments of the progressive transmutation of the soul.

Indeed – Tymieniecka exclaims– through the moral and entirely freely chosen work of the conscience, the self-enclosed ontopoietic course may be undone and remolded in a free redeeming course!

The logos of life has lead us to a borderline place between the ontopoietic logos of life and logos' sacral turn toward territory that is beyond the reach of the logos of the vital individualization of beingness.[81] It is here that the Great Metamorphosis takes place, "that completes life's meaning in a transition from temporal life to a-temporality, or better, hyper-temporality".[82]

At this point Anna-Teresa Tymieniecka can undertake a radical metaphysical re-elaboration, suitable for the needs that spring from the decline of the modern theoretical paradigm. In fact, philosophical inquiry into the principle of all things, that phenomenology of life set off again, now engages the field of being no longer in its generic and static wholeness, which embraces all-that-is, but also and above all in its continual concrete becoming and proceeding, by incessant auto-articulation: therefore, responding to the ancient need to "save the phenomena" means undertaking a research of *philosophia prima* directed at the objective of "theorizing" the overall phenomenon of the new "fullness of the Logos in the key of Life." Really, what has thrown itself wide open before us is a path of theoretical research that we did not believe existed, on which instead we can adventurously embark, renewing the need of the Enlightenment and Kant to "sapere aude!"(=dare to know!). We now catch sight of a unitary logos leading us, that animates the parmenidean sphere and the same absolute Hegelian Spirit and that, autoindividualizing itself through ontopoiesis, shows it can intrinsically connect phenomena emerging bit by bit from the inorganic to the organic, to the human, weaving a "metaontopoietic" network of innumerable metamorphic passages of transcendence, that open it in the direction of the divine, newly risen to sight, according to the perspective of *philosophia perennis*, already delineated by G. W. Leibniz, when, to rationally understand the truth of the propositions of fact, he introduced the principle of sufficient reason, which, while establishing a foundational dynamic tending toward the infinite, made it possible to construct a solid ladder of truth in order to always better suit the fullness of the logos.[83]

University of Macerata, Macerata, Italy
e-mail: itcalz@tin.it
Translated by Sheila Beatty

NOTES

[1] Cf.: H. Jonas, *Life, Death and the Body in the Theory of Being*, in: Id. (ed. by), *The Phenomenon of Life. Toward a Philosophical Biology*, Northwestern University Press, Evanston, Illinois, 2001, pp. 10, 25.

[2] Ibid., p. 10.

[3] Cf.: R. Descartes, *Discourse on the Method of Rightly Conducting the Reason and Seeking Truth in the Sciences*, tr. by E. S. Haldane and G. R. T. Ross, Dover, Mineola, NY, 2003, pp. 5–6. Furthermore: http://www.marxists.org/reference/archive/descartes/1635/discourse-method.htm

[4] Descartes, *Discourse on the Method*, op. cit., p. 9.

[5] Ibid., p. 19.

[6] In this regard, see: G. Crapulli, Mathesis universalis. *Genesi di un'idea nel XVI secolo*, Edizioni dell' Ateneo, Rome 1969.

[7] Descartes, *Discourse on the Method*, op. cit., p. 36.

[8] Ibid., pp. 37–38.

[9] Ibid., p. 37. Cf. in this regard the performative interpretation of the *cogito* proposed by J. Hintikka, *"Cogito ergo sum": Inference or Performance*, in: W. Doney (ed. by), *Descartes. A Collection of Critical Essays*, Notre-Dame University Press, Notre Dame, 1967, pp. 108–139, taken up and valorized by K. O. Apel, *Transzendentaler Semiotik und die Paradigmen der ersten Philosophie*, in: *Globalisierung-Herausforderung für die Philosophie*, Bamberg Universität Verlag, Bamberg, 1998, pp. 21–48. See also: K. O. Apel, *Selected Essays I. Toward a Transcendental Semiotics*, Humanities Press, Altantic Highlands, New Jersey, 1994. Id., *A Spekulative-Hermeneutic Remarks on Hintikka's Performatory Interpretation of Descartes' Cogito ergo sum*, in: R. E. Auxier and L. E. Hahn (eds. by), *The Philosophy of Jaako Hintikka*, "The Library of Living Philosophers", Vol. XXX, Open Court Publisher, Chicago and LaSalle, Illinois, 2006, pp. 357–367. See furthermore: http://philmat/oxfordjournals.org, In addition: A. Rossi, *Possibilità dell'io. Il cogito di Descartes e un dibattito contemporaneo: Heidegger e Henri*, Mimesis, Milan, 2006; S. Nicolosi, *Il dualismo da Cartesio a Leibniz*, Marsilio, Padua, 1987.

[10] Descartes, *Discourse on the Method*, op. cit., pp. 38–39.

[11] Ibid., Part IV, § 5, pp. 39 and following.

[12] Jonas, *The Phenomenon of Life*, op. cit., p. 13.

[13] Ibid., p. 17.

[14] Ibid., p. 11.

[15] E. Cassirer, *Leibniz'System in seinen wisseschaftlichen Grundlagen*, Olms, Hildesheim, 1962, pp. 11–12. Cf. in this regard: D. Carloni, *Pensée mathématique et génération chez Descartes*, "Analecta Husserliana", L (1997), pp. 143–154, where a brief review of the youthful writings of Descartes permits a glimpse of the initial suggestion that the experience of generation, as a universal and cosmic phenomenon, had an influence on the young philosopher and his aspiration to a *mathesis* conceived as origin of the sciences, including those of the phenomena of life.

[16] Jonas, *The Phenomenon of Life*, op. cit., p. 18.

[17] Cf.: C. Wolff, *Philosophia prima sive Ontologia*, Francofurti et Lipsiae, 1736, § 55 and note.

[18] Ibid., § 29.

[19] Ibid., § 70.

[20] Ibid., § 74.

[21] Cf.: I. Kant, *The Critique of Pure Reason*, tr. by J. M. D. Meiklejohn, eBooks Adelaide, 2009, *Introduction*, vi: "metaphysics must be considered as really existing, if not as a science, nevertheless as a natural disposition of the human mind (*metaphysica naturalis*). For human reason, without any instigations imputable to the mere vanity of great knowledge, unceasingly progresses, urged on by its own feeling of need, towards such questions as cannot be answered by any empirical application of reason, or principles derived therefrom; and so there has ever really existed in every man some system of metaphysics. It will always exist, so soon as reason awakes to the exercise of its power of speculation. And now the question arises: 'How is metaphysics, as a natural disposition, possible?' In other words, how, from the nature of universal human reason, do those questions arise which pure reason proposes to itself, and which it is impelled by its own feeling of need to answer as well as it can?". See also: I. Kant, *Prolegomena to Any Future Metaphysic*, tr. by P. Carus, Open Court Publisher, Chicago-London, 1902, § 57: "But metaphysics leads us towards bounds in the dialectical attempts of pure reason (not undertaken arbitrarily or wantonly, but stimulated thereto by the nature of reason itself). And the transcendental ideas, as they do not admit of evasion, and are never capable of realization, serve to point out to us actually not only the bounds of the pure use of reason, but also the way to determine them. Such is the end and the use of this natural predisposition of our reason, which has brought forth metaphysics as its favorite child, whose generation, like every other in the world, is not to be ascribed to blind chance, but to an original germ, wisely organized for great ends. For metaphysics, in its fundamental features, perhaps more than any other science, is placed in us by nature itself, and cannot be considered the production of an arbitrary choice or a casual enlargement in the progress of experience from which it is quite disparate".

[22] Jonas, *The Phenomenon of Life*, op. cit., p. 19.

[23] Ibid., pp. 19–20.

[24] Ibid., pp. 20–21.

[25] Ibid., p. 21.

[26] Cf.: M. Scheler, *Versuche einer Philosophie des Lebens*, in: M. Scheler and M. Frings (eds. by), *Gesammelte Werke*, Francke, Bern-Munich, 1972, III, pp. 314–339. We can find the Schelerian expression: "eine Philosophie aus der Fülle des Erlebens heraus", at p. 314.

[27] Thus , V. Costa, *La verità del mondo: giudizio e teoria del significato in Heidegger*, Vita e Pensiero, Milan, 2003, p. 81.

[28] Cf.: A.-T. Tymieniecka, *Phenomenology of Life (Integral and "Scientific") as the Starting Point of Philosophy*, in: "Analecta Husserliana", L (1997), pp. ix–x.

[29] Cassirer, *Leibniz' System*, op. cit., pp. 13, 11–12.

[30] Cf.: G. W. Leibniz, *Considérations sur les Principes de la Vie et sur les Natures Plastiques*, in: C. J. Gerhardt (ed. by), *Die philosophische Schriften von G. W. Leibniz*, Vol. VI, Weidmann, Berlin, 1885, pp. 539 and following.

[31] Tymieniecka, *Phenomenology of Life (Integral and "Scientific") as the Starting Point of Philosophy*, op. cit., p. xi.

[32] E. Cassirer, *The Philosophy of the Enlightenment*, tr. by J. P. Pettegrove, Princeton University Press, Princeton, New Jersey, 1969, p. 29.

[33] Cf.: Leibniz – Chr. Wolff, *Briefwechsel zwischen Leibniz und Wolff*, in: C. G. Gerhardt (ed. by), H. W. Schmidt Publisher, Halle, 1860, p. 139. Reprint by Georg Olms, Hildesheim, 1963.

[34] Cf.: G. W. Leibiz, *Letter to Bossuet*, in: Foucher de Careil (ed. by), *Oeuvres de Leibniz publiées pour la première fois d'après les manuscripts originaux*, I, Ladrange Publisher, Paris, 1854, p. 363. Reprint by Georg Olms, Hildesheim, 1975.

[35] Jonas, *The Phenomenon of Life*, op. cit., p. 63.

[36] Cassirer, *The Philosophy of the Enlightenment*, op. cit., p. 34.

[37] Cassirer, *Leibniz' System*, op. cit., p. 411. Cf. also the reference to: A. Weismann, *Essays Upon Heredity*, Claredon Press, Oxford, 1889 (Ibid., p. 411, note 1); U. Maturana and F. Varela, *Autopoiesis and Cognition. The Realization of the Living*, Reidel Publishing Company, Dordrecht, 1980. Original title: *De maquinas y seres vivos. Una teoria sobra la organización biológica*, Editorial Universitaria, Santiago de Chile, 1972.

[38] Thus R. Cristin, "Presentazione" at: E. Husserl, *Meditazioni Cartesiane*, it. tr. by F. Costa, Bompiani, Milan, 1994, pp. viii–ix, xv. Cf.: E. Husserl, *Cartesian Meditations. An Introduction to Phenomenology*, tr. by D. Cairns, Martinus Nijhoff Publishers, The Hague/Boston/London, 1982[7]. At § 62, Husserl faces and rebuts the objection against phenomenology, according to which phenomenology is incapable of solving the problems that concern the possibility of Objective knowledge, unless it uses "an unknowledged metaphysics, a concealed adoption of Leibnizian traditions" (p. 148).

[39] A.-T. Tymieniecka, *The Logos of Phenomenology and the Phenomenology of Logos*, in: "Analecta Husserliana", LXXXVIII (2005), p. xv.

[40] Ibid., pp. xiv–xv.

[41] A.-T. Tymieniecka, *Tractatus Brevis. First Principles of the Methaphysics of Life Charting the Human Condition: Man's Creative Act and the Origin of Rationalities*, in: "Analecta Husserliana", XXI (1986), p. 3.

[42] Cf.: D. Verducci, *The Development of the Living Seed of Intentionality from E. Husserl and E. Fink to A.-T. Tymieniecka's Ontopoiesis of Life*, in: "Analecta Husserliana", CV (2010), p. 33; furthermore: Id., *La questione dello sviluppo in prospettiva ontopoietica*, in: "Etica ed Economia", 1 (2007), pp. 45–58; Id., *Examining Development from the Ontopoietical Perspective*, in: "Phenomenological Inquiry", 31 (2007), pp. 17–22.

[43] Tymieniecka, *The Logos of Phenomenology and the Phenomenology of Logos*, op. cit., p. xv.

[44] A.-T. Tymieniecka, *A Note on Edmund Husserl's Late Breakthrough*, in: A.-T. Tymieniecka (ed. by), *Phenomenology World-Wide. Foundations, Expanding Dynamics, Life-Engagements. A Guide for Research and Study*, Kluwer, Dordrecht, 2002, p. 685a.

[45] Cf.: D. Verducci, *Disseminazioni fenomenologiche e innovazioni teoretiche*, in: Id. (ed. by), *Disseminazioni fenomenologiche. A partire dalla fenomenologia della vita*, Eum, Macerata, 2007, pp. 11–27.

[46] Tymieniecka, *A Note on Edmund Husserl's Late Breakthrough*, op. cit., p. 685 a.

[47] A.-T. Tymieniecka, *Die Phänomenologische Selbstbesinnung*, in: "Analecta Husserliana", I (1971), p. 10.

[48] Cf.: D. Verducci, *La trama vivente dell'essere di A.-T. Tymieniecka*, in: A. Ales Bello and F. Brezzi (eds. by), *Il filo(sofare) di Arianna. Percorsi del pensiero femminile del Novecento*, Mimesis, Milano, 2001, pp. 63–89; Id., *The Human Creative Condition Between Autopoiesis and Ontopoiesis in the Thought of Anna-Teresa Tymieniecka*, "Analecta Husserliana", LXXIX (2004), pp. 3–20; Id., *La meta-ontopoesi di A.-T. Tymieniecka come teoresi di solidarietà tra spirito e vita*, in: "Annali di Studi religiosi", 5 (2004), Edizioni Dehoniane, Bologna, pp. 315–335; Id., *The Ontopoiesis of Life: A Theory of Solidarity Between logos and life*, in: "Phenomenological Inquiry", 31 (2007), pp. 23–28.

[49] A.-T. Tymieniecka, *Impetus and Equipoise in the Life-Strategies of Reason*, "Logos and Life", Book 4, Kluwer, Dordrecht/Boston/London, 2000, p. 4.

[50] Cf.: E. Husserl, *Teleologie in der Philosophiegeschichte* (in three chapters, the first two dated June/July 1937 and the last the end of August, 1936): it is a text from cover K III 29 and from pp. 5 and 9 of cover K III 28; now included in volume XXIX of "Husserliana", in: R. N. Smid (ed. by), *Die Krisis der europäischen Wissenschaften und die transzendentale Phänomenologie. Ergänzungsband. Texte aus dem Nachlass 1934–1937*, Kluwer, Dordrecht/Boston/London, 1993, n. 32, pp. 362–420. In the *Anmerkungen des Herausgebers*, it is R. N. Smid who defines the text in question as Husserl's "last philosophical testament" (p. 362). See also: N. Ghigi, *Introduzione a* "La teleologia nella storia della

filosofia", in: E. Husserl (ed. by), *La storia della filosofia e la sua finalità*, it. tr. by N. Ghigi, Città Nuova, Roma, 2004, pp. 11–55.

[51] E. Husserl, *Ideas Pertaining to a Pure Phenomenology and to a Phenomenological Philosophy*, Second Book, tr. by R. Rojcewicz and A. Schuwer, Kluwer, Dordrecht/Boston/London, 1989, § 51, pp. 209–210.

[52] Husserl, *Cartesian Meditations*, op. cit., § 55, p. 120.

[53] Husserl, *Teleologie in der Philosophiegeschichte*, op. cit., p. 364.

[54] Husserl, *Cartesian Meditations*, op. cit., § 60, pp. 139–140.

[55] A.-T. Tymieniecka, *From the Editor*, in: "Analecta Husserliana", I (1971), pp. vi–vii.

[56] A.-T. Tymieniecka, *Phenomenology as the Inspirational Force of Our Times*, in: Id. (ed. by), *Phenomenology World-Wide*, op. cit., p. 3a.

[57] Ibid., p. 2b.

[58] Tymieniecka, *Impetus and Equipoise*, op. cit., p. 4.

[59] Tymieniecka, *Phenomenology as the Inspirational Force of Our Times*, op. cit., pp. 2a, 3a.

[60] Ibid., p. 3a.

[61] Ibid., p. 2a.

[62] A.-T. Tymieniecka, *The Great Metamorphosis of the Logos of Life in Ontopoietic Timing*, in: Id. (ed. by), *Timing and Temporality in Islamic Philosophy and Phenomenology of Life*, 3, Springer, Dordrecht, 2007, p. 20.

[63] Cf.: U. Maturana and F. Varela, *Autopoiesis and Cognition. The Realization of the Living*, Reidel Publishing Company, Dordrecht, 1980. Original title: *De maquinas y seres vivos. Una teoria sobra la organización biológica*, Editorial Universitaria, Santiago de Chile, 1972.

[64] A.-T. Tymieniecka, *Creative Experience and the Critique of Reason*, "Logos and Life", Book 1, "Analecta Husserliana" XXIV (1988), p. 6.

[65] Ibid., p. 28.

[66] Cf.: M. Kronegger and A.-T. Tymieniecka (eds.), *Life. The Human Quest for an Ideal*, in: "Analecta Husserliana" XLIX (1996), p. 15: "I call it [the becoming of life], going back to Aristotle's *Poetics* a 'poietic' process: onto-poietic. In brief, the self-individualization of life is an ontopoietic process."

[67] A.-T. Tymieniecka, *Human Development Between Imaginative Freedom and Vital Constraints*, in: "Phenomenological Inquiry", 31 (2007), p. 8 and following.

[68] Tymieniecka, *The Great Metamorphosis of the Logos of Life*, op. cit., p. 18.

[69] Ibid., p. 19.

[70] Ibid., p. 20.

[71] Cf.: F. W. J. Schelling, *Aphorismen über die Naturphilosophie*, in: K. F. A. Schelling (ed. by), *F. W. J. Schellings Sämtliche Werke*, Vol. VII, J. C. Cotta Publisher, Stuttgart-Augsburg 1861, p. 236.

[72] Tymieniecka, *The Great Metamorphosis of the Logos of Life*, op. cit., p. 20.

[73] Tymieniecka, *Life's Primogenital Timing: Time Projected by the Dynamic Articulation of the Ontogenesis*, in: "Analecta Husserliana", L (1997), p. 4.

[74] Tymieniecka, *The Logos of Phenomenology and the Phenomenology of Logos*, op. cit., p. xix.

[75] Tymieniecka, *Impetus and Equipoise*, op. cit., p. 99.

[76] Tymieniecka, *Life's Primogenital Timing*, op. cit., p. 4.

[77] Tymieniecka, *The Great Metamorphosis of the Logos of Life*, op. cit., p. 31.

[78] Ibid., p. 32.

[79] Ibid., p. 33.

[80] Ibid., p. 35.

[81] Ibid., p. 60.

[82] Ibid., p. 67.

[83] A.-T. Tymieniecka, *The Case of God in the New Enlightenment*, "The Fullness of the Logos in the Key of Life", Book I, "Analecta Husserliana" C (2009).

ELLA BUCENIECE

CRITIQUE OF REASON PROJECTS WITH REFERENCE TO ANTIQUITY: I. KANT AND THE PLATONIC IDEAS, E. HUSSERL AND THE MNEMOSINEAN ENTICEMENT, A.-T. TYMIENIECKA AND THE DYONISIAN LOGOS

Motto The particular Greek manner of reasoning and expounding of thoughts I take to be the distinctive feature of philosophy. The question of philosophy is first and foremost a question of language.

Emmanuel Levinas

ABSTRACT

Every new movement in the phiolosophical thought is not only new, but is always a repetition, a returning to the roots, to the Antiquity. New is not only something that takes place for the first time: it is conscious or unconscious meeting, a short-circuit with what has already taken place; it marks a break in the straight forward movement, but it is also a circularity, a returning back to the by-gone. This feature, in turn, enhances the poignancy of the actual situation, it imparts an ontological dimension to life

The paper deals with the analysis of the notions of "ideas" (Plato, I. Kant), "memories" (E. Husserl) and "the logos of life" (A.-T. Tymieniecka), which serve in modern philosophy, especially phenomenology, for the enhancement of the topicality both of the heritage of the Antiquity and of the issues of present-day relevance.

Indeed, are you able to imagine a philosophical discourse – even any of the most modern ones – without the use of Greek notions and words? At least the word *philosophy* itself has to be present. This goes to show that philosophical enterprise is never likely to be turned into pure analysis, for philosophy by definition is incapable of avoiding its self-designation and even analytical philosophy has to refer to its Greek origins.

Every new movement of philosophical thougt is always a repetition, a returning to the roots of Antiquity. New is not not only something that takes place for the first time: it is conscious or uncoscious meeting, a short-circuit with what has already taken place;it marks a break in the straight forward movement, but it is also a circularity, a returning back to the by-gone. This feature, in turn, enhances the poignancy of the actual situation, it imparts an ontological dimension to life. (Is it possible for philosophical cogitation to be void of ontological significance – that could be posed as the next question).

A.-T. Tymieniecka (ed.), Analecta Husserliana CX, 39–49.
DOI 10.1007/978-94-007-1691-9_4, © Springer Science+Business Media B.V. 2011

Returning is always the same and always a different one. Eternal Returning. A touch of Eternity and also of Non- Being.

I intend in the present essay to expose some meeting-points and to investigate some of the new insights that have appeared through pondering of such notions as *idea*, *Mnemosine* and *Dionysian logos* in the conceptions of Immanuel Kant, Edmund Husserl and Anna-Teresa Tymienecka.

I D E A S

Idea is one of the most abstract, most all-embracive and also one of the most widely known *termini technicus* designed by Plato. It is also a notion most difficult to comprehend. It concerns Platonic metaphysics, ethics, philosophy of nature, cosmology, epistemology and mythology. The notion of *idea* leads to the understanding of essences (ousia, substantia), to he distinction between the essential and the non-essential. For example, in the dialogue *Parmenides* Parmenides asks Socrates as to the possibility of such things as mud, hair, dirt and other despicaple entities having respective ideas or forms, and receives an answer to the effect that such things as these are just what they seem to be and that there are no ideas behind them. This is a very significant, if not the most significant element of Platonic approach, in other words – it leads to the conviction that evil and baseness are not representations of ideas, they are void of ontological foundation. It is the mind, the intellect, the faculty of understanding (*logos, nous*) that perform the separation of the essential from the inessential, thus releasing the energy for creative ordering of the world. This arrangement will remain essentialy unchallenged till Kant.

In starting the investigation of ideas I. Kant (at the opening of the first part of Transcendental Dialektic of his Critique of Pure Reason) begins with the consideration of the notion of the idea itself – its content and habitual meaning – so as to decide about the further use of the word: either to stick to the existing one or to abandon it in favour of a new term. Of course, Kant turns to Plato in order to reveal the semantic field of "idea" and comes to the conclusion that no new term is needed, only the existing one should be augmented and further developed. Kantian reading of Plato accentuates those features which go to form the bases of his own significant contribution, namely – the practical aspects, the whole gamut of notions which refer to human freedom and the laws of corporate existence. In a way Kant approves of the very idea of Platonic "ideas" to the effect that "ideas obtain of their own reality and that they are not just a dream".[1] Yet, for Kant – as distinct from Plato – ideas belong to the reason itself and consequently – they serve as markers of the transcendental field, the latter remaining unattainable to our experience. Yet, without these markers of boundaries our experience would turn out to be a mess of contradictions and a delusive dream.

It is noteworthy that for Kant ideas are responsible for the generation of doubts as to the ability of reason to grasp the essentiality of things, and even more – inability to approach the ideas themselves; and this – in spite of the fact that for Kant – in distinction from Plato – ideas are not confined to some specific sphere but are viewed as a kind of polarity placed at some distance from the things. Thus, Gilles Deleuze writes: "To take an example of Kant. Kant, of all philosophers, discovered the lofty sphere of the transcendental. He is like a brilliant investigator, yet his concern is not with some other kind of world, but with the summits and the depths of our present world".[2] Ideas are correlative with "I think"; they are objects related to all three synthetic functions of reasoning. Reason becomes aware of the existence of ideas through paralogisms, antinomies and ideas turn into problem-fields of reason giving rise to "deliberations in thinking".[3] This is why the ideal of the reason itself remains the first problem of reason, a task to be tackled before all other problems are approached. A certain amount of idealism is a precondition for the ontology of reason; ideas are a matter of reason, not only of things.

E. HUSSERL AND THE MNEMOSINEAN ENTICEMENT

Mnemosine – memory was an epic muse for the Greeks. E. Husserl has also succumbed to her charms.

In continuing the Augustinian and Cartesian tradition of thought E. Husserl has encompassed the phenomenon of memory within several thematic zones. First, Husserl thematisizes memory as belonging to one of the basic forms of consciousness in the form of a specific type on intention *in concreto* (this research is to be found in Vol. XXIII of Husserliana – Phantasie, Bildbewusstsein, Erinnerung – Zur Phänomenologie der Anschaulichen Vergegenwärtigungen. Texte aus dem Nachlass (1898–1925, Herausgeben von Eduard Marbach). It is worth noting that Husserl discusses the phenomenon of memory (Erinnerung)[1] alongside imagination and image-consciousness, not only within the inner-time dimensions. According to E. Marbach, who has arranged the collection and is the author of a substantial Introduction to it, Husserl had not worked out a systematic theory of memory. The theme of memory is to be found disperssed among other phenomenologically significant issues. It is also to be found elsewhere in Husserl's works, as I intend to show later on. Husserl groups memory together with the perceptible-again-presentification notions, with the view of developing a "phenomenology of perceptible presentification" (Phänomenologie der anschaulichen Vergegenwärtigungen). Under Franz Brentano's influence Husserl considers intuitive perception as a kind of inner perception in distinction from outer perception. Yet, according to E. Marbah, this distinction is not to be completely separated from the empirical tradition of European philosophy.[4]

By empiricism Husserl means only the concreatness of intentionality that takes place in the passive syntheses of Ego, where Ego is *constantly surrounded* (Husserl's italics) by things (Cartesian Meditation § 38). However, this is not an empiricism

---

[1] Although German *Erinnerung* is translated into English as "recollection", Husserl uses the word in a wider sense – so as to include the notion of memory (Gedachtnis), therefore it is translated here as memory.

of facts, for in § 39 of the same meditation Husserl brilliantly makes the point concerning the irrationality of empirical facts. This is transcendental empiricism, as I
have observed on other occasion in discussing Husserlian themes.[5]

As usual, Husserl embarks on the phenomenological description through demarcation – he distinguishes between the acts of experiencing of perceptible notions
and those of experiencing conceptual notions. In distinction from conceptual notions, where a thing or an oder of things is thought of, a thing or an image in
perceptible notions appears (*erscheinen*). This shift of attention to a lower level the
experience of mind (*erfahrungslogischen Vernunft*) had been intended by Husserl as
a kind of critique of reason.[6] Memory in the capacity of the again-presentification
(*Vergegenwartigung*) of the perceptible is radically different from the directly perceptible consciousness – i.e. – perceptions or becoming-present (*Gegenwärtigung*),
because in the case of again-presentification such elements as time, belief, intuition
come into play, and perform modification of mind. Although memory, imagination, expectations belong to one and the same group of the acts of mind, yet
there is also some difference between them: memory and expectations are again-
presentification of the established (*setzende Vergegenwärtigung*), while imagination
is the again-presentification of the non-established (*nicht-setzende*). Both groups are
to be distinguished from reproduction, which is the "pure" again-presentification:
"Memory is reproductive modification of perception".[7] This aspect is stressed by
Husserl also in Cartesian Mediation II, §. 19, by saying that in any given memory
the same is repeated in a modified manner, while each actual perception always
contains the past horizon as potentiality of memories, ready to be awakened. The
objectified sense may be revealed also through memory, which is only implicitly marked in the actual cogito, or the act of mind. Which means that the sense
is not to be conceived as a finished giveness of the objectified, but it is always
accompanied by intentionality of the horizons. Memory is constituted by double objectification, but these are not ready-made things. The objectification is performed
also by perceptions, yet perceptions themselves are part of the mind, *Selbstda*, or,
to use Husserl's formulation – "Consciousness consists entirely of consciousness".[8]
So it turns out that the sense is as it were located in memory (*Setzende*), and at
the same time it has to be brought out into reality. Thus the role of memory in
the acts of remembering turns out to be the revitalization of the things themselves
or of the past perceptions in their concreatness. These acts are not used as a material for the formation of conceptional notions, but they obtain of autonomous
significance within the general relations of consciousness and they generate anew
something that has already been in existence. Thus we may agree with J. D. Caputo,
who characterizes Husserl's approach as proto-hermeneutical and relates Husserlian
constitution to existential repetition. "Husserlian constitution is optimistic parallel to existential repetition, a repetition which pushes forward and produces what
it repeats".[9] Bernhard Waldenfeld also speaks of the creative force of repetition
("*noch einmal*"; *palin* – Greek; *iterum* – Latin).[10] Actuality and belief, according to Husserl, are the memory-determining modalities, in distinction from, for
example, imagination which is determined by non-actuality and neutrality. Belief
which is present in mind as actuality (*das Bewusstsein als "Wieder" bewusstsein*)[11]

is involved in modification of mind and is to be distinguished from positive belief as "non-modified intentionality". It is (possibly) in order to draw attention to
this difference, that Husserl often uses the English word "belief". It may seem
paradoxical, yet the use of "belief" in this sense enhances the clarity of memory (der Erinnerungs – und Erwartungsgewissheit): "Belief, however, is not a new
intention, it is nothing but the modal character of clarity as against appearance
and assumption (presumption) (Der belief ist aber nicht ein Hinzutretendes, nicht
eine neue Intention, sondern nichts weiter als der modale Charakter der Gewissheit
gegenüber der Charakteren der Anmutung, Vermutung".)[12] Memories in the capacity of acts of again-presentification are connected not only with the objectification
of perception and the belief-clarity of what is remembered, but also with the formation of "I" identity, as it is clear from the etymology of the German word:
Er – innern (penetrating inside): "Ich erinnere mich an die Erinnerungen Selbst".[13]
Constitution of the identity of the subject in connection with the horizons of memory in Husserl's phenomenology (in the form of Mit-erinnerung, Wieder-erinnerung,
Selbst-erinnerung), as collected in vol. XXX of Husserliana, has been studied by
R. J. Walton un J. V. Iribarne.[14] Yet memory as repetition is not only subjective; it is
also inter-subjective. Husserl has noted this already in Ideas I, §. 29 by saying that
the fields of actual perception and the fields of memory – different for each person
as they are – are at the same time also intersubjective due to the common *Umwelt* of
people living in community.

A special and separate question concerns the phenomenon of memory within the
context of the inner perception or subjective time consciousness. This problem has
justly served as a point of interest for the researchers of phenomenology and continues to do so.[15] This is why I intend to touch only upon some aspects of the theme
bearing directly on my conception of the problem of memory.

One kind of criticism directed at Husserl's teaching in connection with
Zeitbewusstsein and memory holds that Husserl (1) affords unjustified privilege
to the present and to the active Ego within the continuity of the inner time and
(2) consequently fails to obtain the sense of the past itself and the specific manner of its difference from the present. Such kind of criticism is exemplified by
"Bergsonianism" of G. Deleuze and by Michel Henry with his material phenomenology and ontological monism, etc.[16] The fact that memory also "reproduces
itself" not only in the present activity of Ego, but also within the passive acts of synthesis, was pointed out by me earlier (though this is not only immanently affective
subjectivity as the passivity of pure life, as in the view of M. Henry).

Analytical defense of Husserlian analysis and understanding of the consciousness
of inner time has been performed by Michael R. Kelly. He points out that Husserl's
"distinction between the passive synthesis of retention (or primary memory according to Husserl's revised interpretation), which presents time's passage, and the active
synthesis of memory, which represents a past temporal instant, will reveal that consciousness's double-life in the living present establishes both a sense of the past, i.e.,
the past in general, and a consciousness of succession".[17] Passive synthesis does not
objectivize, i.e. – it does not turn the past into an object, but affords opportunity for
its self-revelation. The fact that what is remembered takes place Now, does not rule

out the fact that I consider also the Not – now. It is the *clearness* (mine italics. – E. B.) of memory that permits to speak not only of the living present, but also of the "living past". This is lucidly shown in Husserl's example (from "The Lectures on the consciousness of internal time from the Year 1905," 27. §.) about the remembering of the illuminated theatre, about its re-presentation: "This re-presentation of the perception of the theater must not be understood to imply that, living in the re-presentation, I mean the act of perceiving; on the contrary, I mean the being-present of the perceived object"[18] and "Memory is the re-presentation of something itself in the sense of the past".[19] In a similar way it is possible to remember the present (memory of the Present, §. 29); and this is not to be taken just as a metaphor, it is a real act of consciousness, because consciousness is not a sum of single points, but a continuous fulfillment.

There is one more important question that concerns the temporal character of consciousness – is it at all right to place consciousness on the same level with the modes of time and their manifestations? In this connection Russian phenomenologist V. Molchanov advances a very pertinent and well-substantiated (to my mind) proposal: "It seems that Husserl did not feel at ease with the total identification of consciousness and time. This is seen from the fact that according to Husserl the very deepest layer of subjectivity – the absolute stream of consciousness is in itself a-temporal. Husserl holds that it is only by way of a metaphor that we can call it a stream".[20] Thus we may conclude that consciousness as time is remembering eternity (like in the case of Augustine) because time without eternity is not time at all but a succession of material forms.

And in the end one more significant (and beautiful) addition to Husserlian understanding of time, connected with "narrative technology" approach. Memory was for the ancients the Muse of epics, and thus it is the highest epical faculty – reminds W. Benjamin.[21] Story, narrative lies at the bases of the profoundest relations with one's experience, with the depth and refiguration of time (P. Riceour). Story as an expounded life is a kind of poiesis (from Greek – creation, formation) because by telling a story life expands, grows larger – P. Riceour calls it iconic growth.

It is worth noting that in Husserl's case memory performs its work of passive synthesis by making use of the instruments of narrative, and not those of logical description, thus producing, synthesizing new meanings. Here is how Husserl remembers Mausberg – a location not far from Goetingen: "I was in Mausberg with my children, wonderful sunset. The town illuminated by evening light. Sunlit clouds of steam of the locomotive. Potato field with long diffuse shadows... Dark brown sparkling field. Returning home (Heimkehr)... *Once again* I see these visions before my eyes. These have been 'seen' and seen 'again', though with interruptions".[22] Don't you feel like reading Heidegger when going through passage like this?

Heimkehr.. Returning always means coming home, returning to one's homeland, returning to one's roots, returning to meet oneself, and the others. Returning is always same, and always different. Eternal returning. It is meeting with Non-being and thus also – with Eternity. Such is the force of memory.

A.-T. TYMIENIECKA AND THE DIONYSIAN LOGOS

"...untill we furnish an answer to what is Dyonisic, the Greeks remain unknown and incomprehensible for us." – F. Nietzsche.

A.-T. Tymienecka in her multi-volume work *Logos and Life* presents one of the most fundamental non-reductive approach to life by revealing the ontopoiesis of life or the strategy for subterran manifestation of logos.[23] Notwithstanding the fact that the critique of reason in connection with creative experience is undertaken in Book 1 of the *Logos and Life*, I intend to deal with Book 4 where the Dionysian logos structure is most fully revealed. First and foremost I want to note the conceptual originality of the term, because usually the Dionysian origins and the logoic origins are conceived as a dualistic opposition – even Nietzsche thought so by holding, as he did, that the Attic tragedy originates from both sources, though the tension between Dionysian principle of the instinct of justification of life and the Appolonic principle of individuation and appearance, remains in his conception intact. Tymienecka proposes a new strategy for the unification of the Dionysian the Appolonic principles, because life is not separated from logos, and the ontopoiesis of life is an on-going process in *Logos in Life and Life in Logos* manner. Yet, life and its progression, its self-creative qualities (autopoiesis) retain for the author the status of primacy, and thus her conception may be considered as an engagement in a *critique of reason* project. Dionysian logos is not identical with reason, it streches over a wider field, and is characterized by "uniquely logoic synergies."[24] It may appear at first that A.-T. Tymienecka elaborates on the Husserlian *Lifeworld* conception, yet her approach is marked by significant conceptual and linguistic shifts. Thus, for example, instead of the *Lifeworld* notion she inrtroduces the concept of *World-of-Life*, which obtains of a completely new semantic structure.

It is life and not the world that offers the platform for scientific investigation taking off from the life-world and aiming to install itself in life's workings.[25]

In order to understand the structure of the Dionysian logos and its significance within Tymieniecka's phenomenology of life one can hardly avoid the aforementioned comparison with Nietzsche. Nietzsche was the first one to establish the proper place of the Dionysian pathos in philosophy. This achievement had been facilitated by Heraclitus.: "No one before me has transmitted the Dionysian pathos into a philosophical principle – and that for the lack of tragical wisdom. I entertained some doubts, though, concerning Heraclitus, whose very presence made me feel warmer at heart and was more enjoyable than anything else. His approval of contradictions and of fight, of becoming, while radically rejecting the very notion of 'being' – this is where I recognized the most congenial idea that has ever been entertained."[26]

By performing a kind of reflection on the Dionysian process of becoming in the form of autopoiesis of life, A.-T. Tymieniecka is not denying – as did Nietzsche – the concept of being; just the opposite: for her becoming is creating the full plentitude, the many-sidedness and unity of being, as exemplified by the final chapter of "Logos and Life" "The meta – ontopoietic clousure". Here the author summarizes her position as "recovering the great vision of the all", as revivication of "the great ancient vision of the All". The priority of being in the conception of Tymienecka does not mean only the equation of this concept with the principle of becoming or creativity; it means also non-acceptance of the Husserlian idea about the subject as the transcendental bases of the unity of the world, and other antropocentric gambits of thought: "Within the framework of the phenomenology of life, the human being is envisaged not in antropocentric fashion but as one of the types of beingness that emerges within the evolutionary progress of life as such – not as a crystalized essence, but being in the process of unfolding himself".[27]

If Nietzsche's vision of becoming comes from Heraclitus, Tymienecka's insight starts off "most significantly from the Aristotelian schema of things". This follows from the fact that Tymienecka advances the principle of "Vital Unity-of-Everything-There-is-Alive". Here – in my opinion – we can see a modified structural similarity with Aristotle, because – relying on the opinion of the well-known scholar of Antiquity and Mediewal philosophy V. Tatarkiewitz – we hold that "in his considerations of 'being' Aristotel first and foremost thought about living creatures. The branch of knowledge that he pursued and that formed the bases of his philosophical conceptions was biology."[28]

Here is another significant passage bearing upon our present theme; Tatarkiewitz says: "Plato was the originator of principles, while Aristotel created full-blown theories."[29] In a similar vein we can continue: Husserl was the originator of principles, Tymieniecka is the author of a full-blown theory. The conspicuous presence of Aristotle in Tymieniecka's philosophy is attested both by numerous references to Aristotel's works and by the actualization of the notion of entelecheia – a principle that has been all too neglected in modern philosophy. This principle – Tymieniecka insists – is not a *substantia*: it is self-regulative, itself-adjustive, flexible and changable.[30] In a wider sense the principle of entelecheia represents the sentient logos of life that is one of the profoundest manifestations of the Dionysian logos. Tymieniecka distinguishes between four forms of Logos: the animated (sentient) Logos, the intellectual triadic-noetic Logos, the communicative Dionysian (feeling/sharing) logos and the Promethean (inventive, creative) Logos.[31]

Thus, once again returning to the comparison with Nietzsche, wee see that Tymieniecka is reinterpriting then Dionysian Logos by way of extending the Dionysian/Apollonic distinction and creating a new one – a Dionysian/Promethean juxtaposition. At the same time both pairs of distinction are not inherently contradictory – neither for Nietzsche, nor for Tymieniecka. The difference between the two thinkers appears elsewhere: Nietzsche holds Socrates as being guilty of destroying the Greek tragical wisdom with subsequent consequences for the Western culture. For him Socrates with his rationalistic self-sufficiency and his optimistic "logical totalitaianism" appears as a third deity – a kind of redundant deity standing between the Dionysian and the Apollonic principles.[32] For Tymieniecka exaggeration of the role of reason is also unacceptable, yet she entertains no ideas about the redundancy or abatement of any "deities", seeing that the intellective logos represents the principle of creativity and is of outstanding significance for the various manifestations of the self-individuation of life's antropoiesis.

Sharing-in-life is yet another of the most significant matrixes of the phenomenology of life. Life is a stream and Logos expands itself and differentiates through life in inumerable ways – from the pre-life realm, through living-beingness-in-becoming to the Promethean direction as dialectics of the embodiment and freedom.[33] The various impeti of life, the "driven moves of the logos, call for appropriate measures if they are to be balanced against each other, to be negotiated in their pluridirectional tendencies".[34] The logoic life device is intentionality: "There can be no doubt that the intentional act is in its fulness the implementation of the Dionysian logos that surges with the human creative condition, and yet if wee look closer, consciousness is also the prerogative of animals, even if it be of degree less developed or more rudimentary".[35] And again: "Dionysian logos excels and attains the greatest heights of logoic achievement".[36] In distinction from most of the modern conceptions linking the technological progress with the victory of the formalized instrumental *ratio* over the living life, Tymieniecka considers the technological progress as a specific impetus for the evolution of logos in the direction of human freedom: "The impetus of the Dionysian logos does not stop at any step reached with technological invention. So-called technological progress is nothing other than the impetus of the Dionysian logos in its Promethean aspiration to set the human being free, to make him master of not only his destiny but also of his very own ontopoietic course as set down by the system of life".[37]

By stressing the contiguity of the activity of logos with various forms of intelligibility and also with the sentient significance of mind, Tymieniecka has advanced – in my opinion – a completely novel appraoach to intentionality. And again, she looks for the substantiation of such an approach in the direction of the Greeks, this time – to the Stoics, by drawing attention, in particular, to the distinction between the "unspoken" logos, logos endiathetos and the "spoken" logos, logos prophoricus, seeing that the first one is concerned with rationality in the entelechial code, while the other one gets expression in thinking and in articulated sound.[38] Tymieniecka takes note of this significant distinction and at the same time she stresses also that her conception differs from that of the Stoics: "they did not seem to discern the uniquely sentient attunement of the ways and modes in which the animus in living being binds and then puts assunder, fuses, prompts, diffuses, etc. The ties between and among individuals, convival undertakings, enterprises, projects, cooperative work, through which attunement plays a leadinfg role in society."[39]

These are – in my opinion – some of the most significant spheres and strategies of the manifestation of the Dionysian logos in Tymieniecka's phenomenology of life. General, finalized evaluation of her achievement could be characterized in the following way: first and foremost the concept of the Dionysian logos and of the logos as such, from which the present-day philosophy, with few exceptions (M. Merlo Ponti) is shying away – has been reinterpreted and its topicallity enhanced.

Next – by using the concept of the Dionysian logos Tymieniecka revises and widens the phenomenological understanding of intentionality, bringing to the fore such elements as feelings, emotions, passions and the enjoyment of life.

And thirdly – by way of refusing to separate life and logos and by holding both elements as integral parts of the structure of autopoiesis that is developing in te course of the life-processes – reason, logos does not become encompassed within boundaries (as is the case with the grandiose Kantian system), for it is not limited by forms of our understanding– on the one hand, and by undisscursiveness of ideas – on the other. Thus Logos is given an opportunity to undergo changes, to become transformed through evolovement in infinity. This, of course, is not the mechanically extended infinity, but the infinity of creative processses. This is why Tymieniecka at the end of her book introduces the notion of "other infinity" and of the "novel logoic sphere in its 'other' infinity".

In conclusion: all themes touched upon in the present article – concerning ideas, memory and the Dionysian logos – are engaged in tackling – in my opinion – of the over-all general question about the unbounded capacity of reason to balance and to harmonize the sphere of the world and the sphere of the human effort, or – to take the cue from the Greeks once again – to obtain a measure (*metron*) amongst the various "things" – between the contradiction -stricken human being and the equally heterogenious development of the world habitually designated as "progress".

Latvian University, Riga, Latvia
e-mail: e.buceniece@gmail.com

NOTES

[1] Immanuel Kant, Critique of Pure Reason, Riga, 1931, p. 258. (in Latvian).
[2] Gilles Deleuze, Difference and Repetition, Saint-Peterburg, 1998, p. 170. (in Russian).
[3] Ibid., p. 174.
[4] Edmund Husserl, Phantasie, Bildbewusstsein, Erinnerung. Zur Phänomenologie der anschaulichen Vergegenwärtigungen. Texte aus dem Nachlass (1898–1925). Husserliana, Bd. XXIII, Herausgeben von Eduard Marbach, Martinus Nijhoff Publishers, 1980, S. XLVII.
[5] E. Buceniece, "Sensuous Experience and Transcendental Empiricism (F. Brentano, E. Husserl, P. Dāle)". in: Analecta Husserliana, XCV, 2008, pp. 333–342.
[6] Edmund Husserl, Phantasie, Bildbewusstsein, Erinnerung. Zur Phänomenologie der anschaulichen Vergegenwärtigungen. Texte aus dem Nachlass (1898–1925). Husserliana, Bd. XXIII, Herausgeben von Eduard Marbach, Martinus Nijhoff Publishers, 1980, S.XLII.
[7] Ibid., S.305–306.
[8] Ibid., S.265 f.
[9] J. D. Caputo, Radical Hermeneutics, Indiana University Press, Bloomington, IN, 1987, p. 34.
[10] B. Waldenfels, "Die verändernde Kraft der Wiederholung," in: Zeitschrift für Ästhetik und Allgemeine Kunstwissenschaft, 46/1, Feliks Meiner Verlag, Hamburg, 2001, S. 5.
[11] Edmund Husserl, Phantasie, Bildbewusstsein, Erinnerung. Zur Phänomenologie der anschaulichen Vergegenwärtigungen. Texte aus dem Nachlass (1898–1925). Husserliana, Bd. XXIII, Herausgeben von Eduard Marbach, Martinus Nijhoff Publishers, 1980, S.287.
[12] Ibid., S.297.
[13] Ibid., S.204.
[14] R. J. Walton, "Horizontes de la memoria I la identidad", in Escritos de Filosofia 19, Buenos Aires, 2000, S. 299–317; J. V. Iribarne, "Memoria y elvido en relacion con la identidad personal", – Idem, S. 195–213.
[15] J. Brough, "Husserl and the Deconstruction of Time," Review of Metaphysics, 1993, 46: 503–536; A. Al-Saji, "The Memory of Another Past: Bergson, Deleuze and New Theory of Time", Continental Philosophy Review, 2004, 37(2): 203–239. etc.

[16] M. Staudigal, "Umsturz der Phänomenologie? Zu Michel Henrys Kritik an Husserl", Phähomenologische Forschungen, Hamburg, Feliks Meiner Verlag, 2002, S. 87–109.

[17] M. R. Kelly, "Husserl, Deleuzean Bergsonism and the Sense of the Past in General", Husserl Studies, 2008, 24: 15–30.

[18] Edmund Husserl, The Lectures on the consciousness of internal time from the Year 1905. In: Husserl E. Collected Works, Editor: Rudolf Bernett, Vol. IV – On the Phenomenology of the Consciousness of Internal Time (1893–1917), translated by J. B. Brough, Kluwer, 1991, p. 61.

[19] Ibid.

[20] V. Molchanov, Introduction. in: E. Husserl, Collected Works, Vol. I Phenomenological Consciousness of Time, Moscow, Gnosis, 1994, p. XIV (in Russian, Молчанов В., Предисловие – Гуссерль Э. Собрание сочинений, т. I. Феноменология внутреннего сознания времени. Москва, Гнозис, 1994, С. XIV.).

[21] Walter Benjamin, Story-teller, Iluminations, Riga, 2005. (in Latvian, Benjamins V. Stāstnieks. – Iluminācijas.Riga, 2005).

[22] Edmund Husserl, Phantasie, Bildbewusstsein, Erinnerung. Zur Phänomenologie der anschaulichen Vergegenwärtigungen. Texte aus dem Nachlass (1898–1925). Husserliana, Bd. XXIII, Herausgeben von Eduard Marbach, Martinus Nijhoff Publishers, 1980, S.287.

[23] Anna-Teresa Tymienecka, Impetus and Equipose in the Life-Strategies of Reason. Logos and Life, Book 4, in: Analecta Husserliana, Vol. LXX, Kluwer, Dordrecht/Boston/London, 2000, p. 25.

[24] Ibid., p. 369.

[25] Ibid., p. 20.

[26] Friedrich Nietzsche, Kritische Studienausgabe., Bd.6., S. 613.

[27] Anna-Teresa Tymienecka, Impetus and Equipose in the Life-Strategies of Reason. Logos and Life, Book 4, in: Analecta Husserliana, Vol. LXX, Kluwer, Dordrecht/Boston/London, 2000, p. 597.

[28] Vladislav Tatarkiewicz, History of Philosophy, University of Perma Press, New York, NY, 2000, p. 162. (in Russian).

[29] Ibid.

[30] Anna-Teresa Tymienecka, Impetus and Equipose in the Life-Strategies of Reason. Logos and Life, Book 4, in: Analecta Husserliana, Vol. LXX, Kluwer, Dordrecht/Boston/London, 2000, p. 86.

[31] Ibid.

[32] I. Ijabs, "Afterword" to: F. Nietzsche, Die Geburt der Tragödie oder:Griechentum und Pessimismus, S.173. (in Latvian).

[33] A.-T. Tymienecka, Impetus and Equipose in the Life-Strategies of Reason. Logos and Life, Book 4, in: Analecta Husserliana, Vol. LXX, Kluwer, Dordrecht/Boston/London, 2000, p. 436.

[34] Ibid.

[35] Ibid., p. 519.

[36] Ibid., p. 521.

[37] Ibid.

[38] Ibid., p. 337.

[39] Ibid.

TANSU ACIK

WHAT WAS A CLASSIC UNTIL THE BEGINNING OF 20TH CENTURY?

ABSTRACT

While in the 17th and 18th centuries, the concept of "classic" and its derivatives denoted superiority or competence especially attributed to Roman Antiquity writers, with the revolutions of romantic and then historicist thoughts, they started to qualify a style parallel to romantic conception or challenging the opposing styles. That happened through 19th century in connection with the canon formation in various European languages.

Democracy, social class, mass, national consciousness, civilization, culture, progress, standard, art, literature, education, and humanism. While we use these concepts to explain or to describe this and that, we tend to forget that all of them gained their current meaning in European languages through 19th century. Raymond Williams analyses the changes in the meanings of about hundred and ten of such concepts.[1] The concepts we have listed above are selected for being related to the concept "classic". We are going to add here "classic" and "classicism" to R. Williams' list. Let's survey the semantic field of the "classic/classical" historically until the beginning of 20th century.

Before our discussion, we have to examine the concept of canon. The concept of classic and its correlative canon do not only have a meaning within their proper contexts, but they have recently become scientific concepts carrying both analytical and heuristic powers, thanks to the works of Jan Assmann and Adeila Assmann.

A research path opened in the last thirty or fourty years, discovered fundamental differences between orally transmitted culture and written culture in terms of thought structures.[2] J. Assmann's research which has brought forth the normative and formal structures of the classic and canon concepts in Antiquity, provides important clues for our inquiry.[3] J. Assmann explores the concept of canon in contrast with the concept of classic, in the high written cultures of Mediterranean Antiquity. According to Assmann's decoding, the meaning of the word "canon" evolved from meaning ruler, scale, meter, towards the meanings example (b), table and list (d) derived from the meaning concrete scale; and criteria (a) principle, norm, rule (c) derived from the meaning abstract scale. The canon of text is based on the meanings (c, d); the canon of text, or canon in common words is binding and official at the highest level. He establishing that canon originated independently and separately in the Torah and Buddhist religious texts. He explains the canonization of the Greek classics of secular nature in Alexandria, and the canonization of Christian, Confucian, Taoist texts referring to those initial, original examples. He

51

A.-T. Tymieniecka (ed.), Analecta Husserliana CX, 51–57.
DOI 10.1007/978-94-007-1691-9_5, © Springer Science+Business Media B.V. 2011

asserts that transition from ritual coherence based on repetition to textual coherence based on interpretation occurred in Antiquity within close dates; that occurred not because of writing as a tool, but through the canonization of tradition, through disruption of tradition in a renovative way. Cultural memory is the highest concept above and including all ramifications of Assmann's type-genre criteria classification, below are tradition and canon, and within the canon branch are classic and canon. The distinctive feature distinguishing tradition and canon, is that the criteria for determining canon is the exclusion of the options, the determining of the boundaries of the chosen. The difference between classic and canon is that in the classical concept, the excluded is not worthless, the classic choice is not binding; while the discrimination between classic and not classic is also based on the distinctions between authority, connectivity, measurability. Assman defines canonization as the emergence of new teaching, and not as the strengthening of tradition nor as the existing culture becoming sacred.

Disruption and not continuity causes the "Ancient" to rise to the throne of unsurpassable excellence. The classic emerges through the interruption that makes it impossible for the traditional to continue to exist and that fixes the relation to the ancient, and on the other hand, with the identification that transcends this interruption and which considers the past as their own past and the ancient masters as their own masters. The past should remain in the past but not be estranged.

If we roughly classify the reference and dictionary data,[4] we can identify three primary meanings of the words "classic" and "classical". The first refers to a certain grandeur, stability, an important text, a standard text –and the meanings of conventional or stereotyped are derived from the reverse of this first set of meanings-; the second one refers to Greek-Latin literature, for example the plural of word in English when alone means this; and last to classic in opposition to romantic in literature and this meaning has in fact emerged from literary discussions about romanticism. The word classic became obsolete in Middle Ages Latin, thus there is no continuity between its derivatives in European languages and its use in Latin. "Classicus" in Latin meaning tax group and class in the sense of classroom, marked the first uses of the word in European languages. The same evolution is roughly observed in the English, French, German and Italian meanings. The adjective "classic" meaning first class is first encountered in the 16th century; in the 17th century add on the meanings important, model, criteria; the adjective is used in the 19th century to denote a certain stage of a language or a culture. In the 17th century, the name "classic" means both a first class thing, and the sum of Greek and Latin literature. The concept of English classics is derived from the latter in the 18th century. Also "classic" means suitable to the Greek-Roman style in the 18th and 19th centuries. In Italian, whereas the adjective means first class in the 17th century, in the beginning of 19th century, it means criterion, measure for works of art. In French, the word is first encountered in the 16th century, and means emulating model, authority or material taught in classrooms in the 17th century.[5] The same meaning is carried on the famous *Encyclopedia* published by Diderot and his friends in the middle of 18th century. The comprehension of the concept encompasses some authors contemporary of Augustus, some from the 2nd century Roma, and authors like Racine,

Molière, Corneille and La Fontaine. If we examine the extension of the concept, elegantly described, proper writing is the measure, with no other criterion. The list that comprises no Greek authors, is an enumeration of authors and not works. Among them, feature secondary authors such as Valerius, Maximus or Frontinus who only provide material for historical research.

In European literature, qualifying a product of living language as classic started in mid 18th century France, upon a retrospective look at their own literary tradition.[6] Some authors from living languages throughout centuries have been remembered as "great" here and there, but we cannot speak of a common concept to qualify them altogether until 19th century. In this sense, the adjective is used in this sense, for the first time to qualify a certain period in French literature. Whether there are similar classical works in other languages or not will be questioned from 18th century on. For example Thoulier d'Olivet says in the second quarter of 18th century Italy has its classical writers we (French) never have. Nietzche asks the same question for German and gives a negative answer in *The Wanderer and his Shadow* (the third part of *Human, All Too Human*). It is not before 19th century that Dante, Shakespeare, Goethe, each start to be considered European classic writers.[7] Goethe and certain writers around him for the first time have used the concepts of classic and classicism in a sense close to today's. Schlegel brothers refer to classic and classicism within the context of discussion on romanticism –we are not talking here about the distinctions and fluctuations between the German forms *classik, klassik, klassizmus, klassische*. Thus, the meaning of the word ceased to be a value term to become a style current, fashion or the name that refers to a period and which can imply diverse qualifications within itself.

The expression is more rarely used in 19th century English (where it appeared within literature debates) than in French; whereas in German its use is originally spread in the second half of 19th century by the historians of German literature. The word is often resorted to in the beginning of 20th century, by literature critics from various backgrounds but who are all opponents to romanticism. To sum up, while in the 17th and 18th centuries, the concept of classic and its derivatives denoted superiority or competence especially attributed to Roman Antiquity writers, with the revolutions of romantic and then historicist thoughts, they started to qualify a style parallel to romantic conception or challenging the opposing styles.

J. J. Winckelmann, the founder of history of art, gives the first definition of classic in arts; while he classifies Greek statues as classical and archaic according criteria that he makes up, he characterizes Greek sculpture masterpieces most common and distinctive features as "a noble simplicity and a quiet grandeur" *(edle Einfalt und stille Grösse) – Gedanken über die Nachahmung der Griechischen Werke in der Mahlerey und Bildbauer-Kunst, 1755 – Thoughts on the Imitation of Greek Works in Painting and Sculpture / Reflections on the Imitation of Greek Works in Painting and Sculpture* (translation by Elfriede Heyer and Roger C. Norton) in terms of stance and expression; this characterization remained as the only criteria in the field of classical art for so long.

Here Winckelmann, uses "Laocoon", that we know today as the product of Hellenistic era, as an example to Greek masterpiece criteria. After affirming that these criteria feature also in the Socratic era prose, he identifies the same features in Rafaello's work. Winckelmann was the founder of modern scientific archaeology and first applied the categories of style systematically to the history of art. As H.- G. Gadamer succintly put, it was originally at the time of Winckelmann a normative concept, it was creative anachronism transformed into a period label, along with such terms as Archaic, Hellenistic, and so on, by historicist scholars: "The concept of the classical now signifies a period of time, a phase of historical development but not a suprahistorical value" (*Truth and Method* 287); With the rise of historical reflection in Germany which took Winckelman's classicism as its standard, a historical concept of a time or period detached itself from Winckelman's sense of the term, it denoted a quite specific stylistic ideal and, in a historically descriptive way, also a time or period that fulfilled this ideal. So the normative side of the term and the historical descriptive side of the term has been fused. When german humanism proclaimed the exemplarity of first Greek then Roman antiquity, the concept of classical came to be used in modern thought to describe the whole of "classical antiquity".

If we continue our investigation exclusively in the field of literature, we observe that the expression is more rarely used in 19th century English (where it appeared within literature debates) than in French; whereas in German its use is originally spread in the second half of 19th century by the historians of German literature. In the beginning of 20th century, the normative side of the concept has often been invoked by literature critics from various backgrounds but who are all opponents to romanticism.

Sainte-Beuve a leading critic of his time wrote his famous essay named *Qu'est-ce qu'un classique?* "What is a Classic?" in 1850 (Christopher Prendergast, *The classic: Sainte-Beuve and the nineteenth-century culture wars*, Oxford, 2007).

This text is not only the oldest, the most detailed written on the subject, it constitutes also a reference in every discussion on the subject. While dwelling on the Greek-Roman tradition, Sainte-Beuve expands the application field of the concept. Through discussing Louis XIVth time as an example, he suggests that this characterization requires a constant and stable resource that is formed slowly and transmitted from generation to generation. Even though he consults and discusses Goethe's "the king of critics" views as a standard, he doesn't consider him as classic.

Yet Homer, Dante, Shakespeare are considered classic even if they don't meet the criteria of Louis XIVth era, the only classic age. By criteria, we don't mean a consistently elaborated measure, but some qualities referring to a style, because he thinks in terms of oppositions introduced by Romantics, such as the one between those who control their inspiration and those who abandon themselves to theirs. In the meantime, of course he mentions the famous quarrell between old and new in 17th century France (*Querelle des anciens et des modernes*). Actually, the biggest part of his essay simply consists in enumerating groups of old and new writers worthy of entering the Pantheon of classics; his list comprises names of authors and not the works. Among them are the Indians Valmiki and Vyasa, Job, Solomon (he does qualify those last two as Prophets), the Iranian Firdevsî, and Confucius. Let's put aside the judgments of Homer that he quotes from others, and his judgments that Sophocles and Aiskhylos are insufficient, crippled, debris, garbage; comedy writer Menandros is part of the list even though at that time complete copies of his texts were not yet available.

German thinkers from consecutive generations have been determinant in the formation of the current meaning of classic as a concept. Let's mention some areas to explore for an extensive study on the subject. J. J. Winckelmann gives the first definition of classic in arts. One should mention and The big picture includes, the review with a new eye of the Ancient Greek and its appropriation by scholars of philology initiated by F. A. Wolf, and by writers surrounding Winckelman's friend Goethe; philosophy of history started in 19th century by German thinkers; and the first secular secondary and higher education institutions achieved by Goethe's friend Wilhelm von Humboldt in Prussia.Even though Goethe doesn't know Greek, and even though his Latin is limited to reading Spinoza as he confesses in the beginning of his *Travel in Italy*, he had proposed higher education based on Greek and Roman texts, because these texts provide an education both ethical and aesthetic.

After that rough survey Instead of giving some conclusion I would like to make some remarks concerning education based on classical texts be it literature , scientific or philosophical, because the concept of classic gave rise to many institution in the fields of education in the 19th century. Modern secondary education and higher education in the West have been heavily influenced by the work of Wilhelm von Humboldt. In the first decades of 19th century in Berlin Goethe's friend W. Humbolt, processes the idea of *Bildung*, self-formation, put forth by Enlightenment thinkers since Herder, and creates the "gymnasium" a secondary school based on studying of Greek-Roman texts in their original language, of math and history; and in 1810, the University of Berlin, namely the first example of modern university.[8] We owe him many key concepts and their applications: PhD based on original research, academic autonomy, innovative scholarship, especially his conception of *Bildung* or cultivation. By the end of the 19th century every state had, more or less aligned its educational system with the Prussian one, even the rival French model. Many universities emphasized a version of the Humboldtian *Bildung* and called it liberal education in English and *culture générale* in French. That approach gave rise to many higher education models such as liberal arts college, core curricula. Those are aiming at imparting general knowledge and developing general intellectual capacities, in contrast to a professional, vocational, or technical curriculum. Rooted in language and dependent in particular on writing, the humanities are inescapably bound to literacy. From reading great works of literature, history, and philosophy, or the symbolic texts of music and the visual arts, humanists proceed to elaborate their insights through language.

A second wave of transformations within the university system followed between the world wars in favor of technical education. The last transformation is the one we had been experiencing, namely the corporatization of the university. Conscientious scholars and teachers must, now and then, ask themselves the basic question of what it means to be educated. Who would object to an education based on direct experience of classical texts themselves, which, as M. Arnold famously formulates are "the best that has been thought and said in the world"? Besides providing common, shared ground of higher education experience for all students humanistic studies could be the most suitable candidate to interrelate the humanities, social sciences, science, and technology. That should have been self-evident, but it is not the case. So defending liberal education against the excesses of professionalism and against the utilitarian academic bureaucracy is a priority.

University of Ankara, DTCF, Faculty of Letters, pk.218 Sıhhiye Ankara 06100, Turkey
e-mail: tansuster@gmail.com

NOTES

[1]   R. Williams, *Culture and Society 1780/1950* (Columbia University Press, 1958); R. Williams, *Keywords, A Vocabulary of Culture and Society* (Fontana, 1976).

[2]   Walter J. Ong, *Orality and Literacy. The Technologizing of the Word* (London, 1982).

[3]   J. Assmann, *Das kulturelle Gedächtnis: Schrift, Erinnerung und politische Identität in frühen Hochkulturen* (C. H. Beck, 2007, pp. 87–129; 272–280).

[4]   *The Oxford English Dictionary*; *Dictionnaire de L'Académie Française*; *Le Petit Robert*; C. Zolli, *Dizionario Etimologico della Lingua Italiana* (Bologna, 1979); P. P. Wiener ed., *Dictionary of the History of Ideas* (New York, NY, 1968) in it René Wellek, "Classicism in Literature". *Historisches Wörterbuch der Philosophie*, ed. J. Ritter, K. Gründer (Stuttgart, 1976); H. Fricke ed., *Reallexicon der Deutschen Literatur Wissenshaft* (Walter de Gruyter, 2000); Ernst Robert Curtius, *European Literature and The Latin Middle Ages*, trans. by W. R. Trask (The Bollingen Library, 1953, [1948]).

[5]   M. L. Clarke, *Classical Education Britain 1500–1900* (Cambridge University Press, 1959).

[6]   Hans Ulrich Gumbert, "Phoenix to Ashes or: From Canon to Classic", trans. by R. Norton, *New Literary History*, 20:1 (1988) pp. 141–163. The Author discusses also a vital short text on the subject by Voltaire here; he argues that the French classics composed mainly by drama started to be regularly staged in France, in the second half of 19th century; he discovers the concept of classic close to its contemporary meaning, through the mediation of the German romantics, in a book on Germany Madame de Staël wrote in the 19th century. Among the arguments suggested in a convincing manner, there is one however of crucial importance, that literature still meaning in 18th century massive, diverse, deep knowledge, and erudition, became transcendental and autonomous through the concept of classic. Also Gumbert arguments that 18th century literature shifted from writing towards reading and interpretation, because it lost its function of socialization and gained an educational value through writing in 19th century.

[7]   E. R. Curtius, *European Literature and the Latin Middle Ages*, trans. by W. R. Trask (The Bollingen Library, 1953) pp. 348–350.

[8]   David Sorkin, "Wilhelm von Humboldt: The Theory and Practice of Self-Formation (Bildung), 1791–1810", *Journal of the History of Ideas*, 44:1 (1983) pp. 55–73., Matthew Arnold's *Culture and Anarchy* published in 1869 is the most famous modern advocacy of high culture and high humanism. According to Arnold culture is "the best which has been thought and said in the world". A generation before Arnold, poet and writer Coleridge, while translating German ideas into English, has first used the form *cultivation* for the concept *bildung*. In his sentences, the concept ceased to be a natural tendency for development and started to mean a certain state of general consciousness in conflict with the concept of civilisation in the sense of general material progress: R. Williams, *Culture and Society 1780/1950* (Columbia University Press, 1958) pp. 49–70. The meaning of "culture" will gradually expand during a century, from individual perfection, to society's overall development, then to arts as a whole, and last to a way of life in both material and intellectual senses (R. Williams, ibid.). The real issue in Arnold is education reform; he discusses

and criticizes the attitude towards education of almost every group, namely liberals, aristocrats, middle-class bourgeoisie, in the context of current political events. He considers each group deficient in terms of understanding education. He advocates "Unification of Education" which would be implemented as late as 1902 in Great Britain, and the superiority of culture and criticism, seen as the individual's efforts for perfection in all aspects against narrow specialization. He debates "Hellenism and Hebraism" as the two main components of British thought. Humboldt's is the only private name – praised – outstanding in contrast to the politics mentioned in current events, the abundance of people of religion, and to the fact that there is absolutely no reference to any example naming a writer new or old. The educational ideals put forward by Humboldt and continued by Arnold and the like, are in a way ideal and supranational regarding their content and purposes, despite otherwise defended opinions in Germany (the above mentioned article by David Sorkin). We will not deliberate here on the connection of this education bill with the ideal of a new citizen, and the training of public officials; we will just point that this education doesn't aim at training experts, but at general education. Arnold's "sweet light", the common must-have that he attributes to the educated, is based on acquaintances with "that which is thought and written in the best way".

SECTION II
LOGOS AND LIFE

SIMEN ANDERSEN ØYEN

THE EXISTENTIALISTIC SUBJECT TODAY

– Jean-Paul Sartre's philosophy in a context of consumerism and individualism

ABSTRACT

This article examines the phenomena of intersubjectivity and freedom in Jean-Paul Sartre's philosophy in the context of our individualistic consumer society. The idea of the situated individual's responsibility introduces the problem or aporia of inter-subjectivity. The early philosophy of Sartre must consequently be seen in relation to a problematized structure in which questions related to bad faith and an authentic life, freedom and anxiety and the aporetic aspects of the intersubjective dimension collaborate in forming an understanding of the historically, physically and socially situated subject. This is the foundation for an individualistic view of life where self-realisation derived from Sartre's concept of freedom will be central. This has some clear parallels to today's consumer society. The article then problematizes whether Sartre's philosophy can be said to be a theoretical justification of processes of individualization or, alternately, whether aspects of this philosophy can have an emancipatory function in regard to the more deterministic aspects of the consumer society.

The central assertion in this article is that the concepts of freedom, responsibility and intersubjectivity as they appear in the early philosophy of Sartre can illuminate current tendencies in our society such as processes of individualization and consumerism. This will entail an internal theoretical discussion, especially of the relationship between the concepts of freedom and intersubjectivity. In addition, the article will contain a critical analysis of late modernity's consumer freedom in light of Sartre's understanding of freedom. The primary focus will be *Being and Nothingness*, which expresses a conception of the subject as absolutely free, and where freedom, as a structure of consciousness, both constitutes the world as well as our understanding of it. At the same time the encounter with other people is presented as a conflict where the constitution of our life-world implies the Other, functioning as a limitation on an egocentric perspective of the world. These aspects of Sartre's philosophy constitute a paradox, making it difficult to deduce normative implications from his thinking. Nonetheless, there remains an existentialistic, individualistic intuitional philosophy which has a particular resonance for contemporary individualization-processes and consumer-based society.

A.-T. Tymieniecka (ed.), Analecta Husserliana CX, 61–74.
DOI 10.1007/978-94-007-1691-9_6, © Springer Science+Business Media B.V. 2011

ABSOLUTE FREEDOM

The concept of freedom in Sartre's philosophy is many faceted and requires analysis on various levels. First and foremost, the term must be seen in relation to the human mode-of-being. One's being is torn loose from what is, and one lives in expectation of something, in relation to something. Consequently, the human mode-of-being is characterized by absence, negation and nothingness due to the function of negating what he is conscious of. Therefore freedom is not something one has in the way one has qualities. "We are freedom", "we are condemned to freedom" or "we are thrown into freedom" are all varieties of the status of freedom in Sartre's philosophy.[1]

Sartre expresses further, according to the phenomenological principle of intentionality, that actions are always intentional in the sense that they are addressed towards a future goal which always occurs in the context of an absence within, or a negation of, the actual situation. Every action assumes that I transcend what is, towards a goal which is yet to be realized. Consciousness is therefore a subject's manner of tearing loose from the past; it is a freedom to break with the causal series which are characteristic of an object's mode-of-being. In this situation, where nothing is given except the external laws of nature, one has to make a choice and these basic ontological conditions entail that consciousness is essentially connected with choice.[2] The individual, in relation to being, is free to conduct himself according to his own wishes. Freedom establishes reality; the subject must give reasons for himself and become his own foundation. This means that one is responsible for oneself, and in Sartre's subject-ontology the subject is therefore defined as both independent and, to some degree, isolated. The term angst is central here. Angst is freedom's reflective understanding of itself. I am filled with angst when I realize that to write this article, keep deadlines, be precise with references, etc. are some of my many opportunities in my immediate circumstances of life, opportunities which owe their justification of existence to me, and are maintained only by me. No one or nothing forces me to write this article. The anxiety about this article is angst if I am anxious, not about whether or not I will complete the article, but about choosing to put it away – to stop maintaining the opportunity of finishing this article.

THE AUTHENTIC CHOICE

The responsibility freedom carries with it leads to questions about the status of one's authentic choices and the opposing existential structure: bad faith. Bad faith is a state in which one rejects the responsibility that freedom demands and avoids the responsibility of transcending one's facticity. Sartre describes bad belief as either a retreat into transcendence or a retreat into facticity, the being-state of objectness. Bad faith rests on the duality of transcendence and facticity, where either the subject denies the one and identifies with the other, or tries to synthesize the two. Bad faith is therefore a self-delusion. In contrast to bad faith, an authentic life requires the acknowledgement of freedom and facticity while being willing to acknowledge ones contingent existence.[3]

This is the basis for Sartre's development of an individualistic and personal philosophy of life, where the subject maintains total responsibility for his actions. An extensive literature[4] has attempted to deduce a normative theory from this concept of responsibility, thereby synthesizing the concepts of freedom and responsibility with theories of the Other as they appear in Sartre's philosophy. This literature seeks to develop an existentialistic ethics based on the idea of authentic existence and choice. Here, choice has universal implications in which responsibility is a responsibility for the Other. Through the subjective project one chooses how one wishes to conduct oneself in relation to the Other and establishes therefore a norm which reaches beyond the subjective realm itself. Sartre's own *Existentialism is a Humanism* must also be understood as a similar experiment in the sense of developing a foundation for an existentialistic normative theory. It is equally possible to locate other normative tendencies from other periods of Sartre's philosophy. According to Thomas C. Anderson there are at least two tendencies like this in addition to the perspective which is presented in Being and Nothingness.[5] The one is from the period after *Critique of Dialectical Reason* and can be described as materialistic, while the other can be located in Sartre's work from the 70's, represented by the title "Power and Freedom".[6] Still, I will argue that there are certain perspectives in *Being and Nothingness* – in particular the interpretation of intersubjectivity – which make it hard to extract an ethics based in the ontological concept of freedom. Consequently, Sartre's ontological concept of intersubjectivity will not, with deductive stringency, be able to be connected to a moral precept, but neither will it exclude it. One might say that Sartre allows for more than just an inference of the connection between ontological theory and moral philosophy. Even though the concept of intersubjectivity excludes a complete ethical system, it is possible to locate certain normative implications in Sartre's work. These implications have and can be made the foundation of a rudimentary ethics. This represents one possible direction of inquiry. However, Sartre's subject can be characterized as isolated, and thus the epistemological integration of the Other will appear problematic – something that I will later claim to be a paradox and aporia.

THE PENETRATING LOOK

Sartre describes social constitution through what he describes as *the look*. Because of *the look* I can experience the Other. To be seen by the Other is the basic existential relation between humans. *The look* is the subject who sees me as an object. Existentially this intersubjective relationship and the presence of the Other are doubly or ambiguous faceted. There is an interplay between subject oriented and object oriented attitudes. But there is also an explanation as to how the self experiences the Other as object amongst other objects in the world, and is himself experienced as object amongst other objects. The Other does not only appear in my experience as object, but actively reduces me to an object. The relationship is a mutual objectifying and negating of the Other's transcendence. Intersubjectivity must thereby be considered as conflict. This is rooted in an understanding that the subject is forced into self-consciousness through social dialectics. The subject can only assert himself through being in opposition to another subject, and thereby make this other an object. Sartre would say that in the Other's *look* I experience that my freedom is threatened and challenged. Through my being-for-the-other I become an object, who can be integrated in his freedom and be made use of in his existential projects.[7] The Other's *look* can make me an instrument, dependent on his being. My being is therefore to a great extent developed because of the Other's freedom, and this implies a partial alienation of my opportunities. Because of the Other, a great portion of self-knowledge is located outside of ourselves. This, which is called the other part, is still me, but out of reach, outside my radius of action, outside my sphere of knowledge. Sartre exemplifies this through the feeling of shame and how being ashamed necessitates the Other. It is through shame and similar experiences that the Other is constituted for me as one different from me, being-in-itself, and in similar ways I am constituted for the Other as a being-for-the-other.[8] This intersubjectivity is constituted as an alternation between object- and subject-orientation. How the term intersubjectivity expresses ambiguousness or an aporia in proportion to the epistemological status of freedom is explicitly seen here.

An analysis of *the look* demonstrates the paradoxical and aporetical in Sartre's theory of the Other. It is paradoxical because consciousness is defined both as freedom and as sovereign in its understanding. Even if Sartre says that the subject is always free to transcend what he stands face to face with, the intersubjective dimension has determining epistemological implications that are difficult to neglect. This paradox can be formulated like this: How can the subject be in an already socially arranged world and how can the Other objectify the subject's being when the subject is at the same time torn lose from everything outside himself? One can therefore discuss whether Sartre succeeds in proving an actual decentring of the subject's sovereign epistemological position.

It is a further problem and paradox that consciousness is seen from the outside as if it was an object, but at the same time comes into view as behaviour and embodied intentionality. The experience of another human is therefore to be understood as this paradox: that the Other in front of me is an object, but still exists for himself, as another consciousness. The Other's existence lies within the contradictions of the subject-object relation. The Other's experiences are radically removed from me and are an eternal synthesis of unrevealed qualities. But it is only because the unfamiliar subject in this way escapes my direct experience that he is experienced as the Other. This duality warns of an epistemological problem because the subject cannot be an object to himself. The Other can consequently never be understood purely as an object among the world's objects, but more what one might call a privileged object or an ecstatic relation.[9]

From this explanation one can, on the one hand ask oneself how Sartre, by outlining a demarcation between sovereign consciousness and human commonality can escape an abstract rationalism. Sartre's subject-ontology is in danger of becoming an abstract rationalism where the self is self-sufficient, the free consciousness is *sui generis*, and in the end, consciousness defined as freedom is the main condition and the sovereign instance for all knowledge. On the other hand, it is a question about how he can locate commitment in the Other. This is still problematic. Sartre's perspective on intersubjectivity will to a certain extent exclude human interaction and will also partly exclude the understanding of how phenomena appear or arise in relations between individuals. In my opinion, Sartre's understanding of the relation between subjects on one side, and the relation between the subject and the situation on the other, is too individualistic and ahistorical: The self's freedom is totalized and the social dimension disappears to certain degree. Here can Sartre's subject-oriented philosophy be accused of ruling out collective political actions, since it complicates engagement in something bigger than oneself. The problem of intersubjectivity is therefore not only a question about to what extent the subject can experience and know something about other humans and their experiences. It also concerns human coexistence and the possibility for interaction.

It is consequently difficult to derive anything but a personal and individualistic philosophy of life from Sartre, one in which self realization rooted in the concept of freedom is central. The intersubjective dimension is too unsettled and the aporetical aspects of this dimension can only partially ground a normative structure. Even though *Being and Nothingness* includes several value-oriented commitments, due to the fact that the concepts of bad faith, alienation and responsibility have normative implications, it seems that an ethical system based on this philosophy is excluded. Nonetheless, as we will see, Sartre's early philosophy might still have relevance when it comes to understanding current social processes.

NORMATIVE IMPLICATIONS – A CONSISTENT MORAL OR A PHILOSOPHY OF LIFE?

A possible or tentative ethics based on Sartre's ontology will contain a more extensive question about judgement and to a larger degree, consideration of situation than traditional ethical systems such as deontological ethics. It is based in the situation, where the content of ethics is variable and where it acts to derail system-building in ethics. To have to adjust to a new situation every time is a trait of this normative theory of value, but its contents are still open and variable. There is no objective knowledge or objective universal ends to guide our conduct other than our truthfulness through authenticity and our consistency in our choices.[10] It is a situated ethics, without any objective criteria to guide us between right and wrong or precepts for the good life – anchored in the ordinary life-world. However, it is – considering the aporetic aspects of the concept of intersubjectivity – difficult to transfer or adapt the generalised reciprocity which is characteristic of face-to-face ethics to an existentialistic ethics based in Sartre' philosophy. While face-to-face ethics is based in the intimacy of the Other's presence and the moral relevance of these circumstances, those same aspects are absent in the normative implications of Sartre's philosophy and can consequently be said to represent the actual problem or obstacle to developing an existentialistic ethics.

Based in Sartre's philosophy, an existentialistic ethics can be described as a tentative philosophy of life where the responsibility for one's own choices and a realisation of the individual are central. There are, of course, normative implications here. However, there are no established norms previous to the choice, and it is therefore a problem for Sartre to give reasons for a normative relevant difference between, for example, a nun and a torturer.

The relevant aspects of intersubjectivity are characterized by a theoretical aporia which is seemingly incompatible with the moral implications located in a concept of responsibility. Consequently, it is problematic to extract something more from the different perspectives of *Being and Nothingness* than an individual philosophy of life. The dimension of intersubjectivity and its aporetic structure makes the foundation for normative system-building difficult, perhaps impossible – which is in general accordance with Sartre's philosophy taken as a whole. *Being and Nothingness* contains – as mentioned – several value orientated determinants. However, given the paradoxal nature of the concept of intersubjectivity and given the priority and sovereign position of freedom, an interpretation of these different concepts can proceed in multiple directions. An attempt to synthesize the ontological concept of freedom and the dimension of intersubjectivity may give some indications of a normative theory, but an ethical system is excluded. In accordance with this, two interpretations seem especially reasonable. The first one involves an individualistic philosophy of life critical to hypostatic values, or any attempt to give reasons for norms outside the individual's existential projects. However, even the rejection of an ethical system is a normative position. The other interpretation makes possible a rudimentary existentialistic ethics based on the idea of authentic existence and choice, where choice may have universal implications. A precondition for this interpretation is that the concept of bad faith can not be ethically neutral, thereby excluding an ethical pluralism. In this case, the idea of the individual's responsibility in the situation announces a radical situational based ethics. Choice is subjective but through engagement in a project individual chooses to engage in committed forms of living.

At the end (with basis in the second interpretation) we are left with a rudimentary ethics which contains elements of virtue ethics, deontological ethics, discourse ethics and a face-to-face or situated ethics. In Sartre's philosophy it is decisive or conclusive as to what kind of person I am through my choices and my way of living. The individual finds values in those activities which he is insolvably and inseparably engaged in. This has a clear parallel to virtue ethics' concept of "praxis". Values are constituted through our praxis. At the same time, this Sartrian ethics has deontological traits in the manner that this ethics is universal in its form and that we commit others by our own choices. The universal aspects to an ethics of freedom are the irreducible position of freedom, the choice's committing status and that we are condemned to act upon a situation. It has a similarity to discourse ethics because it seeks the intersubjective preconditions and presuppositions for an ethics. At last, it is a situated or face-to-face ethics because its contents are open and only constituted in the situation and in everyday interaction with the Other.[11]

THE EXISTENTIALISTIC INDIVIDUAL

Important aspects of Sartre's early philosophy can shed light on central tendencies of our time, particularly modern and late modern individualization processes. This applies specifically to the concepts of freedom and intersubjectivity, as well as what can be described as an individualistic life-philosophy. This approach to Sartre's philosophy is also an examination of the extent in which the theoretical perspectives which appear in *Being and Nothingness* include a diagnostic of society and an anticipation of our contemporary understanding of the individual. This obviously concerns ideas that were closely connected to the circumstances at the time Sartre wrote *Being and Nothingness*. But Sartre's philosophy of subjectivity can also be seen as being ahead of its time, legitimating theoretically the current zeitgeist and our individualistic social paradigm.

There are, however, problems in identifying Sartre's early philosophy of subjectivity as an individualistic philosophy. This is a philosophy that primarily examines the subject on an ontological level, and not on a social or moral level. Freedom must be understood as being of ontological character. The integration of Sartre's subject in a social and political context and the transition to an individualization which is contextualized in society must be considered problematic. Still, there are many aspects to this philosophy of subjectivity that agree with an individualistic philosophy where the freedom of the subject and the position of choice can be located within the meaning of individualization. At the same time, this philosophy of subjectivity has a special resonance in our western individualistic culture – not only in relation to questions about freedom and choice – but also seen in relation to the question of authenticity and responsibility. The ontological concept of freedom can consequently function as basis for this kind of socially founded concept of the individual. Accordingly, it can be claimed that the non-social individual is the most central figure in Sartre's philosophy, and that the perspective that appears is a kind of methodological individualism.[12]

THE ANTAGONISTIC INDIVIDUAL

One can say that our age's distinctive individualization processes began in the 1960s and 1970s, with for instance the 68-rebellions, and was strengthened by the neoliberal ideological turn in the United States and Great Britain around the 1970s and 1980s.[13] This is a tendency that makes it possible to claim that the individual in today's western society is the fundamental unit in social reproduction – at the expense of the family and other collective structures.[14] In contrast to the individual in so-called pre-modern societies, the modern individual does not have a given permanent identity or social function – other than perhaps that of a consumer. The modern identity is open, unfinished and differentiated in that it is not determined by socially given roles, but is constituted in a plurality of divided spheres of value and culture. This development must be seen as a consequence of the neutralization of tradition and existing social ideologies on one hand, and the differentiating of social functions on the other.[15] These aspects of modernity and the contemporary society are to be found in Sartre's philosophy: the status of the choice and the subject's possibility to project his freedom towards an open future. They are present in the sense that the individual creates meaning and identity where tradition and socially given roles no longer have privileged positions.

If our late modern age consequently can be characterized as a time in which collective ways of action or collective identities are considerably weakened compared to earlier periods, the individual can – according to this development – be described as independent, with more options available. With this development the individual is made increasingly responsible; responsibility for the self and one's own body is held solely by oneself. The individual is left to define his life, and identify his own projects.[16] Freedom has become more subjective, as Sartre describes in the beginning of the 1940's in *Being and Nothingness*. It has been emancipated, in the sense of the boundlessness we find in Sartre's understanding of consciousness. This is a conception of individual autonomy: the individual is free to conduct himself within being as he wishes. Accordingly, Sartre's understanding of freedom and consciousness gives, in a social context, an understanding of the individual as free in the sense of being released from restraint. With this freedom comes the total responsibility for oneself.

However, the individualization processes are complex. It is a complexity that makes it problematic to locate certain development patterns and dominant tendencies. But individualization processes can be understood as the individual's expansion of his own autonomy, as an expansion of individual roles and lifestyles. Rights, education, career and expectations concerning mobility are individualized in the sense that interests and actions to an increasing extent are understood as singularized terms.[17] As a consequence of individualization processes a development towards autonomy means that individuals become more and more isolated due to the network of anonymous social contacts expanding. Individuals are increasingly concerned with their own interests independent of other people.[18] This understanding of the individual is also to be found in liberalistic political theory. Here I first and foremost refer to liberalistic contract theory with roots in the theories of John Locke and Thomas Hobbes. In this tradition it is assumed, though simplified, that individuals abstain from the unlimited freedom of the state of nature and support a political and legal authority that shall guarantee the individual's life, security and property. The motivation to agree on a contract that protects the individual's basic rights is enlightened self interest. This is an ideal-typical understanding of the state where the state's legitimacy is dependent on its protection of individual rights. This approach contains an understanding of mankind as isolated individuals who are concerned with maximizing their own interests in competition with other individuals. One underlying condition of this understanding is the comprehension of the human being as a rational individual best suited to define his own interests alone. Further, the individual takes precedence over institutions and communities where individual freedom and individual rights are basic political and moral imperatives. In addition, freedom of choice is closely related to the individual's behaviour in a market. Last, it is a condition that the political sphere is an arena for the protection of individual rights.[19] In this approach to the individual's position in the political, I will claim that the individual's most important interests are strongly secured by negative rights, for example the right to property or the protection from random governmental involvement in private life and family. These rights apply to every individual and protect the individual's autonomy against encroachment by the state or by other individuals. The individual consequently holds a position as an independent, self sufficient monad or unit, with individuals having an antagonistic relationship to one another.[20] This conception is based on the understanding that destructive relations between individuals who compete are dissolved in a collectively positive and functional maximization of benefits for society. This is a conception that also can be found in liberalistic theories' meta-narrative about the invisible hand.[21]

The liberalistic understanding of the individual as shown above is – though slightly simplified – the one we find in liberalistic political thinking. This view does partly correspond to Sartre's concept of intersubjectivity, in the way he describes social dialectics and the subject's partially isolated position. According to Sartre, the individual takes precedence over institutions and communities. Subjective freedom is a basic existential imperative, prior to any formation of association such as society or social grouping. Further, social dialectics are described by Sartre as antagonistic; the subject asserts himself by standing in opposition to the Other, and the subject can be made an instrument to be exploited in the Other's existential projects. In the same way as in the liberalistic understanding of the individual the Other's freedom is reduced by my freedom. However, seen from Sartre's perspective, the position on subjectivity that liberalism is based on will risk being accused of being essentialistic in the sense that the individual and his identity are strongly related to economic interests and behaviour in a market. The subject chooses himself, and the choice is subjective, but not necessarily instrumental, if instrumental means that the individual utilizes himself after given rules and is only oriented towards his own benefit. To define the subject as rational and instrumentally oriented, like liberalism seems to do, is incompatible with Sartre's resistance to claiming anything at all about human nature or essence. The subject is never identical with himself, and identity cannot be understood as substance, but as a self creating and self justifying process in which the individual is his own foundation. Existential freedom and authentic choice consequently include something more than the well-informed, forward looking and planned choice in a market. Sartre's subject is open-ended and without essence, better corresponding to, or able to adjust to other approaches to sociality and the role of subjectivity in the social.

Mouffe and Laclau criticize the theory of the subject as a self-transparent, rational agent. They argue that every position on subjectivity is a discursive position, part of a discourse's open character.[22] Accordingly, the social is described as an irreducible plurality, meaning that it cannot be reduced to an underlying homogeneous principle or essence. In other words, we have, regarding both the individual and the social arena, no fully unified identities. This can be seen in light of Mouffe and Laclau's concept of "antagonism". An antagonistic relation is a relationship between, for instance, two subjects where the presence of the one subject prevents the other in achieving full presence. Here the antagonism does not take place under conditions of a competition with given regulating principles, but in a non-transparent social and political arena. This can be seen in relation to Sartre's concept of intersubjectivity. On Mouffe and Laclau's understanding of the social, the Other's presence will prevent me from being totally myself. Relations do not occur as totalities, but from within the impossibility of their full constitution. The Other's presence cannot be subordinated as a positive differential element in a causal chain. I cannot be a complete presence to myself, but the force that antagonizes me cannot be a complete presence either. Accordingly, Sartre describes the Other as an inexhaustible synthesis of non-revealed qualities. At the same time, an important part of the knowledge about ourselves is located outside ourselves. The Other will therefore be a constituting exteriority in which identity is created in relation to others and where the subject sees himself and inscribes his own identity. Likewise the structure of *the look* will be a constituting exteriority in the sense that the subject's being is developed due to the Other's freedom, entailing that the subject is dependent on the Other revealing himself.

The consequence of this view of intersubjectivity is that society and the social is infused with antagonism and will consequently never become transparent and totally present. According to Mouffe and Laclau, the subject's and individual's co-existence cannot be shaped according to an objective and understandable pattern. Rather than being a transparent arena where social agents consider their interests rationally in competition with others, the social is an arena consisting of balance and aggregation between different groups and individuals, or constellations of institutions and power which take place in a cultural and historical development. This description of sociality can function as a continuation of Sartre's understanding of intersubjectivity where the social antagonism that Mouffe and Laclau describe can be founded in Sartre's antagonistic concept of intersubjectivity. The paradoxical in Sartre's interpretation can consequently better explain the complexity of the dynamic processes that characterize social phenomena and human relations than, for example, the liberalistic approach, which reduces the subject to a rational and benefit-oriented being.

SARTRE'S SUBJECT IN A CONSUMER SOCIETY

These individualization processes must also be seen in relation to new and developing governance structures and liberal market freedoms, at the expense of government regulation. We see a new form of governing rationality that moves towards less direct political central planning to the advantage of more undefined and individualistic forms of politics, where the political expands into the private. This development has, as mentioned, its basis in the development of advanced liberal democracies where a change has taken place from state sovereignty to governance techniques based in economic structures.[23] Each of us must govern ourselves, and this freedom is also – according to Sartre – the responsibility of creating oneself. This identity development often takes place as a participant in a market through the role of consumer.[24] The individual must govern himself – something which is consistent with how, for example, commercial industry manufactures consumption as a creating of the self. Every choice in the market is presented as a new start (and this does not only apply to diet products), proof that we freely create our life through choices and actions.

In an individualistic consumer society, consumption is considered, according to Bauman, as a calling, a profession, or universal human right that does not recognize the exception in that it is the individual's skills as consumer that defines one as a person.[25] Consumption is inseparably related to individualism and identity. The choice of the right market-based products or services is regarded as a responsibility that rests upon the individual. The individual is constituted as a consuming agent whose goal is the maximizing of his own well being through his actions in a world of products. Products embody a seemingly personal meaning that reflects the individual and that person one wishes to become.[26] Like Sartre's subject, the individual has to choose between a number of possibilities and project meaning onto the objects, thereby constituting himself through the selection. Consumer ideology can be illuminating for how Sartre's total freedom and total responsibility relates to the individualization processes of our time. Sartre's view of responsibility lacks substantial values and should seemingly be compatible with the understanding of consumer responsibility. We create ourselves and are responsible for who we are. This perspective on Sartre's responsible individual is based on the view that individuals in a market have actual freedom and are made responsible as reflective, participating actors.[27] The individual's right to pursuit one's own values and interests in a market creates social processes in which organizations and actors wish to influence market processes through consumer behaviour. This can be seen in campaigns for the protection of, for example, nature and the climate or solidarity with the poor.[28] This development includes something more than a fundamental view about the subject being hedonistic. The role as consumer can be understood in the sense that consumption represents new roles for citizens that supplement and replace classical political roles through consumer movements. It is however difficult to see that environment problems or the climate threat can be every person's responsibility, for example. It is more legitimate to ask whether *it is* every individual's and consumer's responsibility to reduce global pollution or address threats to the climate. Is this not a responsibility that stretches beyond the individual's possibilities of action? Here Sartre's philosophy could be criticized for supporting our tendency towards making social problems the individual's problem. This individualization has a concealing function for the identification of the social origin of general problems.

However, the existential responsibility cannot be reduced exhaustively to buying Max Havelaar or other fairtrade-products. Sartre's individualistic philosophy of life cannot unconditionally give reasons for the subject's expression of freedom that is consistent with consumerism because Sartre's subject is not synonymous with consumerism's comprehension of the subject. The subject is a nothingness and not the complex of desires that consumerism seems to depend on. Individualized consumption cannot function as compensation for basic existential projects, and consumption as the basis for identification with a social role can also be understood as a variation of bad faith. Rather, existentialistic authenticity could be understood as the rejection

of identifying our needs with consumption and our personality with certain products. Our self-realization, projects and identity construction must be understood as
preceding a society of consumerism where the possibilities of expressing individual
freedom through consumption are related to a standardized production of symbols
determined through, amongst others, the design- and marketing-industry.[29] We are
not condemned to market freedom or the materialistic market choice in the same
way as "we are freedom".[30] Consciousness is so defined as to break free from the
causal series which are characteristic of being – also for the market-based being.
In this context one can understand the emancipating aspects of Sartre's philosophy
confronted with what can be understood as deterministic market liberalism. At the
same time this perspective is a comment on the ongoing intellectual debate which
concerns alternatives to the capitalistic system which can be summarized by Fredric
Jameson's famous quote:

Even after the "end of history," there has seemed to persist some historical curiosity of a generally systemic – rather than a merely anecdotal – kind: not merely to know what will happen next, but as a more
general anxiety about the larger fate of destiny of our system or mode of production as such – about
which individual experience (of a postmodern kind) tells us that it must be eternal, while our intelligence suggests this feeling to be most improbable indeed, without coming up with plausible scenarios
as to its disintegration of replacement. It seems to be easier for us today to imagine the thoroughgoing
deterioration of the earth and of nature than the breakdown of late capitalism [. . .].[31]

Market imperatives and the lack of alternatives to sovereign market mechanisms
can be understood as a forced colonization of existence. Freedom is made instrumental and is coded towards consumption; individualization occurs as an atomizing
process. In this context the market and its fictive freedom implies an alienation
where Sartre's philosophy can be revealed to represent an alternative. The subject is
not a reflection of reality, but conducts himself freely in regard to it – also to market
mechanisms that can seem fatalistic and absolute.

Dominant economic mechanisms and processes are consequences of globaliza-
tion. That more and more social functions and values find their expression through
market arrangements can seem alienating – a form of bad faith. Economic con-
junctures, the hierarchical labour market, large international corporations, consumer
products' advantageous position in society, the commoditization of culture and art
and market competition has an alienating function in which the Other's intentions
and plans are realized at the expense of the subject's own.[32] Consumer society is
also divided into layers and classes where large groups are prevented from following emancipatory projects within a market or as a modern liberation project. Sartre's
early philosophy can therefore dissolve the understanding that social relationships
seem so determined that they are considered being of the same character as natural
phenomena. One is always free to transcend what he is opposed to, also seemingly
determined society and market structures. What is authentic in this case would be to
acknowledge that one has freedom when it comes to structures outside oneself, and
that the structures are therefore changeable. Freedom as an ontological structure of
consciousness comes before every other determination of human characterizations
and therefore denies that there exist structures and essences in society that will give
these determinations a privileged position.

CONCLUDING REMARKS

Sartre's early philosophy and its normative implications must be seen in relation to a problem in which questions surrounding bad faith and authenticity, freedom and angst and the aporetic aspects of the intersubjective dimension intersect within the conception of the historical, embodied and socially situated subject. The subject must justify himself, becoming his own foundation and is thereby ultimately responsible. However, despite that the subject can acknowledge that he is free in relation to the Other and despite that the subject, by virtue of his actions, gives meaning to life, intersubjectivity's acknowledgeable structures always haunt the subject. These aspects of the dimension of subjectivity in Sartre's philosophy are partly consistent with current individualism and the modern consumption-oriented, selfish individual. The extreme individualization processes of our age that are coded towards consumption parallel Sartre's philosophy of subjectivity: the individual is absolutely free and has total responsibility for himself. At the same time, this perspective illuminates liberating elements in Sartre's philosophy. There are aspects to this philosophy that challenge how liberalism and consumerism understand the concept of the individual and provide a foundation for alternative understandings of contemporary individualized market society.

The Centre for the Study of the Sciences and the Humanities, University of Bergen, Bergen, Norway
e-mail: simen.oyen@ua.uib.no

NOTES

[1] Sartre (2001).
[2] Vestre (1993).
[3] Daniels (2005).
[4] Anderson (1993), Rendtorff (1993), Kerner (1990).
[5] Anderson (1993).
[6] Ibid.
[7] Sartre (2001).
[8] Ibid.
[9] Oliver (2005).
[10] Kerner (1990).
[11] For more on this theme see Øyen (2010).
[12] Østerberg (2009).
[13] Beck (1992).
[14] Madsen (2006).
[15] Honneth (2004).
[16] Rose (1999).
[17] Madsen (2006).
[18] Simmel (2004).
[19] Habermas (1996).
[20] Ibid.
[21] Laclau and Mouffe (2001).
[22] Ibid.

74 SIMEN ANDERSEN ØYEN

23 Negri and Hardt (2000).
24 Rose (1999).
25 Bauman (2007): p. 231.
26 Madsen (2006).
27 Jensen (2007).
28 Ibid.
29 Madsen (2006).
30 Sartre (2001).
31 Jameson (1994): pp. xi–xii.
32 Jensen (2007).

REFERENCES

Anderson, Thomas C. 1993. *Sartre's two ethics: From authenticity to integral humanity*. Chicago: Open Court.
Bauman, Zygmunt. 2007. *Forbrukersamfunn in Schelderup, Gerhard Emil & Knudsen, Morten William: Forbrukersosiologi: makt, tegn og mening i forbrukersamfunnet*. Oslo: Cappelen akademisk forlag.
Beck, Ulrich. 1992. *Risk society: Towards a new modernity* (trans: Ritter, Mark). London: Sage.
Daniels, Michael. 2005. Camus' Meursault, and Sartrian irresponsibility. In *Analecta Husserliana*, ed. A.-T. Tymieniecka, Vol. 85. Dordrecht: Kluwer/Springer.
Habermas, Jürgen. 1996. *Between facts and norms*. Cambridge, MA: MIT Press.
Honneth, Axel. 2004. Organized self-realization – some paradoxes of individualization. *European Journal of Social Theory* 7(4):463–478. (London, Thousand Oaks, CA and New Delhi: Sage).
Jameson, Fredric. 1994. *The seeds of time*. New York: Columbia University Press.
Jensen, Thor Øivind. 2007. Gerhard Emil Identitet og forbruk. In *Forbrukersosiologi: makt, tegn og mening i forbrukersamfunnet*, eds. Schelderup and Morten William Knudsen, 81–113. Oslo: Cappelen akademisk forlag.
Kerner, George C. 1990. *Three philosophical moralists: Mill, Kant and Sartre*. New York, NY: Oxford University Press.
Laclau, Ernesto and Chantal Mouffe. 2001. *Hegemony and socialist strategy. Towards a radical democratic politics*. London: Verso.
Madsen, Ole Jacob. 2006. In psychology we trust.... *Vardøger* 30:157–187.
Negri, Antonio and Michael Hardt. 2000. *Empire*. Cambridge, MA: Harvard University Press.
Oliver, Kelly. 2005. *The phenomenology of intersubjectivity*. In *Analecta Husserliana*, ed. A.-T. Tymieniecka, Vol. LIV, 193–207. Dordrecht: Kluwer.
Østerberg, Dag. 2009. *Hinsides liberalismen* (Klassekampen 02.06.2009).
Øyen, Simen. 2010. Multiple Demokratier in Ole Jacob Madsen and Simen Øyen. *Markedets Fremtid. Kapitalismen i Krise?* Oslo: Cappelen Akademiske Forlag.
Rendtorff, Jacob Dahl. 1993. *Frihed og etik i Jean-Paul Sartres filosofi*. Oslo: Nordisk sommeruniversitet.
Rose, Nikolas. 1999. *Governing the soul*. London: Free Association Books.
Sartre, Jean-Paul. 2001. *Being and nothingness – A phenomenological essay on ontology*, 2nd ed. (trans: Barnes, H.E.). New York, NY: Citadel Press.
Simmel, Georg. 2004. Schriften zur Soziologie: Eine Auswahl, in Honneth, Axel: Organized self-realization – some paradoxes of individualization. *European Journal of Social Theory* 7(4):463–478. (London, Thousand Oaks, CA and New Delhi: Sage)
Vestre, Bernt. 1993. Introduction in Jean-Paul Sartre. In *Væren og intet: i utvalg*, 11–89. Oslo: Pax.

ROBERT SWITZER

RE-TURNING TO THE REAL: PHENOMENOLOGICAL APPROPRIATIONS OF PLATO'S "IDEAS" AND THE ALLEGORY OF THE CAVE

ABSTRACT

This paper focuses on the way one thinker in the phenomenological tradition, Martin Heidegger, has appropriated, re-worked and radically re-cast what is arguably the great founding vision of Western metaphysics, the cave allegory at the heart of Plato's *Republic*. I take as my text the long, detailed and, inevitably, somewhat idiosyncratic interpretation of the cave story and Plato's "theory of forms" presented by Heidegger in the first half of his winter 1931–32 lecture series in Freiburg, entitled *The Essence of Truth* and published as volume 34 of his *Gesamtausgabe*. Through the lens of a close reading of Heidegger's analysis, I articulate two distinct themes of continuing concern within the broader phenomenological movement: the place of eidetic essences—paradigmatic structures of intentionality which, in Heidegger's term, "pre-model" the transcendent objects which come forward for us, 'as' what they are, in experience—and secondly, the place of the "quest" archetype, the dream of liberation from the shackles of the ordinary and, through philosophical questioning, the turn (or return) to the "essence" of human existence. My wider goal is to show that phenomenology has served not merely epistemological but also broadly "ethical" ends: its aims—in the work of both Heidegger and Husserl, I argue—have been not merely to justify, but to transform, both our claims to truth, and our very lives.

I suspect, at least among those of us who have dedicated ourselves to the study of philosophy, that there are very few who do not vividly recall the first time we encountered Plato's allegory of the cave. No text or tale is more central to philosophy than this story of a shackled prisoner, for whom a play of shadows is all of reality, finally liberated to the light of truth—and arguably no single conception has been more interpreted and debated than that of the ἰδέα (the abstract eternal essence or "Form") which, Plato tells us, the prisoner sees and recognizes as the truly real upon his ascent out of darkness and confusion. Plato's story from *The Republic*, as this suggests, contains two main elements which, though related, are fundamentally distinct: an account or *logos* of truth and the Forms, and as such we might say of the λογοξ itself, and a *logos* of life, or philosophy as the true life or liberation to life. My goal here is to examine both in terms of their influence upon, and appropriation by, one of the key figures in the phenomenological movement, Martin Heidegger. I will principally examine his detailed reading of the cave allegory and related issues

A.-T. Tymieniecka (ed.), Analecta Husserliana CX, 75–90.
DOI 10.1007/978-94-007-1691-9_7, © Springer Science+Business Media B.V. 2011

in the first half of his winter 1931–1932 lecture series in Freiburg, entitled *The Essence of Truth.*

The first broad themes I wish to consider concern truth and our encounter, in knowledge and perception, with the objects of our experience; this will be my primary focus here—detailing the rather unexpected connections that arise between Plato's theory of ideas and Heidegger's own thinking. These are important concerns not only to understanding Heidegger's ontology but also, given the consistent (though evolving) Platonism of Husserl's thought (not least in its efforts to establish transcendental noetic essences constitutive of the intentional correlates in lived experience), to highlighting how Heidegger remains more deeply in the "wake" of his teacher's lasting influence than many would recognize.

Along with, though as much as possible distinct from, the discussion of Plato's "ideas," I want to examine an even more perennial theme: the transformation and realization of human existence in what Plato spoke of as "the ascent from the cave" or, let us say, the turning of the soul. Such themes may seem, at first glance, less native to phenomenology "as such;" in my opinion, however, concerns with the possibility of a transformation to a fully realized life, resting on but going beyond the Socratic ideal of the "examined life" as the only life "worth living," pervade Husserl's thought—as does the conviction that phenomenology is the last and best hope of achieving it. As a founding part of phenomenology, these concerns have also remained close to most of the subsequent thinkers in the movement. Clearly, I cannot establish this here in detail; two quick citations from Husserl's work will have to suffice.

Let me turn first to the *Cartesian Meditations.* Here, Husserl holds that phenomenology, while excluding "every naïve metaphysics," does not exclude "metaphysics as such;" rather, all the traditional philosophical questions remain, including those concerning "the possibility of a 'genuine' human life," but freed from the old errors and grounded instead on "an all-embracing self-investigation," understood not in terms of an isolated Cartesian ego, but as universal and "intermonadic."[1] Earlier, in *Erste Philosophie*, Husserl held that the philosopher "necessarily requires an individual resolve which, originally and as such, makes him a philosopher, an original self-causation, as it were, which is an original act of self-creation." For Husserl, this resolve is of course precisely to effect the phenomenological reduction, as "radical world-denial"—which is for him the necessary means to "viewing an ultimate and true reality, and, therewith, for living an ultimately true life." Such is simply not possible in everyday human life, lived in the "natural attitude" and in "kinship with the world:" that is, "a life carried out as an entirely primordial and thoroughly necessary surrender to the world and as a being lost in the world." Instead one needs the wholly "unnatural attitude" of a life "of radical and pure self-reflection upon the pure 'I am,' upon the pure life of the ego and upon the ways in which something gives itself within this life as being in some sense objective, and how it achieves just this sense and this status as something objective solely through the inner and own-most achievement of this life itself."[2]

This reference to "something ... objective"—a measure to life that emerges within life but is in some sense beyond it—returns us of course to the Platonic meditations on the Forms. My claim here, however, is this: For Husserl, and we shall see, for Heidegger as well, the aims of phenomenology are not merely to justify, but to transform, both our claims to truth, and our very lives. But let me now turn more directly to the Platonic "ideas."

There is a fire in Plato's cave, and one can well imagine that the first emergence of "virtual worlds," the imaginary realms that now so dominate our leisure hours, was in the dream-like state induced by story-tellers as our earliest ancestors crowded around this dancing, artificial light, as the dark of night closed around them. But it was Plato who first vividly brought home to us the notion that the everyday world around us, plain as day, can itself be seen as "mere show," a tissue of illusion. The appeal of this is as much mystical as philosophical, but Plato's own concerns seem to be centered on how it is that we are able to perceive and give an account of things in the world in terms of stable formulations, and, in general, aspire to knowledge—despite the continually shifting nature of the experienced world.

Heidegger's own abiding philosophical question, of course, is the question of being: the actuality of the actual. We encounter the actuality of things every day— most simply when we are not stopping to reflect, but busy with our work, as in his well known example, in *Being and Time*, of the carpenter at his work-bench.[3] The hammer in our hand hardly seems "shadowy," though when our attention is drawn to it—when it breaks, for example, or a philosopher like Plato interrupts us with his questions—it can suddenly seem uncanny, questionable. For Plato, of course, that the hammer breaks, gets thrown away, and that the wood and iron then slowly decompose in the land-fill, are arguments against its true hammer-being; what we must catch sight of instead is the hammer "as such," the Form of hammer, the ἰδέα. But, Heidegger asks, "*what kind of* seeing is this, in which ideas come into view?" Clearly it cannot be with "our bodily eyes, for with the latter we see precisely the beings that Plato calls shadows," and the Forms or "ideas" are, for Plato, emphatically "*other* than these beings."[4]

Heidegger's response to the suggestion that the ideas have nothing to do with bodily seeing is emphatic. "Not so fast," he cautions his students, "Do we see beings with our bodily eyes? Doubtless we do!" With this, he launches into his own, distinctive kind of phenomenological account of seeing. This account rejects the "traditional" approach of locating the ideas in a "world beyond," and places them rather in "the between," as we might call it: the zone of contact and differentiation between perceiver and perceived, knower and known, subject and object (though recognizing, of course, the inadequacy of these "metaphysical" terms and oppositions already for Heidegger in 1931). More broadly, Heidegger offers the reader a hermeneutic interpretation of the ideas as that which allows "what is" to come forward in our experience *as* "what it is."

Heidegger begins his account with the suggestion that to see, or to hear, is to "hold ourselves in a perception," to "register something that is presented to us." We hear tones, we see colors and, coextensively, shapes; we also see "glowing, sparkling, glittering"—the brightness of illumination. (36) But in fact, we are rarely aware of such elemental perceptions; instead, we hear the phone, we see the book. Seeing, especially—which Heidegger later acknowledges is the privileged access to the real for the Greeks (74)—gives us the look, the form of the things before us. Or does it? Is this—the form—something we can sense, can there be a "sensation of form"? For Heidegger, assuredly not: "What is sensed with our eyes is not the book," but merely, for example, the reddish brown of its cover; indeed, he goes on to clarify, "as such," as sensory organs, the eyes do not even give us colors. Instead, the sense of sight "sees" colors "with the eyes," which is to say, with their assistance, "by means of them"—but never the book "as such." Thus, "when we say that 'we see the book', we use 'see' in a meaning which goes beyond perceiving the object by means of the sense of sight with the help of our eyes."

We would never see anything like a book were we not able to see in another *more primordial* sense. To this latter kind of 'seeing' there belongs an *understanding* [*Verstehen*] of what it *is* that one encounters: book, door, house, tree. We *recognize* the thing as a book. This recognition registers the look that is given to us: of the book, table, door. We see *what* the thing is from the way it *looks*: we *see* its what-being. 'Seeing' is now a *perceiving*; of something, to be sure, namely this as a book, but no longer through our eyes and sense of sight [but rather] in the sense that we comport ourselves to what is presented to us. (37–38)

Once again, we are generally quite unaware of this—at least, until someone draws our attention to it, which is what Plato did with his discovery of the "ideas;" Plato brought us to begin to recognize what happens, every day, when we without hesitation see or take hold of something as the thing that it is. Heidegger writes,

'Iδέα is therefore the *look* [*Anblick*] of something *as* something. It is through these looks that individual things *present* themselves as this and that, as *being-present*. Presence [*Anwesenheit*] for the Greeks is παρουσια, shortened as ούσια, and means *being*. That something *is* means that it is present [*es ist anwesend*], or better: that it *presences* [*west an*] in the present [*Gegenwart*]. The look, ιδέα, thus gives *what* something presences *as*, i.e. what a thing *is*, its *being*. (38)

The "seeing" of the idea, which is to say, for Heidegger, the "understanding" of the "what-being and how-being" of a thing, is what "first allows beings to be recognized as the beings they are." Hence, "we never see beings with our bodily eyes unless we are also seeing 'ideas.' " (38–39)

This may well give us pause. On the one hand, we are presented with what amounts to a basic phenomenological insight: that eidetic structures of some kind are conditions of the possibility of experience. Specifically, for Heidegger here, the ideas somehow enable the presencing or standing-out of things as "what they are," hence the being of beings. Thus, to repeat his last point, to see the book is also to see the "idea" of the book, as that "in terms of which" let us say, we perceive it. And so it is that Heidegger can also affirm that Plato's discovery of the ideas was not some "far flung speculation" but "relates to what everyone sees and grasps in comportment to being." (38) But relates how? The evident difficulty is that the prisoners in the cave—which is to say, presumably, all of us in our average everydayness—see the book but *not* the idea (or rather, in Plato's terms, seeing the book as we do is seeing a dim shadow *of* the idea—a copy of a copy, at each stage further removed, "flattened," and "dimmed down" from the original). As Heidegger goes on immediately to say, "the prisoners in the cave see only shadow-beings and think that these are all *there are*; they know nothing of being, of the understanding of being." (39)

In addressing this apparent contradiction, which has us "seeing" the ideas but also knowing nothing of them, we will do well to recall that Heidegger's focus throughout his reading of the cave allegory is the place in it of truth as ἀλήθεια (*aletheia*), which he translates as *Unverborgenheit*, unhiddenness. Specifically, what he finds in Plato, and even within the cave allegory passages as they unfold, is a turning away from the "originary" Greek sense of truth as ἀλήθεια to truth as correctness of assertions, ὀρθότης which, as *adequatio* and "correspondence," has dominated Western metaphysical thinking, eclipsing the former despite being a derivative mode of truth "grounded in the particular manner of orientation and proximity to beings, i.e. in the way in which beings are in each case unhidden." (26) From the outset, Heidegger has stressed that there is already truth as ἀλήθεια, unhiddenness, in the shadowy realm of the cave—but the prisoners are blind to it. That is—and this is very much a theme at the core of Heidegger's thinking from first to last—they experience beings but not being, and lack explicit understanding of the ontological difference. Although things present themselves in the cave only as "shadows," they nonetheless stand forth in the light; but the standing-forth itself and as such— in truth, which is to say, in ἀλήθεια—remains occluded. The prisoners see what is present but not its presence, its "unhiddenness." One could say: there is truth here, there is being—but unrecognized. The light, without which there could be no shadows, has not itself been brought to light. So it is that Heidegger follows up his assertion that the cave-dwellers "know nothing of being," with the words, "Therefore they must remove themselves from the shadow-beings" and "make an ascent, taking leave from the cave and everything in the lower region [. . .] for the light and brightness of day, for the 'ideas'." (39)

The reference to "ascent" here, central as it is to the allegory, may also give us pause. In the Platonic context, such talk makes sense: what Plato articulates, not just here but in congruent allegories in the *Symposium* and the *Phaedrus*, is an actual migration of the soul: his tales tell of a movement of the seer, an ascent from the unsteady vision of the ever-fading instance that somehow participates in the eternal Form, to the Form itself, which—problematically, of course, as Aristotle first instructed us—exists at a distance from its particulars, independently, in itself, in some kind of "other place" (e.g., the hyper-uranian "heavens"). But, as we have already seen, Heidegger's interpretation of the "ideas" is far more phenomenological than metaphysical (or mystical). Troubled how something so much a part of everydayness could be still be "won" in the liberation from the cave, we focused on the inherent *elusiveness* of the ideas; structuring the visible and bringing them to vision, they remain themselves *in*visible, like lenses we do not see but see *through*. As Heidegger puts it, we go through our daily lives without once "suspecting" that "in order to see this book, door, and so forth, we must already understand what 'book' and 'door' mean," which, he continues, is "nothing else but the seeing of the look, the ἰδέα." (30) But if, as he goes on to say, this is seeing "the *being* of beings [*das Sein des Seienden*]," not only are we faced with the task of recovering that which withdraws, remains hidden or has, as the later Heidegger often puts it, fallen into oblivion—it is also clear that the place of our doing so can be none other than here, in our confrontation or encounter with the beings themselves. For Heidegger, in short—in contradistinction from Plato—talk of ascent can be only talk, at best a "metaphor;" there is no "higher realm" to ascend to, but "merely" the task of thinking: letting oneself enter more fully into the "draft" or "current" of being.[5]

In Husserlian terms, what is called for is not a geographic displacement but a shift in regard, a refocusing of attention; the increasingly central place of the reduction in Husserl's phenomenology reflects a growing awareness of the need to step back from the worldly entities that interest and consume us, to better grasp the intentional structures constitutive of experience. For Heidegger, however, there is more here than a mere shift of attitude or refocusing of regard; while not literally an ascent, it is something as profoundly shaking and transformative: a "liberation," as Heidegger calls it, or a turning, as from the shadows into the light. This issue, as I indicated, we shall return to; for now, let me follow Heidegger's own analysis of the light, and its relationship to the work of the ideas in bringing forward the things of our concern, in their being.

Heidegger identifies a number of related terms here which, precisely in their inter-relatedness to each other and to the making-visible of the ideas, call out for consideration: brightness, transparency, and light. Although we can see the source of light (for example, the fire, the sun), for the most part we do not see light itself but *by* the light; hence, as Heidegger writes, it is "nothing which can be grasped hold of; it is something intangible, almost like nothingness and the void." (40) This seems fitting; like the nothing, like the "power of the negative," we might suggest, light differentiates, bringing out boundaries and outlining edges, bringing things forward as "standing out" against the ground. Brightness, specifically, Heidegger says, is a word borrowed from the realm of sound: it means "penetrating;" and in the light, brightness is a letting-through that first makes sight possible. "Brightness is visibility, the opening and spreading out of the open," the "originally transparent" that stands, like the ideas themselves, in the between: it lets through the thing "to be viewed" as visible, and also lets the view through *to* the thing. (41)

This bi-directional letting-through of the visible is of course nothing other than a letting-through of the being of beings; Heidegger calls this "precisely the basic accomplishment of the idea." (42)

What is seen in and as the idea is, outside the allegory, the *being* (the what-being and how-being) of beings. ʼΙδέα is what is sighted in advance, what gets perceived in advance and lets beings through as the *interpretation* of 'being'. The idea allows us to see a being as what it is, lets the being *come* to us [. . .]. Only where being, the what-being of things, is understood, is there a letting-through of beings. Being, the idea, is what lets-through: *the light*. What the idea accomplishes is given in the fundamental nature of light. (42)

We said earlier that the idea is a kind of lens; here, this can help elucidate Heidegger's claim that light, the letting-through, is itself let-through by the idea. We see a thing "as a book" only when "we understand its sense of being in the light of its what-being," namely the "idea" we have, beforehand, of book as such.

If there were no light at all in the cave, the prisoners would not even see shadows. But they do not *know* anything about the light which is already in their sight, just as little as someone who sees a book knows that he already sees something more than, and different from, what he can sense with his eyes, i.e. that he must already understand what 'book' as such means. (42)

For Heidegger, understanding [*Verstehen*] is a standing-before something that gives an overview; we "have its measure" (2), we "see its blueprint." (45). We now see that understanding is the fore-going opening that lets-through "what is;" he calls it the "pre-modeling projection of being." (45)

To understand being means to project in advance the essential lawfulness and the essential construction of beings. Becoming free for beings, seeing-in-the-light, means to enact the projection of being [*Seinsentwurf*], so that a look (picture) of beings is projected and held up in advance, so that in viewing this look one can relate to beings as such.(45)

How is light related to freedom? Characteristically, Heidegger invokes the clearing in the forest [*Waldlichtung*]; free from trees, from "encroachments," we might say, the clearing "gives free access for going through and looking through." (44) While perhaps helpful, one feels a fuller elucidation is needed; light, as freedom, needs to be brought more fully to light.

Whenever we take a step back from the immediate, whenever we at last recognize an assumption of our own that we did not know we had, but that we now see holds us back, whenever we shift to a new perspective full of fresh possibilities and pathways of advancement, we say "I've seen the light." The light is the medium of truth in which things come forward as what, truly, they are.

The light, then, broadens our awareness; just so, it can "dawn on us" that we are prisoners. We see by the transparent letting-through of the light, and that which was restricted, held fast within narrow confines (namely, our vision of things in the light), begins to open, to brighten. Thus it is, I would suggest, that we can best understand what is at first a somewhat bewildering claim: it is the light itself, Heidegger tells us, " 'seeing the light,' that gives freedom." (43) This "becoming free for the light" is "to understand being and essence," and hence "to experience beings as such." Therefore, Heidegger writes, "the essence of freedom" is "the *illuminating view*" which lets beings freely be (what they are). Only "from and in freedom" do "beings become more beingful, because being this or that." As Heidegger continues,

Becoming free means understanding being as such, which understanding first of all lets beings *as* beings *be*. Whether beings become more beingful or less beingful is therefore up to the freedom of man. Freedom is measured according to the primordiality, breadth, and decisiveness of the binding, i.e. this *individual* grasping himself as *being-there* [*Da-sein*], set back into the isolation and thrownness of his historical past and future. The more primordial the binding, the greater proximity to beings. (44–45)

Here again we may well pause. What is the source of this light, and so the ground of the "measure" of the ideas by which we see? And, secondly, what is the source, the nature, and above all the aim or purpose ($\tau\varepsilon\lambda o\varsigma$) of the freedom that Heidegger associates with the light? Heidegger himself is well aware of these issues, invoking the familiar assertion of Protagoras as he asks, "What is man, such that he could become the measure of everything? Can the essence of truth be given over to man?" (54) We shall see Heidegger's own response shortly, but we should first note that locating the light of truth as *lumen naturale* in man has a long history in metaphysical thought, perhaps best expressed in Descartes' rationalism, and best lampooned in Nietzsche's image of the "madman" who, using his feeble "lantern" in the bright light of morning, fails to find God and so announces, "God is dead,"[6]

For Platonism, the madman's efforts to locate and illuminate God by the light of human reason would be tantamount to lighting up the sun with a flashlight; it is of course the good, and nothing else, that provides not only the light of intelligibility and grounds the being of all that is, but is also the $\tau\varepsilon\lambda o\varsigma$, the guiding principle and measure by which, and towards which, we navigate as individuals and as $\pi\acute{o}\lambda\iota\varsigma$ (community)—in short, that "binds" human freedom. Heidegger, in contrast, sees freedom as a "binding of oneself for oneself, such that one remains always bound in advance." (43) And to this, of course, our question will be: bound, yes, but to what?

Heidegger's response is vital to our examination precisely in its doubleness: He writes that to be "authentically free" means, "I can acquire power by binding myself to what lets-through;" hence, such binding "is not loss of power, but a taking into one's possession." (44)

That is, I am bound to the thing—presumably, to the being of the beings—but only in simultaneously binding them to me, taking them up in an act of appropriation that makes them mine, takes them as "my own;" and in doing so there is apparently no "giving way" to something higher, but a self-assertion, an enhancement of my own-most power or—let us venture, given the dominance of Nietzsche's thought over Heidegger throughout the 1930s—will to power. The freedom described by Heidegger is not freedom *from* the shackles of unexamined assumptions or narrow thinking, or *from* the tyranny of propaganda or received opinion (we should note that at no point does Heidegger discuss those who shape the shared reality of the cave dwellers by manipulating the puppets and statues that cast the shadows by the light of the fire in Plato's allegory). Nor is it freedom as a "letting-shine" of an extra-human truth. Rather, at this point in Heidegger's thinking, it appears to be the freedom *to* impose our will, to lay out in advance, to bring beings within what he would later come to call the standing reserve, the instrumental matrix in which things are brought to a stand and "de-realized" precisely as endlessly transmutable quanta of power or energy, at our disposal.

Before we can judge the appropriateness of this criticism, we should note that Heidegger's later view of technology is at least hinted at in the account of modern science to which he now turns, as one of three examples of how "such freedom" as we have been discussing, that is, the appropriative "pre-modeling projection of being," actually brings us into closeness to (or distance from) beings. Not surprisingly, Heidegger looks very critically at the rise of modern science—but not because it involved a "projection" which "delineated in advance what was henceforth to be understood as nature and natural process," but because of the reductive nature of that projection, which limited nature to "a spatio-temporally determined totality of movement of masspoints." Thus, though beginning as a bringing-forward into closeness of beings for us, "the projection has forfeited its original essential character of liberation," such that beings are no longer made "more beingful," but less. (45) Nonetheless, he concludes, "this penetration into nature happened on the basis of, and along the path of, a paradigmatic projection of the being of these beings, the beings of nature." (46)

The other two examples Heidegger examines are history and art; of the latter, especially poetry, he affirms that it can happen that "the artist possesses essential insight for the possible, for bringing out the inner possibilities of beings, thus for making man see what it really is with which he so blindly busies himself." He then adds, "What is essential in the discovery of reality happened and happens not through science, but through primordial philosophy, as well as through great poetry and its projections." (47)

At this point, in Section Nine of his lectures, Heidegger returns to his guiding concern, which is laying out as fully as possible the nature of truth as ἀλήθεια. As unhiddenness, truth belongs to beings, not to our assertions or statements; it is the coming-forward of beings as what they are, in the light and for a seeing, according to a projective guiding fore-having or sketching in advance in terms of the paradigmatic "ideas." That these templates, let us say, or as Heidegger calls them, "blueprints"—that is, the ideas—should themselves come to awareness or come-forward in unhiddenness, is precisely the accomplishment symbolized in the story of the "ascent" out of the cave. Since "the unhiddenness of beings *originates* in them," Heidegger tells us, the ideas, once recognized in what is presumably a new level of seeing, become "the most beingful beings, the primordially unhidden." (48, cf. 51) In this, Heidegger appears to be preserving the "degrees of reality" doctrine associated with Plato's "theory of Forms;" the ideas (Forms) are not only more true than their instances, the particulars, they are also more real (since they are, for Plato, perfectly and fully what they are). But in fact, this endorsement of Plato is merely apparent; although more "beingful" than the beings salient in everyday experience, for Heidegger the ideas do not have, contra Plato, any self-subsistent (even less, "eternal") independent existence.

As we have seen, the ideas on Heidegger's account "are" only as "sighted" in and by the "pre-modeling perceiving" of things by human beings, within the "coming to light" of truth as ἀλήθεια (unhiddenness). They thus have no existence "in themselves;" how, after all, could one conceive of a "look" (ἰδέα) that is not seen? Heidegger writes:

What might ideas be 'in themselves'? Idea is what is sighted. What is sighted is so only in seeing and for seeing. An unsighted sighted is like a round square or wooden iron. 'Ideas': we must at last be serious with this Platonic term for being. 'Being sighted' is not something else in addition, an additional predicate, something which occasionally happens to the ideas. Instead, it is what characterizes them as such. (51)

Heidegger recognizes his divergence from Platonism at this point, but characteristically insists it is Plato who held back and could go no further—"with the consequence that the whole problem of ideas was forced along a false track." (51–52)

For Heidegger, the true path involves a return to the problem we touched on earlier: bringing the Forms down to earth, re-situating them in the "between," as I have put it, as formative of the human encounter with "what is"—even if this runs the risk of relativizing them, of reducing them to the "merely subjective." In Heidegger's words, "The problem of ideas can only be posed anew by grasping it from the primordial unity of what is perceived on the one hand, and what does the perceiving on the other hand." The ideas are the "look;" in the light, they let the being "be seen." But this, Heidegger tells us, is "a looking in the sense of per-ceiving [*Er-blickens*]," which is to say, not a passive taking-in but "projection," an active, primary "*form-ing*" of "what is looked at *through* the looking and *in* the looking, i.e. forming in advance, modeling." (52)

At the origin of the unhiddenness of beings, i.e. at being's letting-through of beings, the perceiving is no less involved than what is per-ceived in perceiving — the ideas. *Together* these constitute unhiddenness, meaning they are nothing 'in themselves', they are never *objects*. The ideas, as what is sighted, *are* (if we can speak in this way at all) only in this perceiving seeing; they have an essential connection with perceiving. (52)

This does not, however, mean the Forms (ideas) are "merely in our heads;" rather, Heidegger asserts, they are "neither objectively present nor subjectively produced." That is, he continues, "Both, what is sighted as such, and the perceiving, *together* belong to the origination of unhiddenness, that is, to the *occurrence* of truth." (53)

Before we come to focus on this question in more detail, let me briefly sketch the final elements of Heidegger's account here: The perceiving of the idea, which we have characterized as projective, binding pre-modeling, Heidegger now names as "de-concealing [*Ent-bergen*];" (53) it is this which brings together viewing, freedom and light in their unity. It is also what properly defines the "liberation" of turning from the shadows into "the light of day" beyond the cave.

To be deconcealing is the innermost accomplishment of liberation. It is *care* [*Sorge*] itself: becoming-free as binding oneself to the ideas, as letting *being* give the lead. Therefore becoming-free, this perceiving of the ideas, this understanding-in-advance of being and the essence of things, has the *character of deconcealing* [*ist entbergsam*].

Deconcealing, in short, "belongs to the inner drive of this seeing," this "looking-into-the-light." (53) It can even be said of deconcealment, Heidegger continues, that it "first creates the perceivable in its innermost connection," for only in and through it does the "unhiddenness of beings" come to pass.[7]

Just as there are no ideas without man, so with truth: "the essence of truth qua $\dot{\alpha}\lambda\dot{\eta}\theta\varepsilon\iota\alpha$ (unhiddenness)," Heidegger writes, "is deconcealment, therefore located in man himself." Would not such a reduction of truth to the "*merely* human" serve simply to annihilate it? Do we descend here into nihilism? Heidegger's response is that the charge of relativism is too easy; it rests on countless unexamined presuppositions, most notably that the essence of "the human" is a given and well understood by everyone. Heidegger then asks, "From where are we to take the concept of man, and how are we to justify ourselves against the objection of an attempted human-ization of the essence of truth?" [54] It is the cave allegory itself that provides the answer, Heidegger holds, for it gives "precisely the history in which man comes to himself as a being in the midst of beings," a history in which the "decisive" oc-currence is nothing other than "our" projective de-concealment; it is the essence of truth, as unhiddenness, that first discloses the essence of human existence. The allegory of the cave, as we shall see, shows us an individual who, in the fundamen-tal occurrence of his Dasein, is "*set out into* the truth" [*in die Wahrheit ver-setzt*]. Heidegger continues:

Truth is neither somewhere *over* man (as validity in itself), nor is it in man as a psychical subject, but man is '*in*' *the truth*. Truth is something greater than man. The latter is in the truth only if, and only in so far as, he masters his nature, holds himself within the unhiddenness of beings, and comports himself to this unhiddenness. (55)

We are perhaps left with the feeling that Heidegger has sidestepped the real question of the "relativism" of truth. On the one hand, he has just said that truth is greater than man; on the other, one of his concluding points in this part of the lecture course is that "that truth itself is not ultimate, but stands under an empowerment." (82) Earlier, we recall, in words that might seem to anticipate the views of the so-called "later Heidegger," he suggested that "binding oneself to the ideas" is "letting being give the lead" (53)—but the question, of course, is: lead to where? This question remains resolutely unanswered. Instead, Heidegger tells us that the real question of the essence of the human, echoing Nietzsche, is not identifying what we are but "becoming what we can be;" for this, we must "come to a decision" on ourselves, on "the powers that carry and define" us. Man can only be understood, in other words, "as a being bound to his own possibilities, bound in a way that itself frees the space within which he pursues his own being in this or that manner." (55) As this makes very clear, we are offered no hint of a $\tau\varepsilon\lambda o\varsigma$, no clear sense of what the good for man, or the realization of our Dasein, would be.

This is not to say that Heidegger ignores "the idea of the good" in his analysis. It is there, but its role is ontological, and explicitly *not* normative. That is, Heidegger cautions us, we must free ourselves from the outset from "any kind of sentimental conception of this idea," for "it is not at all a matter of ethics or morality." Rather, this "highest idea," which lies out "beyond" all ideas, is the enabling ground of both seeing and being-seen, both the capacity of vision and the visibility of the visible, at once in themselves and in their connectedness; the good ($\dot{\alpha}\gamma\alpha\theta\acute{o}\nu$) is the light which makes both possible and is also their common link or bond—in Plato's language, the "yoke" ($\zeta\upsilon\gamma\acute{o}\nu$) under which both are harnessed. But the light of the good does not merely facilitate, let us say, knowing on the one side, understood in terms of sight, and the known in its truth, as unhiddenness or becoming-visible, on the other; rather, for Heidegger, it is for each and in their unity the enabling power ($\delta\acute{\upsilon}\nu\alpha\mu\iota\varsigma$).

The highest idea, although itself barely visible, is what makes possible *both* being and unhiddenness, i.e. it is what *empowers* being *and* unhiddenness as what they are. The highest idea, therefore, is this empowering, the empowering for *being* which as such *gives* itself simultaneously with the empowerment of *unhiddenness* as *occurrence*. In this way it is an intimation of $\alpha\acute{\iota}\tau\acute{\iota}\alpha$ (of 'power', 'mastery'). (72)

That is, the good, power, gives and sustains not just the visibility (intelligibility) but the existence of what is, beings, and also of ideas (which bring into the light of unhiddenness beings in their being). "In so far as being-as-idea means empowerment for being, the making manifest of beings," Heidegger writes, it follows that the idea of the good surpasses both "being as such and truth." (79) Heidegger supports his interpretation of the good by citing Plato's *Sophist* where, at 247 d-e, the Eleatic Stranger suggests that, in Heidegger's words, "the essence of being is found in $\delta\acute{\upsilon}\nu\alpha\mu\iota\varsigma$, i.e. in empowerment and nothing else." (80)

Against the horizon of Greek thought and certainly, for us, in the light of Levinas' later criticisms of Heideggerian ontology and his far-ranging meditations on Plato's "the Good" and ethics as "first philosophy," this may well seem like a disturbing "hollowing out" of the good into mere "usefulness" or naked power. We note, for example, that on Heidegger's reading, Plato's holds that "the power of the good is to be valued even more highly than the ideas." For Plato, Heidegger continues,

When we ask about the essence of being and unhiddenness, our questioning goes out beyond these, so that we encounter something with the character of empowerment and nothing else. Empowerment is the limit of philosophy (i.e. of metaphysics). Plato calls that which empowers ἀγαθόν. We translate: the good. The proper and original meaning of ἀγαθόν refers to what is good (suitable) for something, what can be put to use. 'Good!' means: it is done! it is decided! It does not have any kind of moral meaning: ethics has corrupted the fundamental meaning of this word. What the Greeks understand by 'good' is what we mean when we say that we buy a pair of good skis, i.e. boards which are sound and durable. The good is the sound, the enduring, as distinct from the harmless meaning suitable for aunties: a good man, i.e. respectable, but without insight and power. (77)

Despite his scorn for the traditional views, Heidegger does grant that "*what* this empowerment is and *how* it occurs has not been answered to the present day." We no longer even ask the original Platonic question—yet, "in the meantime," Heidegger continues, the idea of the "highest good" has "almost become a triviality." He then concludes, in ominous-sounding riddles:

For whoever asks in a philosophical manner, Plato says more than enough. For someone who wants only to establish what the good is in its common usage he says far too little, even nothing at all. If one takes it merely in this latter way, nothing can be done with it. This clarification of the idea of the good *says* anything only for a philosophical questioning. (80)

What this reference to "philosophical questioning" returns us to is the question of transformation, of a "turning of the soul," which is of course at the very heart of Plato's allegory of the cave. Clearly there is a special conception of "the philosophical" being developed, both by Plato and by Heidegger: not simply philosophy as insight into knowledge and reality, as embodied for example in a "theory of Forms," but philosophy as a way of life, a mode of human existence—and not just any mode, but a most "essential" and "authentic" possibility; let us say, not a mode merely but a *model* of the "realized" human. That is, philosophy has a doubleness here: it is both the process or means of human transformation, and that to which we aspire, the turning itself and that to which we turn.

Having traced in detail the careful articulation of Heidegger's reading of Plato's "theory of ideas," I want briefly to sketch Heidegger's peculiar vision of what I am calling "the turning" in these lectures on the *Republic*. We gain access to this issue through the question just touched on: philosophical questioning itself, as Heidegger pictures it.

Heidegger tells us that "understanding the cave allegory means grasping the history of human essence, which means grasping oneself in one's own-most history." To do this is to question philosophically—and it is precisely in this that the transformation is enacted. As he continues, in words strikingly evocative of Husserl's "reduction,"

This demands, when we begin to philosophize at any rate, putting out of action diverse concepts and non-concepts of man, irrespective of their obviousness or currency. At the same time it means understanding what the clarification of the essence of ἀλήθεια implies for knowledge of human essence. [. . .] Man must first place himself in question, must comport himself to himself as that being who is asked about, and who, in this asking, becomes uneasy. [. . . For] only by entering into the dangerous region of philosophy is it possible for man to realize his nature as transcending himself into the unhiddenness of beings. Man apart from philosophy is something else. (56)

It is precisely this questioning that leads to the liberation from the cave; for the liberated one is a philosopher, one whose own existence is in question. A striking feature of Heidegger's reading of Plato's allegory is his distinguishing a first, failed liberation from a second successful ascent out of the cave. The first attempt fails because the prisoners, though freed of their chains, still do not have what the prisoners as such all lack: an understanding of *difference*—between light and shadow, between appearance and reality (21); presumably, given what we have seen concerning the ideas as "the being of beings," what they lack is ultimately an understanding of the ontological difference itself.

To be sure, the difference between shadows and things announces itself, but the former prisoner does not enact this difference, cannot grasp it as such, cannot bring the distinguished things into relationship. But the difference occurs in the enactment of the differentiation. To bring the differentiation to enactment would be being-human [*Menschsein*], existing [*Existieren*]. (28)

Instead, the prisoners turn back towards the cave wall, fully unaware of the connection, or difference, between the statues and the shadows they cast. But ultimately the failure is not merely one of vision, which could be cured by forcing the prisoner violently the rest of the way from the cave; it is, Heidegger says, a failure of will. And hence what is needed is "a change in the inner man"—precisely "in his willing." This initial liberation or "turning" fails, in short, because the prisoner does not become "free for himself," that is, does not come to stand "in the ground of his essence." (28) This, it seems, requires the liberator; as Heidegger writes, "the liberator is the bearer of a differentiation." (66)

Overall, what has emerged from our discussions as the essence of human existence, is the questioning stance in the midst of "what is;" though he detests "propositions," Heidegger himself puts this into a propositional form as follows: "man is the being who exists in the perceiving of being." But to understand this statement requires something very different than propositional logic:

The truth of this statement (precisely because it says something philosophical) can only be philosophically (as I say) enkindled and appropriated, that is, only when the questioning that understands being in the questionability of beings in the whole takes its standpoint from a fundamental decision, from a fundamental stance towards being and towards its limit in nothingness. (57)

Throughout the long and shifting pathways of his thinking, Heidegger saw philosophy as a transformative undertaking; its task, in the language of his later work, is to take up our place, in the humility of questioning, within the withdrawing mystery of the "gift" of being. This was never pictured as something that would bear practical benefits, advance scientific knowledge or even found an ethics or system of values. Rather, as a return to ourselves, in the essence of our being, and also to that which calls to us—in the things themselves, in originary language and in the worlding of the world—it is intrinsically vital. In this period of his life, however, in the early 1930s, this picture, while still recognizable, has a distinctive—and perhaps, today, somewhat off-putting—tone or flavor. For me, this is signaled here by the words "decision" and "stance." It is also, to say the least, rather atypical for Heidegger to go on to say, stridently, "What this means is not a matter for further talking, but rather for doing." (57)

Heidegger builds on the notion of "stance" in announcing that philosophical enquiry requires, above all, a standpoint: hence,

The right choice of standpoint, the courage to a standpoint, the setting in action of a standpoint and the holding out within it, is the task; a task, admittedly, which can only be enacted in philosophical work, not prior to it and not subsequently. (57)

Evidently, this philosophical work must now, in the current "crisis" (of 1931–32) and faced with the increasing disregard for and "poisoning" of philosophy in the nation, include specifically political activity. Of course, one must keep in mind that his text is the *Republic*, in which Plato announces as his "third wave," his most "laughable" doctrine, the necessity of the "philosopher king." Nonetheless, it is chilling to hear Heidegger pronounce on this. It is not exactly, he sys, that "professors are to become chancellors of the state;" rather,

Philosophers are to become φύλακες, guardians. Control and organization of the state is to be undertaken by philosophers, who set standards and rules in accordance with their widest and deepest freely inquiring knowledge, thus determining the general course which society should follow. As philosophers they must be in a position to know clearly and rigorously what man is, and how things stand with respect to his being and ability-to-be. (73)

As we have seen, however, this kind of knowledge—at least, in the detailed sense that political action would seem to demand—remains highly elusive; there is no counterpart in Heidegger's account to the Forms most sought after in Plato's Socratic dialogues, such as justice, virtue and piety; and the good itself appears, as we have seen, in the guise of quanta of power rather than the guiding quality of goodness.

Heidegger makes it clear, however, that when "the liberator"—the philosopher, though Heidegger insists it is not himself, that he can only "prepare the way for the philosopher who will come" (62)—returns into the cave to help those still in chains, he does not reason with the prisoners. That is, the philosopher-liberator seeks to achieve his aims not by trying to "persuade the cave-dwellers by reference to norms, grounds and proofs," namely, with reference to the "aims and intentions of the cave"—this would merely make him "laughable"—but "by laying hold of them violently and dragging them away." (62) Nonetheless, he takes this drastic action in a spirit of political solidarity, at least with some: "Being free, being a liberator, is to act together in history with those to whom one belongs in one's nature," (92) presumably, also, with what Heidegger earlier spoke of as "strident courage that can also wait, that is not deterred by reversals." (32) The liberator, we read, is

someone who has become free in that he looks into the light, has the illuminating view, thus has a surer footing in the ground of human-historical Dasein. Only then does he gain power to the violence he must employ in liberation. This violence is no blind caprice, but is the dragging of the others out into that light which already fills and binds his own view. This violence is also not some kind of crudity, but is tact of the highest rigor, that rigor of the spirit to which he, the liberator, has already obligated himself. (59)

In doing this, the philosopher is even more heroic in that he faces (as Socrates well knew) the threat of death; this of course recalls Heidegger's account of the resoluteness of authentic being-unto-death in *Being and Time*[8]—but the real death for philosophy we are told is the leveling down and "poisoning" of discourse just mentioned. It is this sorry state of public babble, along with the distribution of "honors" to the unworthy, that is so often repeated in Heidegger's depiction of the cave that one cannot help but feel a pervading bitterness—contemporary, one senses, and quite personal—far in excess of what any retelling of Plato's allegory could justify.

I stress these concerns simply to bring into relief the suggestion that, while Heidegger's break from traditional metaphysical thinking certainly has had tremendous phenomenological impact and influence, in many ways reinvigorating the movement as a whole and influencing even Husserl,[9] it also seems to entail disturbing reminders of the limits of philosophy, particularly as his anti-foundational stance—a stance precisely over the abyss, one might say—was somehow made the basis, during the early 1930s at least, for political intervention. The precise context and setting of Plato's own attempt at such intervention, in Sicily, are for the most part long forgotten, though we know that he barely escaped with his life; in Heidegger's case, sadly, the stench of the historical stage on which he sought to play a role is still quite horrifically pungent.

Let me close on a note more pleasant, I hope, and more lastingly germane to Heidegger's efforts here: I have spoken, in my title and in these concluding comments, of a (re)turn to the real; I want, finally, to highlight the sense of these words that would be closest, I think, to Heidegger's own ultimate goal in these lectures. One advantage of speaking of "turning" rather than ascent is that one can turn away as easily as towards—and, for Heidegger, this is just what Plato did in his cave allegory, in ultimately occluding ἀλήθεια in favor of truth as correctness of assertions; hence the motivation for Heidegger's, as he himself says, often "violent" re-appropriation of it. Nonetheless, by his own account, Heidegger's efforts have been a failure: he has not achieved what he "strove for," namely, a "return into history [*Geschichte*], such that this becomes our occurrence [*Geschehen*], such that our own history is renewed." The reason, he says, is that we are no longer "touched" historically by the occurrence, in Plato, of truth as unhiddenness, ἀλήθεια, but remain at the level of "purely theoretical reflection." This is not, however, our failing—but Plato's:

What already happens in Plato is the waning of the fundamental experience, i.e. of a specific fundamental stance [*Grundstellung*] of man towards beings, and the weakening of the word ἀλήθεια in its basic meaning. This is only the beginning of that history through which Western man lost his ground as an existing being, in order to end up in contemporary groundlessness. (87)

In fact, however, this failure became the basis of efforts far more lasting and important than Heidegger's short-lived political debacles: a continuing effort, throughout his philosophical work, to reclaim and to re-turn precisely into the withdrawn enigma at the origin of Western philosophy, the always veiled-unveiling event of truth as ἀλήθεια.

Acknowledgements The author gratefully acknowledges the support of a faculty development grant from The American University in Cairo, which enabled participation in the World Congress of Phenomenology in Bergen.

The American University in Cairo, New Cairo 11835, Egypt
e-mail: switzer@aucegypt.edu

NOTES

[1] Edmund Husserl, *Cartesian Meditations*, transl. Dorian Cairns (The Hague: Martinus Nijhoff, 1960), pp. 156–57. (Emphasis removed.)

[2] Edmund Husserl, *Erste Philosophie (1923–24) Part II: Theorie der Phänomenologischen Reduktion*, Ed. Rudolf Boehm, *Husserliana*, vol. 8 (The Hague: Martinus Nijhoff, 1959), pp. 19, 166, 123 and 141. Quoted in Ludwig Landgrebe, *The Phenomenology of Edmund Husserl: Six Essays*, Ed. Donn Welton, Trans. R. O. Elveton et al., (Ithaca, NY: Cornell University Press, 1983), pp. 72–73.

[3] Martin Heidegger, *Sein und Zeit* (Tubingen: Max Niemeyer, 1953) sections 15 and 16 (pp. H 66 ff); translated into English as *Being and Time*, trans. J. Macquarrie and E. Robinson (New York: Harper and Row, 1962), pp. 95 ff.

[4] Martin Heidegger, *The Essence of Truth: On Plato's Cave Allegory and Theaetetus*, trans. Ted Sadler (London: Continuum, 2002), p. 36. Subsequent references to this text will be to this edition; page numbers will be enclosed in parentheses and inserted into the text.

[5] Cf. Martin Heidegger, *What is Called Thinking?* Trans. J. Glenn Gray (New York: Harper and Row, 1968), p. 17.

[6] Friedrich Nietzsche, *The Gay Science*, Trans. Walter Kaufmann (New York: Vintage, 1974), section 125, p. 181 f.

[7] Some sense of the profound ramifications of the identification Heidegger make in the just-quoted passage between deconcealing and care (*Sorge*) can be glimpsed if we recall that, in *Sein und Zeit*, it is as care that the "totality of Being-in-the-world as a structural whole" reveals itself. See Heidegger, Being and Time, p. 274 (H231) *et passim*. To consider this in the detail it deserves would far exceed the bounds of the present paper.

[8] Cf. Martin Heidegger, *Being and Time*, Division Two, especially sections 53 and 60.

[9] As Landgrebe, among many others, has attested to; see Ludwig Landgrebe, Op cit, p. 100.

ANDREAS BRENNER

LIVING LIFE AND MAKING LIFE

ABSTRACT

The question "What is life?" has long been a major discussion point in all cultures. Nowadays whilst both Synthetic Biology and the Computer Sciences are trying to create life the question on life is becoming even more important. In oder to answer this question the paper will present the biophilosophy of Humberto Maturana and Francesco Varela. The paper aims to display that this biophilosophy is very close to Husserlian phenomenology. It will be shown that a living system is autonomous and an creation by its own and dependent from its environment which is made by the living entity itself. Living entities cannot be understood without their own logos.

QUESTIONING LIFE

Human beings are beings who are able to scrutinize their own life. Scrutinizing their own life, human beings ask themselves questions such as *"Who I am?"* or *"What is the reason that I am?"* This is the kind of question which is fundamental for the creation of cultural constructions such as religion, philosophy, literature, the arts and music. Culture can be understood as the attempt to give answers to these fundamental questions. One of these fundamental questions is the question "What is life?"

From the ancient tradition we get the answer that life is something that is in motion. But not everything that is in motion is alive, only what is in self-motion is alive, Plato points out. But what makes the moving, move? The moving power cannot be a material one as material matter in general can only come to motion if it is moved by something else. Hence the living is not brought to motion by a material but by a non-material entity. And this entity is called "soul".[1]

Like his teacher Plato, Aristotle also supports the conception of the living's self-motion and appreciates the soul as the basis of the motion. But in contrast to Plato Aristotle acknowledges matter as being on the same level as the soul: Soul without matter is not alive just as material without soul is not alive. Only the interaction of soul and matter makes something alive. So we can summarize that the living is in a self-powered motion. To clarify the term motion we have to understand that motion is not only movement but it is every kind of change. In this way breath, nutritional support and even thinking, as Aristotle points out, are kinds of movement.[2]

If we agree with Aristotle that movement is the characteristic of the living we also agree that the end of the movement is the end of the living's life. This Platonic-Aristotleian concept can be seen as the basis of understanding life.

91

A.-T. Tymieniecka (ed.), Analecta Husserliana CX, 91–102.
DOI 10.1007/978-94-007-1691-9_8, © Springer Science+Business Media B.V. 2011

In the following I would like to present a modern advancement of this ancient theory. I am going to speak about the two Chilean Biophilosophers Humberto Maturana (*1928) and Francisco Varela (1946–2001). The starting point of their work is a hermeneutic turn as it was established by Husserlian Phenomenology. Comparable with Husserl, Maturana and Varela also noticed a lack of awareness in the common scientific way of understanding phenomena. This problem of understanding is discussed by Husserl using the term Lifeworld. Husserl criticized the predominance of the scientific world view as leading to the danger of reductionism and maintained the meaning of the Lifeworld and its acceptance as the precondition of understanding.[3] Similarly Maturana and Varela criticized the dominant approach of scientific research as reductionistic, especially the widely held opinion that it is possible to understand living entities by means of a description from outside. In contrast to this opinion both biophilosophers became more and more convinced that life is something which can only be understood from inside. This became the basis for criticizing objectivism in epistemology and the starting point for a new understanding of life. Objectivism in epistemology is seen by Maturana and Varela as an inadequate way of understanding phenomena which are not objectifiable and which only can be understood under their own laws. Those who make a clear distinction between living and non-living systems will not agree with the opinion that living systems can be fully described from outside. The difference is that which lies between the subjective and the objective position in epistemology. A prominent position of objectivism is Cartesian epistemology which is often seen as reductionistic and mechanistic. It is a kind of reductionism to reduce living entities to qualities which are exclusively typical for mechanical systems but not for living entities. Such reductionism takes place by means of a description from an outside perspective. To have this opinion one sees the deficiency between mechanical and living entities in the difference between the status of both and in this point even the anti-mechanists can follow Descartes and his differentiation between res cogitans and res extensa.[4] No matter what else he was referring to, Descartes pointed out that the living entity is more than matter. Furthermore, the anit-mechanists can agree with Descartes's view that it is completely impossible to reconstruct the spirit.[5] This all together suggests the presumption that living entities are of a level of complexity which forbids any simple explanation. This mechanism is the paradigm of a simple explanation that can be seen in the way of mechanistic explanations. These explanations are characterized by cause-and-effect-chains which are focussed on the parts of a system and not on the whole. However if we understand living entities as wholes we share the conviction that the whole is more than its parts and that does mean that the whole cannot be completely described and understood by its parts. This position is represented, among others, by the position of holism.[6] When criticizing mechanism and avoiding simple cause-and-effect-chains one has to explain what the whole makes a whole. If we are convinced that the whole is more than its parts no explanation can be accepted which is focussed on a phenomena's parts alone. And this is the starting point of the autopoiesis-theory. As made clear before, autopoiesis-theory is denying both mechanism and dualism and the simple thinking in cause-and-effect-chains. Instead of thinking in terms of causes and effects Maturana and Varela promote a thinking in relations. The paradigm of relation which is also important for the theory of holism notes connections between every part of a system and makes clear the changeability of the whole by a change of the different connections. This description characterizes a phenomena that is not static but in motion and the motion is seen as not completely predictable. The reason for this is not the phenomena's complexity alone, as this would be a quantitative question, but also the phenomena's quality. The phenomena we are speaking about are not simple machines but "living machines".[7] Living machines are different from man-made machines primarily not in view of the matter or their complexity but in view of their activity. Activity can be regarded as synonymous with life as William James declared.[8] In this way we can describe man-made machines as passive and describe only living machines as active. But what is the meaning of being active? While passivity is in general understood as a status caused by s.o./sth. else, activity is understood as status caused by the phenomena itself. This makes the difference. To say it in Maturana's and Varela's words: "An autopoietic machine is a machine organized (defined as a unity) as a network of processes of production (transformation and destruction) of components that produces the components which: (i) through their interactions and transformations continuously regenerate and realize the network of processes (relations) that produced them; and (ii) constitute it (the machine) as a concrete unity in the space in which they (the components) exist by specifying the topological domain of its realization as such a network."[9] To put it in a nutshell, you can understand a living machine as sth which "continuously generates and specifies its own organization through its operation as a system of production of its own components, and does this in an endless turnover of components under conditions of continuous perturbations and compensation of perturbations."[10] These descriptions are fundamental for further differentiations between man-made machines and living machines: The first are static the second are homeostatic systems, the first can be completely described, while the second cannot as there will be a remaining. If we ask why we cannot describe living entities completely and what the reason is for the fact that every description will keep sth in the dark we refer again to the idea of autopoiesis: Only autopoietic systems do have sth you could call a self. Therefore autopoietic systems are subjects and not objects as man-made machines. In contrast to an object, the subject and its being a self cannot be understood as sth finished but as sth in a permanent change. The subject's situation is both being the author of its own being and being the origin[11] of its own being. Obviously both descriptions are inadequate to understand an object which is made by someone or sth else, i.e. it is allopoietic. Furthermore an object lacks any kind of subjectivity that is the result of its allopoietic status. Autopoietic systems, living machines or shortly living entities are characterized by the contrary, as Pier Luigi Luisi points out: "The most general property of an autopoietic system is the capability to generating its own components via a network process that is internal to the boundary."[12] The phenomena of living which was described by Plato and Aristotle as being in selfmotion can be understood by Maturana/Varela as *autonomous, having individuality* and *being unities*.[13] Each of these characterizations have an ethical impact as only phenomena which do fulfil these criteria have an intrinsic value and only phenomena with

an intrinsic value can be member of the moral universe.[14] As the purpose of this
paper is not to discuss moral questions we shall ignore these questions and go on
to discuss the ontological questions which arise from the understanding of living
entities as autopoietic systems. In this way we have to scrutinize each of the given
characterizations. Let us start with *autonomy*. When Maturana/Varela speak about
autonomy it is evident that they are not refering to a philosophical understanding of
autonomy as it was argued by Kant. The distinction between the Kantian and the
Maturanian term is obvious in so far as Maturana/Varela neither speak about moral
challenges nor about rationality in an exclusively human manner. If this is the difference between the Kantian and the Maturanian use of the term of rationality there
is also common ground. Both refer to cognition. But there is a big gap between
mainstream philosophy and Maturana and Varela's position: While the majority of
philosophy ascribes cognitive capabilities only to human beings and describes processes of epistemological orientation of other living beings as only quasi cognitive,
Maturana and Varela declare cognition as a conditio sine qua non of being alive
in general. That every living entity is a cognitive entity is both the result and the
precondition of being autonomous. That is the case as only autonomous entities are
able to understand the world because understanding needs perception as well as
intentionality and intentionality refers to an autonomous self. Even this very first
criterion of autonomy exemplifies the way autopoietic processes work as well as explaining the other criteria of *having individuality* and *being unities*: The self which
is the precondition for individuality establishes the unit and is thinkable only in the
context of an individual unit. This simultaneity is the coherence of any autopoietic
process and at the same time makes a further distinction from allopoietic systems.
Their genesis takes place on the chronological table where the latter is the better as
it is closer to its final completion. This is the distinction from autopoietic systems
which are at each time completed as they have at each time their own standing on the
chronological table. For autopoietic systems no time is better than the other, which
can be seen as the proof that every time makes sense. Later we will discuss the concept of sense as a result of an epistemological process, while here we focus on the
meaning of sense for the self itself. Autopoietic systems are self-centered systems
whose self is not fixed but in motion. The never-ending change of the self is the
result of sense-making experiences which all together form the biography of the living entity's self. In this way biography has to be understood as the sediment of these
experiences which are inscribed in a living entity's own history and which make an
entity unique. Here we can see again these peculiar structure which is typical for
autopoietic processes: A phenomenon, for example, a self, generate epiphenomena
which for their part transform the phenomenon. To make it concrete: While the self
is making sense the sense will make the self.

   After these considerations we can summarize the main content of autopoiesis-
theory as follows. The autopoietic structure describes systems which are created by
itself. As the autopoietic structure is the main difference to allopoietic systems we
can regard the autopoietic capability as the decisive fact which makes sth alive. This
position has some important consequences:

1. We do not need any outer position of a creator to understand an living entity. We can understand the living, i.e. its becoming, its growth and its change by itself.
2. The living which is the reason for its own being has to be seen as a *self*.
3. In the autopoietic sense of explanation the self is created by itself.

What autopoiesis means can be exemplified by the phenomena of growth. Growth, and not reproduction as often mentioned is the very characteristic phenomenon of a living being. The process of growth can only be understood autopoieticly: if something growths it is changing its form. The change of the form is not the result of an outer influence but of an inner process. In the process of growth each organism is changing itself permanently. Because the organism's change is caused by the organism itself the organism keeps its identity in each phase of growth. Without the concept of autopoiesis we would have to identify an organism in its early phase and in its later phase as different entities. But this would not make sense. And in fact of this everyone of us would, regarding a child's picture of her- or himself, say "that's me" and not "That is the one I came from." This simple fact of transtemporal identity can be understood by the idea of autopoietic genesis.[15] What Maturana and Varela are going to explain with the autopoiesis-theory was centuries before put in a bright picture when Samuel Taylor Coleridge and William Wordsworth wrote that "The child is the father of the man" and it was as well illustrated by M. C. Escher's drawing "Drawing Hands".[16]

This idea also makes clear that living organisms cannot be fully described from the outer perspective as such a description would objectivate what is subjective and that is the self. In this way we can say that the autopoiesis-theory is on the one hand quite revolutionary, while on the other hand it is embedded in a tradition of thoughts which are represented by Plato and Augustine and which were burried by Aristotele and Aquinas as Stafford Beer points out.[17] This background makes clear that autopoiesis is not only a perspective for understanding life but as well a particular cultural concept. In this way autopoiesis has a lot of in common with cultural concepts such as Romanticism which arises in the 18th century or Holism in the 19th century. Both positions were established as critical responses to a formation of rationalism which reduces our worldview by simplification. In the same way autopoiesis-theory also widens our world view and brings to mind life's inner perspective.

The concept of living as a self was introduced to the discussion by different thinkers and Maturana and Varela were not the first ones. But what is the distinction between their concept and that of the others, let us say the Romanticism?

To put it in a nutshell: The others take the term "self" as a deus ex machina and do not deliver any explanation of how the self comes into being. But only when we understand the being of the self we also understand how the self works. Autopoiesis-theory explains the emergence of the self as a process which is stimulated by the phenomena which is called a self. It is important to see that the self is not any finished entity but a work in process. The processor which is generating the process is at the same time creating itself as the processor. This is the meaning of the self.

Even here the phenomena of growth helps to understand the meaning of the self as growth cannot be understood as a process which is generated from outside: As discussed above we need the concept of an inner process to make the idea of transtemporal identity plausible. In addition only a concept of the self enables us to understand growth as an activity as it is. Without the concept of the self we could only describe a different status of an entity what strictly speaking means that we speak about different entities. Only the assumption of the inner perspective, i.e. the self-perspective, combines the different phases in time to one history of one entity. That makes the fundamental difference clear between the growth of organic entities and the growth of machines in the process of production. Organic entities which are growing are changing *themselves*, in contrast to machines which are changed by someone else when they are "growing". This means that the first retains its identity even when it is changed, the second changes its identity in the process of production i.e. it is permanently becoming something else.

THE EMERGENCE OF LIFE

With the help of the autopoiesis-theory we are going to discuss the question of life's emergence. This question can be discussed in two ways, from the perspective of life as such and from the perspective of the individual life of an individual living being. Both perspectives lead to the same ground, namely to explain how the transformation from the non-living to the living status can be thought. As the theory of emergence has pointed out, only the conditions of life can be formulated but not the way these conditions take place. Consequently it looks to be impossible to explain the genesis from a non-living to a living status. This is a problem which cannot be solved as the distinction between the non-living and the living is a qualitive one and not a quanitive one what means that there is no smooth transition. That means that the becoming of life cannot be seen as a gradual process or as Maturana and Varela point it out: "Either a system is an autopoietic system or it is not."[18] The living and the non-living are from different ontological status.

To clarify more differences between living and non-living entities we can study the relationship both entities have to their surrounding world. In this regard the differentiation between open and closed systems is helpful: Living entities can be seen as open systems which are characterized by open borders, in contrast to non-living systems which are characterized as closed systems with closed borders. This idea was brought into debate by Ludwig von Bertalanffy (1901–1972): "Living systems are open systems, maintaining themselves in exchange of materials with environment, and in continuous building up and breaking down of their components".[19]

But what exactly is the difference between an open and a closed systems? In the first view the difference might not be seen because all systems, even the closed ones are in an exchange with the surrounding world. Even closed systems *react to* their world. Take for example measuring aggregates which collect their data from outside and answer to this situation. What however makes the difference between a closed and an open system is expressed by the category of activity. To make this point clear we can see that open systems *respond* to their surrounding world. Being able to answer is significant for an activity. This can be shown by the open system's status of the border. The border of open systems is not a line where the system ends but part of the system itself. That is not the case with closed systems e.g. a machine. The border of the machine marks the end of the machine that is not the case by open systems e.g. a mammal: The mammal's skin is the outer border of the living being and at the same time an integral part of it. It is interesting to see that the skin has the same important capability as each cell has, that is the bridge function between outside and inside what is called osmosis.

The interesting question now is how open systems are possible i.e. what is the origin of an open system? I am going to discuss this question related to Synthetic Biology's plan of creating life.

LET'S CREATE LIFE

The history of mankind is full of ideas of creating life: Starting with the fall of mankind people were fascinated by the idea of being like God and creating life. Famous projects as Doktor Faustus (by Christopher Marlowe, 1589 and Wolfgang von Goethe, 1808) or Dr. Frankenstein (by Mary Shelley, 1818) give evidence that the idea to be as powerful as God and to create life is never gone. In our times the desire to make life can be studied by the brand new branch of Biology, the so called Synthetic Biology.

Synthetic Biology looks to design and to construct new biological systems which are not found in nature. There is one metaphor you can hear in the debate of Synthetic Biology (SynBio) very often that is the metaphor of "playing God" which is not only used by the critics of SynBio but also by its promoters. In this way Craig Venter who became famous for mapping the human genome declares that he is able to create life and to do God's job.[20]

I won't discuss the hubris of this assertion but will take a look at its logical coherence and ask whether it is possible to create life. I am not going to discuss this question on the level of natural sciences but in theory. Therefore we have to acknowledge that the question of creating life is primarily not a challenge for natural sciences but for philosophy. And the problem which is in consideration is not a scientific one but an hermeneutic one. To make this point clear we will take a look at SynBio's key terms as there are *living machine, construction and the code of life*.

SynBio calls living entities "living machines". It is important to see that the term "machine" is here not used as a metaphor as Maturana and Varela does it but as a description of reality as it is seen by the SynBiologists. To ask what the consequences of this understanding are we will see that the term "machine" includes both the idea of an inner construction as well as the idea of being constructed. And that exactly is what SynBio is planning to do: to construct life. The most important blueprint of constructionism in biology is the idea of DNA as the code of life. If DNA is seen as the life-code the next step is to decode the code for reading the life's text. This kind of research reminds of the metaphor of the "book of life" which is in debate since the Middle Ages. As maintained before, the natural sciences do not think in metaphors even when they are using them. Natural sciences are using metaphors to explain reality without clarifying these metaphors. This is the fact when e.g. DNA is called the software of life and the cell is called the life's hardware, or when it is said that mankind is standing at a breakthrough and the first time humans are able to create real life out of dead matter is coming soon.[21]

THE FOUNDATIONS

As declared before in this paper I am not interested in the ethical and legal background of SynBio but in its theoretical background. In fact of this the paper tried to do both to answer the question "What is life?" and to understand the SynBio's understanding of life. This understanding was shown as reductionistic and mechanistic as it remains on the perspective from outside which describes living entities as closed systems. In contrast, the autopoiesis-theory looks to be adequate for living phenomena as it enables to take the perspective from inside. In the following I would like to put the autopoiesis-theory in a wider context. In the first view, autopoiesis is the name of a biological theory in the second view however, we will see that it has an hermeneutic approach and is a philosophy.

It was even a hermeneutic approach which brought Maturana on the path to his revolutionary research. He was still a student of biology when he realized that the phenomena of the living are of a special kind which makes a special way of speaking necessary: Maturana was convinced that the characteristics of biological phenomena make it obvious that we cannot discuss such phenomena in terms of function or as a means to an end. Therefore we do need an another language for debating living phenomena. As Maturana could not find any alternative language in the realm of natural sciences he started to elaborate his own language. With Maturana and Varela we can summarize the challenge of this new language as follows: This language has to be able

1. to perform the living phenomenon's position in general
2. to describe a phenomenon by excluding the describers' position
3. to exclude every kind of a "means to an end"-thinking

Within the natural sciences it is hard to find positions which fulfil these challenges and you would find no one which is acknowledged by the scientific mainstream. The very few positions which can be found are part of zoology. Most of them are from ethologists, think about scientists such as Konrad Lorenz[22] and Adolf Portmann.[23] What these scientists have in common is their interest in the phenomenon which you could call an "interest without any interest". An "interest without any interest" fulfils the condition that Immanuel Kant formulated for an aesthetic approach.[24] Aesthetic perception differs from any other kind of perception as it is self-sufficient, i.e. it does not look for any further result. In this way if it is not mixed up with other interests a pure aesthetic perception has no interest. However, as Kant points out, the aesthetic position of having no interest can also be the starting point of an interest.[25]

This situation marks the difference between the classical scientific approach and the alternative of Maturana/Varela and a lot of others: If you are perceiving a phenomenon on behalf of s.th. you will get a different understanding of it from the one you would have if you had no interest. In the latter case you will get an interest in the phenomenon, i.e. in the phenomenon by itself.

Now it becomes clear why the mainstream position can be seen as reductionistic: it reduces a phenomenon to function. This way is reductive as every living entity is more than its function. This becomes plausible if you realize that the technical term of function does not describe the phenomenon's way of activity but the observer's view of the phenomenon's activity. The reduction mentioned is not only the reduction to function but also the reduction to the observer's view. For that reason the alternative way of perception is the sine qua non-argument for perception as such, otherwise we only perceive our own interests.

It is obvious that this sine qua non-argument of perception only refers to living phenomena and not to artificial ones: Artificial phenomena such as man-made machines are completely understandable by the observer's interests as these machines lack any interest of their own. In the other way around, living machines are characterized only by the observer's interests and can be completely perceived from an external perspective.

As a result of these considerations we understand living systems as centred on the self and expressed by a self. It is a consequence of this idea to assume for all living systems a self, regardless of its evolutionary stage as Anna-Teresa Tymieniecka points out when she writes that "we do have to take into consideration that animals, even those of the simplest constitution, being endowed with a minimal degree of conscious sentience, do manifest reflexes manifesting the retrieval of past instants in the present, so that urgency of acting in the present becomes apparent to them."[26]

If we ask what makes a self, a self, we tend to characterize the quality of selfhood with consciousness. Doing so we run the risk of an *anthropocentric-rationalistic fallacy* which looks consciousness as exclusively founded on neuronal capacities. It is important to see that the formation of consciousness is necessarily founded neither on any level of the neuronal apparatus (e.g. the human being ones) nor on a neuronal apparatus in general. Therefore we can speak about selfhood also in relation to animals on an very low evolutionary level and even in relation to plants.[27] In order to be a self it is not relevant to have a brain but being able to recognize the inner and the outer world in relation to itself. This way of understanding is the basic condition of living.

What does "understanding" mean here? Understanding is the result of collecting and interpreting data. We can maintain that all living things are able to do so. Who ever fails in this endangers his life and in the long run he will die. In this way a disease e.g. cancer can be understood as a misinterpretation of signs. The living systems' capacity to read signs can be proved by all living systems and is analyzed in the field of Biosemiotics.[28]

Biosemiotics is the result of an interdisciplinary research programme which adapts the approach of linguistics and hermeneutics to the realm of the living. Based on the idea that all living systems are both cognitive and corporal, biosemiotics takes the idea of Jakob von Uexküll's environment-theory and the theory of communication. The theory of environment has made clear that every living thing has an environment which is not static but which is performed by the inhabitant of the environment.[29] The capacity to perform its own environment refers to the other capacity of the living: its cognitive capacity. Recognizing the world is a very challenging process. First of all it makes the distinction necessary, between "self" and "non-self" furthermore a linguistic understanding is asked. The basal linguistic capability has to be superior to a simple sender-reciever-model. Being able to communicate postulates a sense of oneself and the self's world. Incidentally this marks the difference between the living and the non-living: The non-living only reacts to the world in the sense of the sender-receiver-model. Only the living entity which has an understanding of itself is able to answer, in other words, to communicate.

What we have learned from Biosemiotics i.e. the capacity to read and answer signs refers to a general part of logos. As we can now say, living systems have an "ontopoietic sense of the logos of life", as A.-T. Tymieniecka pointed it out.[30] And so the assumption of the logos looks to be consititutive for the understanding of living processes. The importance of this is shown in an empirical way by Biosemiotics which has proved that living entities live in a realm of sense. This means that the living need a cosmos of the logos.

Let us return to the theory of autopoiesis: If the cosmos of the logos is the basis of living, we see that autopoiesis-theory has to presuppose this realm of sense i.e. the logos of life. Without any idea of logos no autopoietic action is possible.

University of Basel, Philosophisches Seminar Nadelberg 6–8, CH-4051 Basel, Switzerland
e-mail: andreas.brenner@unibas.ch

NOTES

[1] Plato, *Phaedrus*, London 1998, p. 245.
[2] Aristotle, *Metaphysics*, London 1976, Chap. 8, Aristotle, *De Anima*, Oxford 1993, 411b.
[3] Edmund Husserl, *Die Krisis der europäischen Wissenschaften und die transzendentale Phänomenologie*. Husserliana VI, Dordrecht 1976, p. 133.
[4] René Descartes, *Discours on the Method*. IV, 2, Cambridge 1986.
[5] René Descartes, *Discours*. V, 10f.
[6] Jan Christiaan Smuts, *Holism and Evolution*, London, Bombay 1926; Jacob von Uexküll, *Der Organismus und die Umwelt* (1931), in *idem: Kompositionslehre der Natur*, Frankfurt 1980, pp. 305–342.
[7] Nowadays the term "living machines" is a trademark for an apparatus for cleaning wastewater. There were probably Maturana and Varela the first which used these term for describing the difference between man-made machines.
[8] William James, The Experience of Activity, in *Essays in Radical Empiricism*, Cambridge, MA 1976, p. 82.

[9] Humberto Maturana, Francisco Varela, *Autopoiesis and Cognition. The Realization of the Living,* Dordrecht 1980, p. 78 f.

[10] H. Maturana, F. Varela, *Autopoieses and Cognition,* p. 79.

[11] The origin of an autopoietic system will be discussed in the chapter about the emergence of life, see below.

[12] Pier Luigi Luisi, *The Emergence of Life. From Chemical Origins to Synthetic Biology,* Cambridge 2006, p. 159.

[13] H. Maturana, F. Varela, *Autopoiesis and Cognition,* p. 80 f.

[14] This idea was brought into debate by Tom Regan as an argument for respecting animals and was used as well for integrating plants into the moral universe, for the first s. Regan, *The Case for Animal Rights,* Berkeley 1983, p. 235, for the later s. Anthony Trewavas, Plant Intelligence. In *Annals of Botany,* Vol. 92, 1–20 (2003). The moral impact of autopoiesis-theory is discussed by Nicholas Agar, *Life's Intrinsic Value. Science, Ethics and Nature,* New York 2001, p. 67.

[15] See the idea of transtemporal identity by Martine Nida-Rümelin, *Der Blick von innen. Zur transtemporalen Identiät bewusstseinsfähiger Wesen,* Frankfurt/M 2006.

[16] Wordsworth, William, Samuel Taylor Coleridge, *Lyrical Ballads,* London 1999 and F. H. Bool, *M. C. Escher: Life and Work,* Amsterdam 1981.

[17] Here I follow Stafford Beer, in H. Maturana, F. Varela, *Autopoiesis and Cognition,* p. 63.

[18] H. Maturana, F. Varela, *Autopoiesis and Cognition,* p. 94.

[19] Ludwig von Bertalanffy, The Theory of Open Systems in Physics and Biology. In *Science,* Vol. 111, 23 (1950).

[20] D. G. Gibson, Complete Chemical Synthesis, Assembly, and Cloning of a Mycoplasma genitalium Genome. In *Science,* Vol. 319, 1215–1220 (2008). Further links: E. Pilkington, I am Creating Artificial Life, Declares US Gene Pioneer, The Guardian, 6 October 2007; J. Kaiser, Attempt to Patent Artificial Organisms Draws Protest. In *Science,* Vol. 316, 1557 (15 June 2007); P. Aldhous, Countdown to a Synthetic Lifeform, New Scientist Magazine, 11 July 2007, pp. 6–7.

[21] M. Schmidt, C. Meinhart, *Synbiosafe. Synthetic Biology and Its Safety and Ethical Aspects,* Dvd 2009.

[22] K. Lorenz, *Behind the Mirror,* London 1977.

[23] Adolf Portmann, *Essays in Philosophical Zoology. The Living Form and the Seeing Eye,* Lewiston 1990.

[24] Immanuel Kant, *Kritik der Urteilskraft,* Werkausgabe Bd. X, Frankfurt/M 1979, § 2, pp. 116–117.

[25] Ibid., Footnote (p. 117).

[26] Anna-Teresa Tymieniecka, *Memory in the Ontopoiesis of Life, Book 1,* Dordrecht 2009a, p. Xii.

[27] Anthony Trewavas, Aspects of Plant Intelligence. In *Annals of Botany,* Vol. 92, 1–20 (2003).

[28] G. Witzany, *Biocommunication and Natural Genome Editing,* Dordrecht 2009; T. Sebeok, Biosemiotics: Its Roots, Proliferation, and Prospects. In *Semiotica,* Vol. 134 (1/4), 61–78 (2001).

[29] Jakob von Uexküll, *Kompositionslehre der Natur,* Berlin 1980.

[30] Anna-Teresa Tymieniecka i.c., p. XV and The Fullness of the Logos in the Key of Life. *Book I, The Case of God in the New Enlightenment,* Dordrecht 2009b, p. 63.

REFERENCES

Agar, Nicholas. 2001. *Life's intrinsic value. Science, ethics and nature.* New York: Columbia University Press.

Aldhous, P. 2007. Countdown to a synthetic lifeform. New Scientist magazine, 11 July 2007.

Aristotle. 1976. *Metaphysics.* London: Clarendon.

Aristotle. 1993. *De Anima.* Oxford: Clarendon.

Bertalanffy, Ludwig von. 1950. The theory of open systems in physics and biology. In *Science,* Vol. 111, pp. 23–29.

Bool, F. H. and M. C. Escher. 1981. *Life and work.* Amsterdam: Abrams.

Descartes, René. 1986. *Discours on the method.* IV, 2, London: Dent.

Gibson, D. G. 2008. Complete chemical synthesis, assembly, and cloning of a *Mycoplasma genitalium* genome. *Science* 319:1215–1220.

Husserl, Edmund. 1976. *Die Krisis der europäischen Wissenschaften und die transzendentale Phänomenologie*. Husserliana VI, Dordrecht: Springer.

James, William. 1976. The experience of activity. In *Essays in Radical Empiricism*. Cambridge, MA: Harvard University Press.

Kaiser, J. 2007. Attempt to patent artificial organisms draws protest. *Science* 316, 15 June 2007.

Kant, Immanuel. 1979. *Kritik der Urteilskraft*, Werkausgabe Bd. X, Frankfurt/M: Suhrkamp.

Luisi, Pier Luigi. 2006. *The emergence of life. From chemical origins to synthetic biology*. Cambridge: Cambridge University Press.

Maturana, Humberto, and Francisco Varela. 1980. *Autopoiesis and cognition. The realization of the living*. Dordrecht: Reidel.

Nida-Rümelin, Martine. 2006. *Der Blick von innen. Zur transtemporalen Identität bewusstseinsfähiger Wesen*. Frankfurt/M: Suhrkamp.

Pilkington, E. 2007. I am creating artificial life, declares US gene pioneer. The Guardian, October 6, 2007.

Plato. 1998, *Phaedrus*. London: Cambridge University Press.

Portmann, Adolf. 1990. *Essays in philosophical zoology. The living form and the seeing eye*. Lewiston: Mellen.

Regan, Tom. 1983. *The case for animal rights*. Berkeley: University of California Press.

Smuts, Jan Christiaan. 1926. *Holism and evolution*. London, Bombay: MacMillan.

Trewavas, Anthony. 2003. Plant intelligence. *Annals of Botany* 92:1–20.

Tymieniecka, Anna-Teresa. 2009a. *Memory in the ontopoiesis of life, Book 1*. Dordrecht: Springer.

Tymieniecka, Anna-Teresa. 2009b. *The fullness of the logos in the key of life. Book I, The case of god in the new enlightenment*. Dordrecht: Springer.

Uexküll, Jacob von. 1980. Der Organismus und die Umwelt (1931). In *idem: Kompositionslehre der Natur*. Berlin: Propyläen

Wordsworth, William, and Samuel Taylor Coleridge. 1999. *Lyrical ballads*. London.

MANJULIKA GHOSH

MAN'S WORLD AND *LOGOS* AS FEELING

ABSTRACT

Man's relationship to the world is a perennial problem of philosophy. The problem is one of accounting for man's experience of the world. Although man shares the world with other living beings, his experience of the world is radically different from theirs. In accounting for his experience of the world he also understands himself, he becomes self-aware, as it were. The experience of the world, it has been claimed is the experience of an articulated, structured reality. Otherwise, the human mind would simply be lost in the maze of the multifarious and discreet perceptions of what it encounters in the world. In the ancient Western philosophy, the source of the fundamental order in the cosmos as a whole has been traced to the *logos*. In modern philosophy, it is resurrected especially in the philosophy of Kant and Husserl. The *logos* or the rationality of there being an ordered world of experience is cognitive rationality. This paper explores feeling as a hidden modality of the *logos*. In feeling we have a fundamental awareness of the object as a unity. This is a primitive experience. Here, an attempt will be made to understand the *logos* from the angle of felt experience imposing order on the world.

Man's relationship to the world is a perennial quest of philosophy. The problem is one of accounting for man's experience of the world. Although man shares the world with other living beings, his experience of the world is different from theirs. In accounting for his experience of the world he also understands himself; he becomes self-aware, as it were. The experience of the world, it has been claimed, is the experience of an articulated, structured reality. Otherwise, the human mind would simply be lost in the maze of the multifarious and discreet perceptions of what it encounters in the world.

In the ancient Western philosophy from Heraclitus to the Philo of Alexandria, the source of the fundamental order, not only in nature but in the cosmos as a whole, has been traced to the animating principle of the *logos*. *Logos* is an idea which not only took hold of the Greek mind, but in the eighteenth and nineteenth century also the Greek spirit reawakened with an enthusiastic upsurge of faith in autonomous reason, a faith in the rationality of all that is. Kant attributed the interconnections and articulation of phenomena to the apriori structures of the human mind. Husserl's phenomenological quest found the ultimate grounding of the world in the transcendental constituting consciousness.

The extraordinary spell of Greek thought on Husserl can be measured from the occurrence and role of Greek words in his works, words like *noésis* and *noéma*, *hylé*, *morphé*, *theoria*, *epistémé*, *entéléchia*, *télos*, *physis*, *doxa*, *nous* as well as *logos* and its derivative, logic. We find the *logos* as the guiding idea running through

A.-T. Tymieniecka (ed.), Analecta Husserliana CX, 103–110.
DOI 10.1007/978-94-007-1691-9_9, © Springer Science+Business Media B.V. 2011

Husserl's enormous corpus of writing. For Husserl, from his early work, *Logical Investigations* to the later *Formal and Transcendental Logic*, the main concern was the reconstruction of knowledge – knowledge which is universally valid. In the *Crisis* the infinite task based on the rationalization of experience is said to be the special telos of Western culture. Transcendental phenomenology is not possible if the *logos* as the unitary telos is not evident at all stages as a functional entelechy. The essence of phenomenology is the "philosophical pursuit of Reason or the *logos*". As Anna-Teresa Tymieniecka observes: "These pursuits of rational structurations, links, articulations of genetic processes, etc., had as their essential reference the cognitive reason of the human mind, especially human intellective cognition".[1] The kinds of phenomena Husserl is interested in are objects of cognition. The cognitive relation to the world is paramount for Husserl. Indeed, we can speak of a *logos* tradition – the ideal of a philosophical culture in the West, permeated by an *ethos of logos*. Plato is taken to be the undisputed father of the *logos* tradition. Plato was searching for timeless truths, which could eliminate the dangers and contingencies that ordinarily seem to vitiate human life. This ideal of reason which stands at the root of Western civilization, born in the works of Plato and Aristotle, is the bequeathing of a tradition, an inheritance or legacy But we have not so far clarified the meaning of "logos". Let us now do it.

"Logos" is a "many-meaninged word". *The Greek-English Lexicon* of Liddell and Scott distinguishes two elements of meaning in it. First, the word by which the inward thought is expressed and second, the inward thought itself. It has been observed that

> This dual nature of its meaning gives 'logos' extraordinary range. Primarily, it refers to those outward sounds that express thought ... It is the ability to give voice to some reasoned thought, word, sentence, talk, speech, explanation, language, discourse, story, argument, rational account – all these function at different times as the proper translation of 'logos'. It is also rendered as thought, reason, rationality, calculation, etc., when it refers to the 'internal talk' that goes on within. 'Logos' thus comprehends all that is verbal and rational within us. The one phrase that begins to capture both these meanings is 'rational account'.[2]

According to Charles Taylor, "What underpinned this connection between saying, words and reason was what one could call a discourse-modelled notion of thought... Because thinking was like discourse, we could use the same word, *logos*, for both".[3] From the perspective of the ancients, there is a third meaning. This logos is not human speech or thought but refers to a "rational structure" which exists outside of the human mind or voice; the rational structure of the world "out there" that can be apprehended by human beings presumably by using their *logos*. For example, Heraclitus begins one of his aphorisms by saying "having listened not to me, but to the logos, it is wise to agree that all things are one".[4] Thus, there is an "ontic *logos*" and a *logos* in the human subject.

At the beginning of *Formal and Transcendental Logic* Husserl makes explicit the meaning of *logos* which owes much to antiquity. It is as follows:

1. In developed language, *Logos* sometimes signifies words or speech itself: sometimes that which is spoken about, the *affair-complex referred to in speaking*: but

also, on the other hand, the propositional thought produced by the speaker either for purposes of communication or for himself: the mental sense, as it were, of the assertoric sentence, that which the speaker means by the expression . . .

2. But, particularly where a scientific interest is active in all these significations of the word *Logos* takes on a more pointed sense, because the idea of a *rational norm* enters into them. *Logos* thus signifies sometimes *reason* itself, as ability and sometimes the action of rational thinking – that is, thinking that has the quality of insight or thinking directed to a truth given in insight.[5]

Husserl's investigations rooted in pure intellective reason led him toward a "formal theory of everything". As Mohanty says: "Towards the concluding portions of *Ideas* I, especially in the chapter on phenomenology of reason Husserl extended the idea of reason from logic to ethics and value theory, to the theory of action (praxis), without developing his detailed views on these matters."[6] Husserl does indeed intend phenomenology as a critique of reason and in this he treaded the Kantian pathway of a critical analysis of reason. In the words of Tymieniecka, ". . . Husserl appears to have stepwise pursued the critique of reason – of human reason – to the point at which the rational chain that had sustained his interrogations, the thread of the cognitive logos, in fact broke down. Despite Husserl's painstaking efforts 'phenomenology of phenomenology' was not accomplished. . ." This makes her wonder "whether the phenomenological pursuit has not ultimately been hiding an ampler conception of rationality than was acknowledged by its founder Husserl and his followers."[7]

What we intend to do in the present context is to show how the confident self-assertion of reason has limits. And the only way to do this is not to make purely theoretical arguments but to base logos in the course of human life; its work in the entirety and the world in which man finds himself. The value and significance of human life, the uniquely beautiful pathos of being human puts a question mark to the claim that investigation of the definite and the stable in human experience, its essential core is the only task of philosophizing. Human life is incomplete, unstable and often unpredictable. Since human life is incomplete there is the urge for creativity. We can make a distinction between construction and creation. We construct a bridge so that we can negotiate the river. Construction serves some utilitarian purpose. In creating something we transcend the given order of things; we create not to achieve some premediated end but because it gives us delight. "Artists create not because art is good but because they are creative".[8] Creative activity is evidence of the fact that we can conjure up other possibilities, and that way we are constantly in the process of making ourselves anew. The authority of reason in defining who we are: rational beings, is a "blasphemy" against the urgings of creativity. Elevating *logos* as the *telos* to evaluate ourselves is evidence of a neglect of life. Equally, the world itself is heterogeneous, filled with objects that differ in kind. There is no world out there safely structured and amenable to the probing eyes of reason. As creativity is the life-affirming attitude, interest in maximizing earthly life brings about cracks, potentially vulnerable openings, in the shield of *logos*. ". . . the world cannot become a determinate and clean subject like medicine or arithmetic."[9] Unlike the realm of number, human world is not fixed; it is electrified by human desires, imagination and feeling. We, human beings, self-consciously alive, cannot be observed, measured and counted. Hence, we hope to discover an alternative conception of *logos* that is compatible with life-affirmation. Feeling, I submit, is that human capacity, that hidden aspect of the *logos* which is crucial for human existence, unfolding the human situation.

Feeling has not enjoyed the same favor of philosophers compared to reason. This is because feeling is alleged to be unstable, hovering between the poles of excitement and depression, waxing and waning. Another reason for the neglect of feeling, as a hidden modality of *logos*, is its confusion with emotion. In common parlance we do not make distinctions between feeling and emotion. But philosophically we must not fail to keep them separate. As Paul Ricoeur says: "Our natural inclination is to speak of feeling in terms apparent to emotions, that is, to affections conceived as (1) inwardly directed states, and (2) mental experiences closely tied to bodily disturbances, as is the case in fear, anger, pleasure and pain."[10] Ricoeur further says: "... both traits come together. To the extent that in emotion we are, so to speak, under the spell of our body, we are delivered to mental states with little intentionality, as though in emotion we lived our body in a more intense way."[11] For him, genuine feelings are not emotion although they may be "embraced" or "surrounded" by it. Rather, they are "negative", "suspensive" experiences in relation to the literal emotions of everyday life. They imply an epoche of our bodily emotions.[12]

However, a further doubt is to be found raising its head. How could it be possible for feeling to function as an aspect or dimension of *logos* – one beyond cognitive rationality? Is not feeling essentially a subjective state considered without reference to an object? As a matter of fact throughout the history of philosophy and psychology feeling is regarded as a state of consciousness without an object. But this is due to our usual interpretation and understanding of feeling. The use of the word "feel" in language is always in reference to an object. Feeling must have an object. In this regard it is like cognitivity of consciousness. When sensing or thinking it is impossible not to sense or think of something. Similarly, in feeling it is impossible not to feel something. Let us try to feel without being directed toward, without feeling some object – it will seem impossible. If feeling is intentional then it is precisely the capacity to enter into relationship to objects. It involves even something more. To clarify what we want to say we may fall back upon Husserl's distinction between objectifying and non-objectifying acts. The non-objectifying act signifies such an act that itself does not possess the mark of being object-constituting but nevertheless aims at an object. Non-objectifying acts mostly refer to a feeling or an act of feeling. Feeling acts are not objectifying acts but are nonetheless aiming at objects.[13] Why we bring in this distinction will be explained below. But before that let us consider one thing. In what does the alleged cognitivity of consciousness consist? In the *logos* tradition, it consists in the absolute validity of the intellective *logos*. But we may speak of knowing a whole range of objects in the sense of, say, recognizing a face, without necessarily describing it; we may speak of knowing a person, a piece of music, of moral good and evil, of knowing the religious dimension of life or God, although unable to make exact true statements about them. Some of the things of which we are aware, may be voluntarily called up, as when we imagine constructively, or they may be voluntarily received, as when we open our eyes upon the sunset. There is hardly any reason to fit in consciousness of objects into the mould of strictly critical cognition. Can consciousness operate in isolation from will and feeling? Is not consciousness colored by these? The English word "feeling" is equivalent in meaning to *rāga*, a concept from Indian musicology. *Rāga*, which is taken as pro-disposition and held as the contrary of *virāga*, contra-disposition, is often interpreted as that which colors the mind or consciousness. That feeling is not so disparagable a candidate in knowledge comes from Russell. Russell describes knowledge of mathematics as having a beauty as cold as marble and comparable to the closing cantos of Dante's *Paradiso*.[14]

Feeling is a kind of experience in which we experience something in a way which is itself fundamental and for which any other reason is not possible. It is an experience which we cannot account for by any other cognition. Feeling is a basic and primitive mode of understanding the world and ourselves. The act of giving meaning to the phenomena is accomplished not by imposing concepts, categories, ideas and principles upon them. Feeling makes the object felt obvious. It is revealed or manifested. It shows itself as itself. "I see a tumbler on the table" is a determinative judgment. Before this determination takes place the being of the thing as naked breaks upon our consciousness. We may call this feeling consciousness before we are concerned with the cognition of the thing in question. In feeling, the object felt is manifested in its totality as a unity, as one. In a subject-predicate judgment this unity is lost and we cannot get it back; getting back the unity will be an endless task. Feeling gives voice to a vision of the world which is not fully accessible to the rational working of the *logos*. Here is an act of awareness that brings consciousness closer to its object while pure intellective reason puts it at a certain distance. The spontaneity and vitality with which feeling relates to its object implies a greater totality of fulfillment than is involved in the cognitive standpoint in which the subject of the cognitive activity is abstracted from its embodiment. We may say that feeling does not constitute its object. In fact, there is no need of that. The object is made our own by our touching it with feeling. The object gives itself over to feeling; it, as it were, "donates" itself.

What has been said above holds true not only of our everyday experience but of all artistic creations. In the latter, the felt unity between the artist and the projected object of his creation is a precondition of artistic activity. In creating something we try to project an alternative vision of the world, of alternative possibilities beyond the habitual pursuits of everyday. It has been held that in aesthetic experience, we are made to realize the world more fully or richly real than we do in normal experience. Art is no less deepening of the world-consciousness than it is a clarification of self-consciousness. Our poet Rabindranath Tagore says: ". . . there is the vast world . . . which is personal to us. We must not merely *know* it, and then put it aside, but must *feel* it – because by feeling it we *feel* ourselves."[15]

The order that feeling bestows on the world is not effected through the mechanism of categories, concepts, ideas, representations, etc. The immediacy with which the felt object is enfolded in the feeling act discovers facts and relation between facts such that facts become meaningful. This interrelatedness and unity of facts is declared to be truth. In feeling we transform facts into human truth. We feel, for example, the *serenity* of the sky, the *placidity* of the lake water, the *gloriousness* of a sunset, the *sublimity* of a mountain and so on; we also feel the appropriateness of a sequence of music, of the positive moral quality of love, or of the religious "numinous". To speak of music; pure music, *mārga saṅgīt* in Sanskrit language, which has no theme outside the musical ones, is a fully developed articulation of meaning which we certainly come to experience through feeling. The world is real when it is known not only by "critical reason", but also when it comes within the range of our feeling. We may here recall again Rabindranath's very acute observation in one of his poems "Śukatārā" which may be rendered in English as "The Morning Star". There is the astronomical reality of the Planet Venus, an "objective" truth of science indeed. And there is the *human* reality of a greater significance to us, of what we call *Śukatārā*, the luminous astral body, appearing like an autumnal dewdrop glistening on the forehead of dawn. The two, Rabindranath avers, deliver to us objectivity, but in different senses; the former in a weaker sense and the latter in a stronger sense. One is "weaker" because it is calculative, and the other is "stronger" because we have made it our own by bringing it to the unitary locus of our feeling, marked by immediacy and intensity. If such a conviction is endorsable then a revision of the received conception of *logos* is called for, by implication. *Logos* understood as cognitive rationality no longer serves as the central concept in understanding the human world. The critical consciousness is not the whole story of man. A man is a full person and not just a cognitive mind and it requires the resources of a full person to understand him and his world. "Our universe is the sum total of what man feels, knows, imagines, reasons to be . . ."[16]

I submit further that feeling establishes the conditions for the possibility of active engagement in a world with others. Feelings of love and sympathy make possible transcendence of the tragic dominance of the self and de-alienated living with others in the world. Society as a community of selves is marked by a we-feeling, that is, a feeling of I with others. It is a model of non-alienated living. Such a society cannot be brought into existence automatically. It is not received as a gift or forced into existence from outside. Society exists because man ever recreates his relations to others through love and sacrifice. The power of love not only sustains life, human and sub-human, it also transforms society into a harmony of persons. This apart, feeling is related to creativity as such. Creativity is very much a part of human life. As Tagore illustrates it: ". . . man by nature is an artist; he never receives passively or accurately in his mind a physical representation of things around him. There undergoes a continual adaptation, a transformation of facts into imagery, through constant touches of sentiment and imagination."[17]

Artistic creativity has a very complex relationship with feeling. The projected object of creation must be made a content of feeling, and sustained by it. A creative genius or any ordinary individual experiencing a creative process is quite conscious about sensing creative ideas that form in his imagination. But the artist intends to look forward to have perspicuous representation. Representation involves, besides will and reason, feelings and sentiments through which the artist in any area nurtures the gestate images. The many modifications that the artist makes on the canvas or the musician in fine-tuning the notes on his musical instrument are shifts from an imperfect state of feeling to a perfect state and finally, result in the feeling states of joy or pride.[18]

The metamorphosis or transformation takes place on another level. The creative art lives in our felt experience unfolding various meanings which undergo metamorphosis according to the way it is appreciated or discarded, or the joy and the satisfaction it provides to us. So both on the level of genesis and the level of appreciation creativity is enmeshed with the feelings of happiness, joy and satisfaction. The creative process is never complete. There is perhaps an element of truth in the lament that the best painting is yet to be drawn or the best poetry is yet to be written. For, there is a continual modification of creative ideas paralleling the refinement, sharpening and deepening of feelings. Creative experiences cannot be limited to human rationality. It might be the case that reason impedes such activities. For example, when trying to improve upon a musical score or creating a new *rāga* within the Indian classical music it is good to suspend the processes of analysis, articulation, justification, etc. So, there are non-reasoning activities like creative activities which suspend the authority of the *logos*.

To conclude: We have stated the first moment of assertion of the *logos* and the later stages of revision. And we have tried to posit feeling as an organizing factor at the very heart of human life and human world at the same time unfolding its relation to creative activity which bestows its own order on reality. Feeling is shown not as a rival of *logos*, subplanting it but as a hidden dimension of it.

Department of Philosophy, University of North Bengal, Siliguri, West Bengal, India
e-mail: ghosh_nbu@rediffmail.com

NOTES

[1] A.-T. Tymieniecka (ed.), "The Logos of Phenomenology and Phenomenology of the Logos", Book 1, *Analecta Husserliana*, Vol. LXXXVIII, (Netherlands: Springer, 2005), Vol. LXXXIII, p. xiii.

[2] David Roochnik, *The Tragedy of Reason: Toward a Platonic Conception of Logos* (New York and London: Routledge, 1990), introduction, p. 12.

[3] Charles Taylor, "Language and Human Nature" in *Human Agency and Language, Philosophical Papers* 1 (Cambridge: Cambridge University Press, 1999 reprint), p. 222.

[4] Heraclitus, Fragment No. 50, from Herman Diels, *Die Fragmente der Vorsokratiker* (Berlin: Weidmann, 1952). Source: David Roochnik. *The Tragedy of Reason, op. cit.*, p. 12.

[5] Edmund Husserl, *Formal and Transcendental Logic*, trans. Dorian Cairns (The Hague: Martinus Nijhoff, 1969), p. 18. Emphasies Husserl's.

[6] J.N. Mohanty, *The Philosophy of Edmund Husserl: A Historical Development* (New Haven and London: Yale University Press, 2007), p. 288.

[7] A.-T. Tymieniecka, "The Logos of Phenomenology and Phenomenology of the Logos", *op. cit.* p. xiv.

[8] Fredrich Nietsche, *Philosophy in the Tragic Age of the Greeks*, trans. Marianna Cowen (Chicago: Dayakrishna, 1962), p. 55.

[9] David Roochnik, *The Tragedy of Reason, op. cit.*, p. 191.

[10] Paul Ricoeur, "The Metaphorical Process", *Critical Inquiry*, Vol. V, No. 1, Special issue on metaphor (1978), p. 156.

110   MANJULIKA GHOSH

Ibid.

Ibid., p. 157.

Edmund Husserl, *Logical Investigations* II, trans. J.N. Findlay, (London: Routledge and Kegan Paul, 1970), p. 743.

Bertrand Russell, *Autobiography* (London: Bantam Books, Allen and Unwin Ltd., 1969), Vol. 3, p. 52.

Rabindranath Tagore, *Creative Unity* (London: Macmillan, 1959), p. 16. Emphasies mine.

Rabindranath Tagore, *The Religion of Man* (London: Unwin, 1963), p. 15.

Rabindranath Tagore, *The Religion of Man*, *op. cit.*, p. 21.

Ranjan K. Panda, "Creative Visualization: A Semantic Analysis", *Journal of Indian Council of Philosophical Research*, Vol. xxi, No. 2 (2004), p. 99.

VELGA VEVERE

THE FEAST OF LIFE OR THE FEAST OF REASON – KIERKEGAARD VERSUS PLATO

"The thought is transparent in the dialogue, and the action in the situation" (Søren Kierkegaard)

ABSTRACT

The article consists of three sections. The first section "Dialogue at the intersection of literature and philosophy" analyzes the fundamental differences between the two modes of human intellectual activity – philosophy and literature on the basis of Anna-Teresa Tymieniecka's philosophy. Nevertheless, the intersection is possible in the form of dialogue. The second chapter "Negative existential maeutics" is dedicated to Kierkegaard's conception of existential maeutics in comparison with the Socratic maeutics. The stress is put upon its negative characteristics – the distance, the interruption, the situation of existential shock. These restrictions are necessary to allow the participants' self-knowing. The third chapter explores they ways how Kierkegaard in his fragments *In Vino Veritas* reenacts Plato's dialogue *Symposium* in order to demonstrante his strategy of negative existential maeutics in practice. If the goal the classical maeutics is the birth of knowledge during the process of conversation, then Kierkegaard's goal is the birth of subjectivity and self-recognition.

The motto by Danish religious thinker Søren Kierkegaard breaks the grounds for the development of the theme "The feast of life or the feast of reason – Kierkegaard versus Plato," as it points towards the special role of dialogue both in philosophy and literature. Of course, it is necessary to take into account the respective differences between these two realms of intellectual endeavors brilliantly disclosed by Anna Teresa Tymieniecka in the book 3 of her monumental work "Logos and Life," entitled "The Passions of the Soul and the Elements in the Onto-Poiesis of Culture." The differences apply also to the dialogue – be it literary or philosophical by its nature, and to the historical sources of the dialogical activity, having in mind, first of all – Socrates' diegmatic dialogue, Plato's intellectual dialogue and Aristotle's theory of drama, and comparison of the antique and modern interpretations of the dialogue and their respective roles in defining personality. All in all the dialogue, according to Kierkegaard, should be viewed as a specific way of conveying essential or existential truth about the world and about the conveyer himself. The form of communication may be almost similar, while the content – as different as it can be speaking of the speculative and the existential mode of philosophizing. Still, the most puzzling question is not so much about the influence of Socrates

A.-T. Tymieniecka (ed.), Analecta Husserliana CX, 111–122.
DOI 10.1007/978-94-007-1691-9_10, © Springer Science+Business Media B.V. 2011

and Plato upon Kierkegaard's mode of thinking (though it is important one), but rather – of the reason why Kierkegaard chooses to reenact one of the Plato's dialogues ("Symposium") in a different setting and with different personages. The aim of the present investigation is to explore the reawakening of certain trends of antiquity in Kierkegaard paying a special attention to the short masterpiece *In Vino Veritas* (part of the longer work "Stages on Life's Way") that appears to be, though not so obvious, the enactment of the Plato's dialogue *Symposium*. The choice of these two particular dialogues (*In Vino Veritas* and *Symposium*) accounts for the title of the present paper, namely, the celebration of reason versus the celebration of life, the intellectual dialogue versus the existential one. Thus the task is at least twofold – first, to explore the influence of antiquity upon the Kierkegaards thought on the basis of the particular example, and, second – to investigate his idiosyncratic conception of the dialogue as the negative existential maeutics that nevertheless bears an imprint of the classical philosophical dialogue.

DIALOGUE AT THE INTERSECTION OF LITERATURE AND PHILOSOPHY

Philosophy and literature are caught in a constant contest as each attempt to absorb each other's task. (Anna-Teresa Tymieniecka)

Despite the obvious similarities between two modalities (literature and philosophy) of the human intellectual activity there are also crucial differences between them. This problem has been discussed in depth in Anna Teresa Tymieniecka's monograph "The Passions of the Soul and the Elements in the Onto-Poiesis of Culture." Both literature and philosophy set their task to present the world, but they differ in the matters of *what* to present and *how* to present it. "... there is an innermost motivation for the writer's urge to write, to communicate something uniquely his own to a public, to the society of his time, and to enrich by his message – or even transform – the culture of his period or of all time even." (Tymieniecka 1990, p. 13) At the same time: "To reveal reasons is, in fact, the main task of the philosophical test." (Tymieniecka 1990, p. 14) As to their relation it falls to philosophy to define and conceptualize the task and the role of literature. Of course, there is no a clear cut demarcation line between those two, and sometimes it becomes quite a difficult, almost impossible task to separate them as they always tend to invade each other's territory. Then could it be possible to speak of the philosophical literature and literary philosophy rather than of literature and philosophy as diverse modes of knowing and presentation? Don't they often have the same concepts in their disposal? Anna Teresa Tymieniecka emphatically insists that despite the similarities and sometimes almost coincidental narrative structures and rhetorical argumentation literature has its own unique vocation that non reducible to any other form of intellectual activity. "The role of literature, that to which it means are geared, is not to explain the world and life as we discover it by positive, universally valid, intellectual means. It is to *recreate* the world and life after we have already lived it and come to know it in the positive sense, to transform what trivial and bare positivism yields through the *creative vision."* (Tymieniecka 1990, pp. 17–18) *Creative vision* is the key concept in speaking about the fate and ongoing development of philosophy and literature in the Occidental tradition. A.-T. Tymieniecka proposes five general distinctions between them on the basis of their respective aims, means of expression (languages) and vision of the underlying structures and laws, attitudes towards the concreteness and abstractedness, and, finally, roles they play in the sphere of human knowing as such. Now let us turn in short to each of these statements of diversity.

If philosophy aims at discovering the most general principles of life and human existence, then literature is concerned with the most unique sand personal vision of the state of affairs within and outside.

The challenge for philosophy, the philosopher's quest is to give the rational and structures explanation of subtle and manifest phenomena of life "in order to provide principles explaining the definite nature of reality according to a most general outline of the vision of each philosopher. . .". (Tymieniecka 1990, p. 19) The writer, in contrary, seeks to fashion his idiosyncratic version of the world and to express it in the most intuitive manner "to give it the most particular, specific, personal incorporation in human life-situations, characters, intertwining of events, etc. in accordance with his deepest feelings, emotions, strivings, and urges – stemming from his own flesh and blood, and spirit." (Ibid.)

This point regards the universality of the language – if the philosopher intends to use the abstract notions, more or less precise and formulated clearly in order to be understood at least within the context of one or another philosophical tradition or school (keeping in mind the Continental and Anglo American divide, for instance), then the writer tries to evoke the most personal feelings, appealing to particularity in order to receive the emotional response.

The role of philosophy is the one of enlightenment "about the parameters of human existence, its nature and prospects, options and limitations; it offers this clarification to all men in all situations and also indicates the proper conduct for their fulfillment." (Ibid.) Whereas literature, accordingly operates on a different level – on the margins of consciousness inhibited by fleeting impressions, vague reminiscences, deeply personal life experiences; in other words, literature tries "*to establish contact* between the living reader and his vision of life." (Tymieniecka 1990, p. 20) And in this sense it promotes the reader's self-understanding and self-inscription on reality, de-ciphering the life-significance and enlargement of the self beyond the limits of the individual ego. These distinctions, Anna Teresa Tymieniecka concludes, allow distinguishing the vocation of literature from the one of philosophy, and at the same time to stress their generic affinities.

Still, in our opinion, the dialogue in the form proposed by Plato is something that could be described as being at the intersection of philosophy and literature as it aims at disclosing the universal structures and forms behind particular appearances. Though, it has to be admitted that Plato himself, in the *Gorgias* has drawn a sharp demarcation line between philosophy and what he calls rhetorical practices. He states that the task of philosophy is truth rather than persuasion as people could be persuaded to believe in untruth by a skillful speaker. Like a sick person can be talked into not following the doctor's orders and this can result in his death; likewise in the legal and political matters. It seems then that philosophy calls for a special form of expression that is not subject to changing opinions and mood swings, the form that could be as transparent, as unvocal as the mathematical equitation, in other words, it calls for a tractate. But what does Plato himself do? He constructs imaginary dialogues between real and/or imaginary characters, inserts comic episodes of mishaps of participants (hiccoughing, drunken behavior), includes everyday expressions and descriptions of daily activities, in short – he creates a story consisting of the beginning, climax and narrative conclusion. All this seemingly (at least on the surface) exhibits the paradigm of the literary expression, not of the philosophical, if we are to believe what Plato has declared in *Gorgias*. Thus we may conclude that all this has been done on purpose. But what purpose? Yes, Plato offers his readers the dramaturgical setting, but at the same time he doesn't create an illusion of illusion of the dramatic action – each speech has to be understood separately. It is to say, that Plato creates a distance between his personal views upon the world and views expressed by different characters, the reader is bound to read and to understand all by himself, without the guidance of the author behind the scene. So by the means of such distancing Plato turns the short literary caprice into the philosophical reflection about the fundamental questions of the world order – be they about the love for wisdom, the highest goodness, the justice, the pre-forms of all existing, and so on. Interestingly enough, the same principle of distance was employed by Søren Kierkegaard in his works, not the least in his *In Vino Veritas*, only his aim is to facilitate the birth of the subjectivity and the subjective truth.

NEGATIVE EXISTENTIAL MAEUTICS

But jus as there is something deterring about irony, it likewise has something extraordinarily seductive and fascinating. Its masquerading and mysteriousness, the telegraphic communication it prompts because an ironist always has to be understood at a distance, the infinite sympathy it presupposes, the fleeting but indescribable instant of understanding that is immediately superceded by the anxiety of misunderstanding – all this holds one prisoner in inextricable bonds. (Søren Kierkegaard)

Of course, Kierkegaard's interest in antiquity was by no means accidental; it has run through his whole authorship. Moreover, he was always apt to use the antique sources for his own purposes (in the development of his philosophical stance), which only rarely complied with the original intentions of the ancient authors. In this respect Kierkegaard's interest is not historical, or rather – not historical in *sensu strictu*. To illustrate this point let us remember his dissertation "On the concept of irony with constant reference to Socrates" (1841). Notwithstanding the scandal in the academic milieu surrounding the process of defense (asking the special permit of the King to write dissertation in Danish contrary to the common practice at the times to submit it in Latin, as well as rendering seemingly non-academic, rather provocative style of narration, etc.) it is a serious research of the concept of irony both in antiquity and modernity. At the same time here, at least in retrospect, it is possible to see some hints of his strategy of existential communication that in our opinion forms the axis of his whole philosophical endeavor. This, in turn, means two things: first, if communication is to be regarded as one of the basic concepts then there appears a possibility of vision of the Kierkegaard's authorship; second, the concept of communication itself functions in two ways – as a certain form of relying information (answering the question *how?*) and exchange of information (intersubjectivity); and as a manifestation of existence that presupposes ethical choice, internalization and self becoming. Thus the use of the concept is quite broad. If the former could be described as an "existential communication" (a form of communication, characterized by distance, gap, understating), the latter – as "existence-communication" (content what is to be communicated and process of communication itself). Perhaps one of the best descriptions of this specific mode of communication is given by Alstair Hannay: "Being 'existential,' such 'communication' differs from that on topics about which people can advise one another, discuss and agree on how to deal with them, or give each other general rules or prescriptions for doing that. An existential matter requires, as it were, a self-provided personal boost on the part of the recipient, something more than the recognition and acceptance of some such rule." (Hannay 2001, p. 12) Such statement, in turn, brings forth the Aristotle's distinction between *techne* and *praxis*, where the latter opens up the possibility to establish harmonious relationship with the word by the means of personal activity; and since the source of disharmony is placed in the outside world, the disharmony can be avoidable. For Kierkegaard, in contrary, the source of disharmony is internal; thus all dialogical activity consists of two steps – towards oneself and only after that – towards others. Therefore, according to Kierkegaard the most important thing in each and every act of communication is the act of self-understanding and self-becoming rather than giving information to someone. Therefore *praxis* for Kierkegaard is mostly inward oriented activity and in order to communicate it a special form of arranged dialogue is of a prime necessity, namely, Kierkegaard stages a situation, that makes it impossible for reader to identify with life positions encoded in the work. He creates a distance between himself and a reader. Of course, such a relationship is asymmetrical, as one of the partners (the initiator) has an advantage – he and only he alone knows possible scenarios of future relations between the author and the reader (this accounts for the term "arranged" dialogue used in the present investigation); he and only he knows that this arranged dialogue won't contribute to the clarification of the matter, but rather – it will make the initial theme less clear, less transparent, and, finally maybe non-important at all. One has to learn that derailing with such provocation in either way – to yield or to resist to it – involves revaluation of the personal attitude towards the text and the tracking those changes within personality which occurred during the process of reading.

To reach the desired effect Kierkegaard uses various rhetorical techniques to stop the dialogue for some time (to stop, not to termination) such as mixing different genres and styles, contrasting life positions and world views within a single book, abandonment of narrative conclusions, narrative ruptures, problematization of the identity of pseudonyms and disclaiming the authorial authority. Regarding this problem Kierkegaardian scholar George Pattison states: "For, like many great works in literature, Kierkegaard's writings themselves construct the role (or roles) that their readers are obliged to assume in the course of their reading. ... we must learn to reflect on how we ourselves are addressed as readers: how we are seduced, how we are abandoned, how we are provoked." (Pattison 1997, p. 292) The interrupted, stalled dialogue, in other words, compels the reader to pay attention to himself first, rather than seek the safety of collective opinion – only understanding the Self could be grounds for understanding others.

Kierkegaard contemplates the concept of maeutics, its meaning and practical applications as early as in the chapter "View made possible" devoted to the Socratic diegmatic (narrative) dialogue of his dissertation. Step by step tracking the outer manifestations and hidden meanings of the Socratic art of questioning Kierkegaard formulates his own principles of the existential maeutics. These idiosyncratic principles received the thorough explication in the reenactment of the Plato's *Symposium* later on in his authorship. The essence of the Socratic maeutics Kierkegaard grasps in the following statement: ". . . thought does not understand itself, does not love itself until it is caught up in the other's being, and for such harmonious being it becomes no only unimportant but also impossible to determine what belongs to each one, because the one always own nothing but owns everything in the other." (Kierkegaard 1992, p. 30) This means that the self-recognition starts with the knowing what other people think of us (we become the co-owners of such information), this leads to the dissolution of all borders between the self and others, and finally to the feel of one's inner poverty since the integrity of the self is lost in the process and the self becomes the source for others to know themselves. In other words, according to Kierkegaard, the true self-recognitions becomes impossible as there is no the sense of the self anymore. The existential maeutics, in contrary, is a process that ensures integrity of the self by maintaining the border between the self and others, nobody can have full knowledge about the other(s), there is always some inner residue left – something unexpressed, untold, withdrawn from the world. "Socrates' questioning was essentially aimed at the knowing subject for the purpose of showing that when all was said and done they knew nothing whatsoever." (Kierkegaard 1992, p. 37) Kierkegaard's maeutics is also directed towards the subject, bet difference lies in the result of the dialogue – not knowing. If the Socratic disciplined (because it presupposes the certain role play, where one person is an interrogator, another – a respondent) dialogue is an attempt to let the thought manifest itself in its objectivity, the Kierkegaardian arranged (there is a role play as well, but the process of interrogation and inner changes while interrogating is much more important than answers received) dialogue results in the birth of subjectivity. Therefore, the Socratic not knowing exhibits the uncertainty about the world and the self in the world, but the Kierkegaardian counterpart exhibits the uncertainty about oneself and the world within this self. For Socrates "to ask questions – that is, the abstract relation between the subjective and the objective – ultimately became the primary issue for him." (Kierkegaard 1992, p. 37) Kierkegaard, in turn, strives to create a situation where the individual could question himself, performing a kind of self-diagnostics that is possible only in the situation of solitude. This self-cognition, seclusion, in turn, is the mandatory condition for making the ethical choice what is the most important for Kierkegaard. This grants the existential status to the dialogue (existential in the sense that the stress is put upon changes within the communicating subject, not to the informative result of the communication). Moreover, in our opinion, for the sake of precision, we can to add also the aspect of negativity to the current description; this aspect characterizes limitations of the dialogue which are applied intentionally by Kierkegaard in order to create a situation of the existential shock for an individual. This initial shock, according to him, is the necessary starting point for self understanding.

Kierkegaard also discusses the difference between two concepts – speaking and interrogating. He believes that only the latter represents the maeutical relation, because "... the subject is an account to be settled between the one asking and one answering, and the thought development fulfills itself in this rocking gait (*altero pede*), in this limping to both sides." (Kierkegaard 1992, p. 35) Asking questions presupposes particular intellectual activity, the absolutely receptive relation to the subject and admittance of not knowing. "Although such a question form is supposed to free the thought from every solely subjective determinant, nevertheless in another respect it succumbs entirely to the subjective as long as the questioner is seen only in an accidental relation to what he is asking about. But if asking questions is seen as a necessary relation to its subject, then asking becomes identical with answering." (Kierkegaard 1992, p. 35) But in the negative existential maeutics such identity is impossible as any relation to the subject is mediated as each and every questioning prompts, first, the self-interrogation of another party and, second, the presentation of the result of this interrogation that only partly accords to the initial question. The main interest lies in the very process of conversing and the respective inner changes within each party during the conversation, rather than in the possible consensus about the matter and objective knowledge about the world and the self. "...intention in asking questions can be twofold. That is, one can ask with the intention of receiving an answer containing the desired fullness, and hence the more one asks, the deeper and more significant becomes the answer; or one can ask without any interest in the answer except to suck out the apparent content by means of the question and thereby to leave emptiness behind... The first is the *speculative* method; the second the *ironic*." (Kierkegaard 1992, p. 36) Irony, in turn, requires particular subjective, indirect style of communication. Kierkegaard's *In Vino Veritas* is an example of such mode of communication especially if we take into account its generic relation to the Plato's *Symposium*.

KIERKEGAARD'S *IN VINO VERITAS* AS REENACTMENT OF PLATO'S SYMPOSIUM

I know very well that I shall not soon forget that banquet in which I participated without being a participant... (Søren Kierkegaard)

The stage for *In Vino Veritas* has been set in the very beginning of the fragment: the time and the place ("So I have deliberately selected an environment on the basis of contrast. I have sought the solitude of the forest, yet not a time when the forest itself is fantastic. For example, the stillness of night would not have been conclusive, because it, too, is in the power of the fantastic. I have sought nature's peacefulness during the very time when it is itself most placid. I have, therefore, chosen the afternoon light.") (Kierkegaard 1988, p. 16), and, the most important, the temporary interior decorations to be put up just for the upcoming event ("The whole setting was to be new creation, and then everything has to be demolished – indeed, it would be all right if even before they rose from the table they were to notice preparation for demolition.") (Kierkegaard 1988, p. 23) Then the mood of the banquet is to be created by the consumption of quite an amount of wine – ". . .no one was to speak before he had drunk enough so that he could detect the influence of the wine or was in the condition in which one says a great deal that one is otherwise not inclined to say – without needing for that reason continually to interrupt continuity of the speech and the thought by hiccups." (Kierkegaard 1988, p. 30) Here we can detect the reference to Plato's dialogue and hiccoughing of Aristophanes after drinking wine. The next resemblance is the theme of the gathering, namely, the one of the erotic love; moreover, the stories told shouldn't be the descriptions of deeply personal stories in their triviality (though they can serve as the starting point for narration), they have to be of the reflective and ironically distanced nature.

The participants of the Copenhagen banquet are various fictional characters from different Kierkegaard's books – Victor Eremita (the editor) and Johannes Seducer of the "Either – Or", Constantine and Young Man from the "Repetition", as well as some previously unknown man – the dressmaker. Hence here the potential reader is confronted by a range of ethical positions, expressed in narratives on various levels: noematical (related to the narrative facts, i.e., the story itself), associative (references to other works and themes by Kierkegaard), and existential (proposition of different life views and existential choices). The very fact that *In Vino Veritas* both structurally and thematically calls on *Simposium*, assigns this fragment a special role in understanding the Kierkegaardian existential maeutics, as playing upon similarities, he makes the differences even more audible. Kierkegaard is interested in the individual rather than the nature of things and the main question he posts is: "What does determine authenticity or inauthenticity of the personality?" Kierkegaard maintains that the individuality can't be reduced to any abstract universal principle; from the viewpoint of the Greek classical philosophy such approach could be regarded as irrational. Kierkegaard wouldn't agree to that since he doesn't oppose the role of reason as such, but rather – the principle of universal objective reason. He strives to enlarge the scope of the notion of truth, placing it outside mere limits of objectivity. Kierkegaard places a special emphasis on the extra-narrative elements in Plato's dialogue such as, Aristophane's hiccoughing, Eryximachus helping him to overcome this misdemeanor, arrival of drunken Alcibiades, Socrates' coming late, arrival of the loud troop of revelers at the very end of the party; each of these episodes are being commented (often if an ironical manner), and all this, according to Kierkegaard serves the purpose to interrupt the dialogue. For example, when belated Socrates arrives, Agathon invites him to lie down besides saying: "I may touch you and have the benefit of that wise thought which came into your mind in the portico, and is now in your possession; for I am certain that you would not come away until you have found what you sought." (Plato, p. 126) The structure of *Symposium* and roles played there by different actors Kierkegaard describes in the following way: "Thus all these speeches are like sliding telescope; the one presentation ingeniously merges into the other and in the process is so lyrically effervescent that it is like wine in crystal so artfully polished that it is not only the bubbling wine in it that intoxicates but also the infinite refraction, the light that blazes forth when one looks down into it." (Kierkegaard 1992, p. 42) This means that every time we cast a glance upon the dialogue, it presents a different facet, a different relation between persons involved; a different set of meanings emerges. It seems that this changing perspective, this rocky gait (*alterno paede*), this unpredictability of the dialogue allows Kierkegaard in his creative reenactment (not imitation, not literal rereading) to practice his own existential maeutics. Kierkegaard makes use also of the Platonic tactics of the double recollection – both dialogues are stories told by people who had heard them from somebody else. Apollodorus repeats the dialogue which he had heard from Aristodemus, and had already once narrated to Glaucon. But even Aristdemus has to rely on other eye witnesses as he falls asleep and doesn't follow the course of events. "Aristodemus was only half awake, and he did not hear the discourse. . ." (Plato, p. 186) The role of Arisodemus in *In Vino Veritas* is being played by William Afham, a silent witness. Certain similarities can be found in the ending of both dialogues – at the day break Socrates leaves others sleeping and goes off at first to the Lyceum and then home. "Thus the dialogue would presumably end without a conclusion, but this 'without a conclusion' is by no means synonymous with a negative conclusion." (Kierkegaard 1992, p. 55) Kierkegaard believes that such conclusion without conclusion is a deliberate step taken by Plato (alias Socrates) in order to leave a reader in the state of not knowing. Whereas in Copenhagen the first rays of sun illuminate the idyllic scene – Judge William and his wife having early morning tea and demonstrated the blissful peace of the married life. No need to remind that Judge William is but one more character populating Kierkegaard's "Either – Or." This fragment ends in the surprise conclusion that contradicts to everything done and said during the banquet. So none of the onlookers "seemed gratified by this outcome, but others were content with making a malicious remark." (Kierkegaard 1988, p. 85) Such an ending, as Kierkegaard sees it, serves as effective instrument to disrupt the unity of the literary piece at hand, and the purpose in doing so is to create a situation where the reader starts to question his own understanding of the material, of the position proposed by the author; such questioning, according to Kierkegaard, is the mandatory precondition for becoming the self, i.e. for actualization of one of the multiple existential possibilities.

The first speaker the Young Man in *In Vino Veritas* presents the scope of problems to be discussed (heterosexual erotic love and marriage) and sets the tone for the further speeches (in general quite arrogant towards women and feminine matters). The opening question is the crucial one – is the erotic love possible at all? According to the Young Man – it is not possible as, on one hand, feelings of love are irrational and from the viewpoint of rationalism they have no sense ("Therefore, you see, in my view Eros is the greatest contradiction imaginable – and comic as well." (Kierkegaard 1988, p. 33)); on the other hand – the idealization of the feeling makes a man unable to fall in love with a real person. The similar contradiction prevails in the very idea of marriage – if the man comes into this world as a whole being then why during his life course does he suddenly feel like a part of that whole? And the child born in this marriage repeats the cycle of being the whole first and then a part later on, thus the tragedy of life is ever growing. The latter statement reminds Aristophane's declaration in the *Symposium*: "Methinks I have a plan which will humble their pride and improve their masnners; men shall continue to exist, but I will cut them in two and then they will be diminished in strength and increased in numbers..." (Plato, p. 144) The similarity of these statements is by no means a coincidence, the reader is to be reminded of the Plato's dialogue in order to have several layers of meaning. The second speaker of *In Vino Veritas* reflects on the feminine inconsistency and dependability upon outer circumstances (women are only relative rather than self substantial beings). He states: "And now for woman, of whom I will speak. I, too, have pondered and have fanthomed her category; I, too, have sought but have also found and have made a matchless discovery, which I now communicate to you. She is properly construed only under the category of jest. It is the man's function to be absolute, to act absolutely, to express the absolute; the woman exists in the relational." (Kierkegaard 1988, p. 48) The most important aspect of the present statement, in our opinion, is the one of doubled reflection, namely, the woman is the construction by help of which the man can carry out his self-reflection. In other words, for the man to understand himself another person (here – a woman) is necessary, but not a real living person, rather – a construct to be used for his own purpose of self-knowing. here again we can see the exposition of Kierkegaard's negative maeutics as the dialogue between these two species of human race is impossible. "Between two such different entities no real interaction can take place." (Ibid.) Victor Eremita turns against marriage as the end of the ideality, because marriage inevitably leads to the philistinism (the woman is not capable of theoretical reflection, her live is the one of everydayness). The Dressmaker, in turn, accentuates that the woman is not worth even to be the object of erotic imagination. It seems that the one and only person in the dialogue to praise the woman is Johannes Seducer. But after criticizing all the previous speakers and their respective positions he gradually comes to the conclusion, that the woman is nothing more than an empty abstraction, a caprice, an instrument for self-reflection, self-construction. "Woman, even less than the god, is whim from a man's brain, a daydream, something one hits upon all by oneself and argues about *pro et contra*." (Kierkegaard 1988, p. 73) Thus after the last speech in our disposal there is a set of quite similar in their attitudes speeches. Doesn't it contradict to the hypothesis of the present paper that Kierkegaard offers different views in order to preclude identification with one single position? Because now the single, it seems, position is being reinforced by multiple repetitions. Yes, it may be so, but then we have to look for the existential content of the dialogue not in the narrative structures of the story itself, but in its generic relation to the Plato's text (playing upon similarities and differences), as well as in the underlying questions: "What does it mean to be the authentic self? Can the authenticity of the self be gained by the means of erotic love?" As to these questions in the end we receive both negative and positive answers – the negative in the speeches of the banqueters, whereas the positive one in the final scene with Judge William and his wife. These contradicting answers compel the reader to make his personal choice on the basis of all material read and changes that occurred within him while reading. "Kierkegaard's unsettling maeutics seeks to keep the individual on the journey to selfhood by preventing the sojourner from sinking roots too deeply in finitude. ... Kierkegaard's Socratic midwifery attends a spiritual rebirth effected by the volitional repetition of transcendent possibility, instead of the cognitive recollection of immanent ideality." (Taylor 1980, p. 104) This description by Mark C. Taylor the most precisely characterizes Kierkegaard's existential maeutics. Kierkegaard prompts the individual to become what he is not. Aesthetical, ethical and religious are not only stages on the life's way but also the steps in self understanding (natural, ethical and religious). The natural stage signifies the emancipation of the self from the non-differentiated status (immediacy) and the beginning of initial self-reflection. The ethical, in turn, comprises the self-realization of the individual in his concreteness, and manifestation of this stage is the ability to make a deliberate decision. However, the authentic self for Kierkegaard is the religious self – "... a person who is fully conscious of the responsibility he bears for his own life constitutes his unique individuality by decisively distinguishing himself from the other selves and by defining his eternal identity in the face of the wholly other God." (Taylor 1980, p. 252) For Kierkegaard the maeutics is first of all the pedagogic strategy to be accomplished only in the indirect manner, i.e. the individual is to be lured in becoming the self. Instead of offering the concrete solutions the author withdraws himself and leaves the reader alone in front of various models of interpretation and existential codes to make a decision on his own. The confrontation (not harmonization) of these models is the place where, as Kierkegaard believes, the self-reflection can start. Nevertheless at first it is only the potency of reflection (immediate existence), the actualization of this potency requires free, unique exercising of the will (reflection) and only after that – the measuring oneself up with the eternity (secondary immediacy). The form of presentation should be suitable for gradually involving the reader in the dramatic dialogue with different personas and the – with himself. In order the reader could be tricked into self-reflection he must understand the text and therefore the author must understand what the potential reader knows and where his interests lie, the suitable form of indirect presentation must be chosen. Kierkegaard compares this maneuver with an attempt to talk sense into the person who is in love and whose infatuation seems ridiculous and unworthy. In case the language is inappropriate for the case, the lover will withdraw in himself and no talk would be possible at all. There wouldn't be a better result in the case of a total identification with a position expressed by the author or some character. In both cases there is no real maeutical relation, in Kierkegaard's view, as there is neither a connection (no ongoing dialogue), nor separation (subject and object become one, thus there is no dialogue as well). The solution to this dilemma, proposed by Kierkegaard is his strategy of existential negative maeutics – a movement towards the self, more precisely, towards self becoming the self.

Kierkegaard's use of Plato in his *In Vino Veritas* is by no means accidental, just a matter of choice, for him the reenactment of the Plato's *Symposium* serves the purpose to promote his own philosophical views in the indirect manner. He deliberately plays upon the similarity of both works (establishing the field of references), as well upon their dissimilarity and incongruity (demonstrating the difference between the classical Greek conception of maeutics and his own conception of the negative existential maeutics).

Institute of Philosophy and Sociology, University of Latvia, Riga, Latvia
e-mail: vvevere@latnet.lv

REFERENCES

Hannay, Alstair. 2001. Something on hermeneutics and communication in Kierkegaard after all. *Søren Kierkegaard Newsletter* 42(September): 8–14.
Kierkegaard, Søren. 1988. *Stages on life's way* (ed. and trans: Hong, Howard W. and Edna H. Hong) Kierkegaard's Writings XI. Princeton, NJ: Princeton University Press.
Kierkegaard, Søren. 1992. *The concept of Irony with continual reference to Socrates/notes on Schelling's Berlin lectures* (ed. and trans: Hong, Howard W. and Edna H. Hong) Kierkegaard's Writings II. Princeton, NJ: Princeton University Press.
Pattison, George. 1997. If Kierkegaard is right about reading, why read Kierkegaard? In *Kierkegaard revisited*, eds. Niels Jørgen Cappelørn and John Stewart, Kierkegaard Studies, Monograph Series 1, 291–309. Berlin, New York: Walter de Gruyter.
Plato. 1977. Symposium. In *The portable Plato*, ed. Scott Buchanan, 121–187. New York, NY: Penguin Books.
Taylor, Mark C. 1980. *Journeys to selfhood. Hegel & Kierkegaard*. Berkley and Los Angeles, CA: University of California Press.
Tymieniecka, Anna-Teresa. 1990. *The passions of the soul and the elements in the Onto-Poiesis of culture,* Logos and Life 3. Dordrecht, Boston, London: Kluwer.

WEI ZHANG

GIBT ES EIN MATERIALES APRIORI?

Mit Moritz Schlicks Kritik An Der Phänomenologie Über Das Verhältnis Zwischen Sprache Und Vernunft Nachzudenken Anfangen

ZUSAMMENFASSUNG

Für die Phänomenologen gibt es freilich ein materiales Apriori, aber diese phänomenologische Behauptung wird zu einer Frage „*Gibt es ein materiales Apriori?*" bei den Opponenten der Phänomenologie, vor allem bei Moritz Schlick. Für Schlick, ist ein materiales Apriori unmöglich. Aber der größte und gründlichste Trugschluss in den Kritiken Schlicks an der Phänomenologie besteht darin, dass Schlick alle Probleme auf einen Satz und ihren Wahrheitswert immer voreilig reduzierte, deswegen können seine Kritiken nicht das phänomenologische eigene Problem treffen. Wahrscheinlich sollen wir Schlick nicht als bloße Opposition ansehen, sondern als Spiegel. In diesem Spiegel kann die Phänomenologie über sich vielleicht besser und tiefer nachdenken. In der Tat kann man finden, dass die Frage „Gibt es ein materiales Apriori?" eigentlich zwei Fragen der verschiedenen Stufen in sich schließen kann, nämlich, erstens: „Gibt es sowohl ein anschauliches Apriori als auch ein grammatikalisches Apriori?" und zweitens: „Welches Verhältnis gibt es zwischen dem anschaulichen Apriori und dem grammatikalischen Apriori?" Während man durch das Prinzip der Selbstgegebenheit oder der absoluten Evidenz auf die erste Frage antworten kann, können wir aufgrund der Lehre der Funktionalisierung der Wesenseinsicht auf die zweite Frage antworten. In diesem Sinne ist es gerade möglich, dass die Struktur des Denkens und die Struktur der Sprache identisch zu sein scheinen und das grammatikalische Apriori auf das anschauliche Apriori fundiert ist.

Es wird als phänomenologisch wichtige Einsicht angesehen, den Gegensatz „a priori – a posteriori" mit dem Gegensatz „formal – material" nicht zu identifizieren, sondern diese zwei Gegensätze zu unterscheiden. Im Gegensatz zu dem materialen Apriori als Kuriosum bei Kant betonen sowohl Husserl als auch Scheler dieses materiale Apriori als reine Tatsache. Scheler führt klarer weiter aus: „Phänomenologie steht und fällt mit der Behauptung, es *gebe* solche Tatsachen –

Wei ZHANG (Dr. Phil.) lehrt Philosophie und Phänomenologie an der Fakultät für Philosophie der Universität Sun Yat-sen (510275 Guangzhou, V. R. China). Seine Forschungsschwerpunkte sind Phänomenologie, Ethik, Konfuzianismus und der deutsche Idealismus. Seine hauptsächliche Publikation ist: *Prolegomena zu einer materialen Wertethik. Schelers Bestimmung des Apriori in Abgrenzung zu Kant und Husserl* (Nordhausen: Verlag Traugott Bautz 2011).
E-mail: renzhizhangcn@gmail.com, renzhizhang@hotmail.com

A.-T. Tymieniecka (ed.), Analecta Husserliana CX, 123–138.
DOI 10.1007/978-94-007-1691-9_11, © Springer Science+Business Media B.V. 2011

und *sie* seien es recht eigentlich, die allen anderen Tatsachen, den Tatsachen der natürlichen und der wissenschaftlichen Weltanschauung, zugrunde lägen, und deren Zusammenhänge allen anderen Zusammenhängen zugrunde lägen." (X, S. 448) Dieser doppelten Behauptung kann aufgrund der ursprünglichen Bedeutung des Apriori die Behauptung, *es gebe ein materiales Apriori*, angegliedert werden. Aber diese phänomenologische Behauptung wird zu einer Frage *„Gibt es ein materiales Apriori?"* bei den Opponenten der Phänomenologie, vor allem bei Moritz Schlick.

Für die Phänomenologen gibt es nicht nur ein materiales Apriori, sondern auch „überall dort materiale Apriontäten", „wo sich Geist in irgendeiner seiner Aktarten aktuiert."[1] Aber für Moritz Schlick, ist ein materiales Apriori unmöglich, sie sprechen immer von dem Irrtum, „der von den Verfechtern des materialen Apriori begangen wird".[2] Diese zwei verschiedenen Positionen stoßen sich von Anfang an ab, so dass es nicht möglich scheint, sie zueinander zu vermitteln. Mit den Worten von E. Tugendhat ist dies ein Streit auf Leben und Tod und nur eine Position kann nur weiter überleben.[3] Deswegen muss man eine Position alternativ einnehmen. Hier werden wir zunächst Schlicks Kritik an Husserl und Scheler kurz umreißen, und dann versuchen wir, für Phänomenologie einzutreten. Wahrscheinlich sollen wir Schlick nicht als bloße Opposition ansehen, sondern als Spiegel. In diesem Spiegel kann die Phänomenologie über sich vielleicht besser und tiefer nachdenken. Am Ende werden Wir einige Ergebnisse dieser Reflexion, vor allem das Verhältnis zwischen Sprache und Vernunft, zu erklären versuchen.

ES GIBT KEIN MATERIALES APRIORI: M. SCHLICKS KRITIK AN HUSSERL UND SCHELER

Wir müssen im Rahmen unseres Themas den Streit zwischen Schlick und der Phänomenologie (vor allem bei Husserl und Scheler) nicht detailliert wiedergeben.[4] Für uns relevant ist die Tatsache, dass Schlicks Kritik an Husserl und Scheler meiner Ansicht nach zwei grundsätzliche Seiten besitzt. Es geht erstens um die Intuition und Wesensschau bzw. Ideation und zweitens um ein materiales Apriori.

Schlick hat das **Erkennen** vom **Kennen** bereits klar unterschieden. Der Unterschied deckt sich mit dem Gegensatz des Nichtmitteilbaren und des Mitteilbaren. Nach Schlick bedeutet „etwas kennen" etwas wesentlich anderes als „etwas erkennen": „kennen" kann man etwas nur durch das Erleben, und dieses ist stets qualitativ; es lässt sich nicht mitteilen, sondern nur im Erlebnis unmittelbar aufzeigen. Dagegen ist „erkennen" immer objektiv und mittelbar, „etwas erkennen" bedeutet, dass sich etwas in einem Urteil oder Satz ausdrückt.[5] Damit ist nach Schlick der große Fehler aufgedeckt, den die Intuitionsphilosophen, z. B. Husserl, begehen: „Sie verwechseln Kennen mit Erkennen. [. . .] Kennen und Erkennen sind so grundverschiedene Begriffe, dass selbst die Umgangssprache dafür verschiedene Worte hat; und doch werden sie von der Mehrzahl der Philosophen hoffnungslos miteinander verwechselt. Der rühmlichen Ausnahmen sind nicht allzu viele. Der Irrtum ist zahlreichen Metaphysikern verhängnisvoll geworden."[6] Das heißt, alle metaphysischen Lehren, z. B. der Voluntarismus, der Bergsonsche Vitalismus und natürlich die Phänomenologie, beruhen nach Schlick auf der Verwechslung von „Kennen" oder „Erleben" und „Erkennen", wenn sie das Transzendente statt das Formale zu erkennen, intuitiv zu erleben versuchen. Aus diesem Grund hielt Schlick alle intuitive Metaphysik für Nonsens, d. h. für eine widersprüchliche Wortverbindung.

Schlick hat den Unterschied zwischen Wissenschaft und Philosophie sowie Metaphysik in Bezug auf das Verhältnis beispielsweise des Satzes erklärt. Man kann sagen: Durch die Philosophie werden Sätze geklärt, durch die Wissenschaften werden Sätze verifiziert, die Metaphysik jedoch hat mit Sätzen nichts zu tun, sondern lediglich mit „Scheinsätzen". Der grundlegende Gedankengang Schlicks lässt sich durch das folgende Schema zusammenfassen:

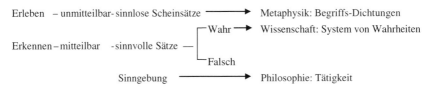

Für Schlick ist dabei zentral, dass der Unterschied zwischen der Falschheit und der Sinnlosigkeit eines Satzes beachtet wird. Durch diesen Unterschied lässt sich das Folgende verstehen: „Der Empirist sagt dem Metaphysiker nicht: »Deine Worte behaupten etwas Falsches«, sondern »Deine Worte behaupten überhaupt nichts!« Er widerspricht ihm nicht, sondern er sagt: »Ich verstehe dich nicht«."[7]

Eben in diesem Sinne gehört die Phänomenologie nach Schlick noch zur intuitiven Metaphysik. Obwohl die zentrale Lehre von der „Ideation" in Husserls Phänomenologie „tatsächlich Richtiges" enthält, fragt Schlick weiter: „Nur ist zur Lösung unseres Problems damit nicht das Geringste geleistet, man hat ihm nur wieder einen neuen Namen gegeben. Wir müssen nämlich weiter fragen: Ist nicht das intentionale Erlebnis als reale psychische Größe von den idealen Gebilden ebenso weit und unüberbrückbar getrennt, wie etwa die Vorstellungen von den Begriffen? Woher weiß ich denn, worauf meine Akte sich richten? Bin ich mit ihnen nicht wieder mitten in der Psychologie, ohne Aussicht, in das Gebiet der Begriffe und der Logik hinüber zu gelangen, wo allein die Strenge und Schärfe herrscht, um deren Möglichkeit wir besorgt waren?"[8]

Das bedeutet: Die Ideation oder Wesensschau in der Phänomenologie und sogar die Anschauung als Prinzip aller Prinzipien werden bei Schlick zum Kennen oder Erleben (nicht zur Erkenntnis), die Phänomenologie wird zur intuitiven Metaphysik (ist also weder Wissenschaft noch Philosophie). Daher kann man die Unmöglichkeit der Phänomenologie und des Intuitionismus betonen. Wie der Schlick-Schüler Julius Kraft sagt, beruht die phänomenologische Methodik der Wesensanschauung „auf einfachen logischen Fehlern" und hält „diese Fehler auf Grund vielfältiger Selbsttäuschungen für Intuitionen".[9]

Um zu demselben Schluss zu kommen, kritisiert Schlick das materiale Apriori in der Phänomenologie. Sowohl die „Wesensschau" bei Husserl als auch „die phänomenologische Erfahrung" bei Scheler werden im Gegensatz zur reinen Anschauung

Kants von Schlick als „Quelle schlechthin allgemeingültiger Sätze" angesehen. „Es wäre natürlich völlig unerlaubt", das Apriori der Phänomenologen als idealen Gegenstand zu bezeichnen. Denn „ein analytischer Satz ist ein solcher, der vermöge seiner bloßen Form wahr ist; wer den Sinn einer Tautologie verstanden hat, hat damit zugleich ihre Wahrheit eingesehen; deshalb ist sie a priori. Bei einem synthetischen Satz aber muss man zuerst den Sinn verstehen, und hinterdrein feststellen, ob er wahr oder falsch ist; deswegen ist er a posteriori."[10] Schlick hat a priori wieder mit der Form verbunden und im Gegensatz zu Scheler betont, dass die Identifizierung des Apriorischen und Formalen bei Kant nicht „Voraussetzung oder Vorurteil" ist, sondern „Ergebnis". In der Tat bedeutet die Form hier für Schlick nur die Form der Sätze. Gerade in diesem Sinne übt Schlick scharfe Kritik an der Kantischen Lehre von den synthetischen Urteilen a priori sowie an der Husserlschen und auch Schelerschen Phänomenologie.

Nach den neueren Entwicklungen der mathematisch- physikalischen Wissenschaften könne es überhaupt keine synthetischen Sätze a priori geben, die nichts als analytische bzw. nur tautologische Sätze a priori seien. Schlick macht klar, „dass alle Aussagen, prinzipiell gesprochen, entweder synthetisch a posteriori oder tautologisch sind; synthetische Sätze a priori scheinen ihm eine logische Unmöglichkeit zu sein."[11] In demselben Sinne hat Schlick ein materiales Apriori abgelehnt. Für ihn ist ein materiales Apriori wie ein synthetisches Urteil a priori logisch unmöglich. Zugleich hat man nach Schlick auch kein irgendwie besonderes Erkenntnisvermögen (z. B. Wesensschau und die phänomenologische Erfahrung), um solche Sätze oder ein solches materiales Apriori gewinnen zu können.

Schlick hat die Phänomenologen damit in ein Dilemma gebracht: Weil es keine synthetischen Sätze a priori oder materialen Sätze a priori gibt, sind Sätze dieser Art in Wahrheit entweder nicht synthetisch oder nicht a priori.[12] Sind sie nicht a priori, sind alle von den Phänomenologen behaupteten Sätze oder Urteile a posteriori; damit gibt es natürlich kein Wesen oder materiales Apriori als idealen Gegenstand usw. Sind die phänomenologischen Sätze oder Urteile nicht synthetisch, sind sie „rein formal- tautologisch", sodass Schlick sogar betonen kann: „Als nichtssagende Formeln enthalten sie keine Erkenntnis und können nicht als Grundlage einer besonderen Wissenschaft dienen. Eine solche Wissenschaft, wie die Phänomenologen sie uns versprachen, existiert ja auch in der Tat nicht."[13]

Aber muss man danach fragen, ob Schlicks Kritiken an der Phänomenologie Husserls und Schelers das phänomenologische eigene Problem treffen können? Kann Schlick daher die revolutionäre Bedeutung der Wesensschau und des materialen Apriori sowie der Phänomenologie selbst anerkennen?

DIE WESENSSCHAU ALS GENUINE METHODE DER ERFASSUNG DES APRIORI

Auch wenn Husserl sich häufig direkt auf Kant bezieht – er bezeichnet z. B. seine Phänomenologie auch als Transzendentalphilosophie – so entfernt er sich doch hinsichtlich des Begriffes „Apriori" entschieden von Kant. Wie oben erwähnt, behauptet Husserl an zahlreichen Stellen, dass Kant der phänomenologisch echte Begriff des Apriori gefehlt habe. Er hat den Begriff „Apriori" bei Kant daher auch als einen „halb mythischen Begriff" bezeichnet,[14] zu dessen Verwendung er sich nicht „herablassen" will. Statt dessen zieht er es vor, den Begriff „Apriori" bei Hume zu erklären, um die Philosophie als strenge fundamentale Wissenschaft zu begründen.[15]

Aber was bedeutet eigentlich für Husserl das Apriori? Kürzer gesagt ist a priori bei Kant hauptsächlich ein Adjektiv, dagegen verwendete Husserl Apriori als ein Nomen. Beispielsweise bedeutet die „Anschauung a priori" bei Kant vor allem eine Art von Form der Anschauung bzw. den subjektiv-strukturellen Charakter a priori, hingegen wird aus der „Anschauung a priori" von Kant bei Husserl „Anschauung *des Apriori*", das heißt, Apriori kann für Husserl in der Anschauung erfasst werden, hat also gegenständlichen Charakter.[16] Aber es ist jetzt zu erklären, wie man eigentlich das Apriori phänomenologisch erfassen kann.

Husserls Bestimmung des gegenständlichen Apriori ist mit seiner Betonung des Seins der idealen oder allgemeinen Gegenstände eng verbunden. In den *II. Logischen Untersuchungen* (LU) analysiert Husserl die Seinsweise der idealen oder allgemeinen Gegenstände. Im Gegensatz zum traditionellen Nominalismus behauptet er, dass die idealen oder allgemeinen Gegenstände wahrhaft existieren. „Es hat evidenterweise nicht bloß einen guten Sinn, von solchen Gegenständen (z. B. von *der* Zahl 2, von *der* Qualität Röte, von dem Satz des Widerspruches u. dgl.) zu sprechen und sie als mit Prädikaten behaftet vorzustellen, sondern wir erfassen auch *einsichtig* gewisse kategorische Wahrheiten, die auf solche ideale Gegenstände bezüglich sind."[17] Damit verteidigt Husserl die „Eigenberechtigung" der idealen oder allgemeinen Gegenstände neben den realen oder individuellen Gegenständen. Aber die idealen Gegenstände haben für Husserl im Gegensatz zum Platonischen Realismus ihnen eigentümliche Seinsweisen. Das heißt, er lehnt sowohl die psychologische Hypostasierung der idealen Gegenstände als auch ihre metaphysische Hypostasierung ab.[18] In der Tat sind die idealen Gegenstände als eigentümliche Seins-Art der Gegenstände uns laut Husserl in einer einsichtigen Ideenschau selbst gegeben. Das gegenständliche Apriori gehört gerade zu solchen idealen Gegenständen.

Diese einsichtige Ideenschau, in der uns das Apriori selbst gegeben ist, wird von Husserl als „Ideation" oder „ideierende Abstraktion" und später „Wesensschau" oder „Wesensanschauung" bezeichnet. Im Unterschied zu der Abstraktion im Empirismus, die die „Hervorhebung irgendeines unselbstständigen Moments an einem sinnlichen Objekte" bedeutet, betont Husserl diese „ideierende Abstraktion", „in welcher statt des unselbstständigen Moments seine »Idee«, sein Allgemeines zum Bewusstsein, zum *aktuellen Gegebensein* kommt."[19] Wie schon gesagt, beruft sich die Phänomenologie als universalster und konsequentester Empirismus auf den erweiterten Begriff der Erfahrung oder Anschauung. Damit hat Husserl bereits in den *Logischen Untersuchungen* den Begriff der Anschauung erweitert: Neben der sinnlichen Anschauung gibt es auch die kategoriale Anschauung.[20] Nach Husserl kann man gemäß der Weise der gegenständlichen Beziehung zwei verschiedene Arten der kategorialen Anschauung, den *synthetischen* Akt und den

abstraktiven Akt, unterscheiden.[21] Der Letztere ist die hier besprochene „ideierende Abstraktion" oder „Ideation" und wird später von Husserl als „Wesensschau"
oder „Wesensanschauung" bezeichnet.

In diesem Sinne kann man sagen: „Die Rechtmäßigkeit des Anspruchs der
Phänomenologie, Wissenschaft zu sein, hängt also davon ab, ob die Methode der
Wesensschau als eine Form der Erkenntnis (d. h. als eine Form der kategorialen
Anschauung) begründet werden kann. Da Husserls Phänomenologie den Anspruch
erhebt, für sich als Methode letzte Selbstbegründung und Selbstrechtfertigung zu
bieten, ist die Klärung der ideierenden Abstraktion ein entscheidendes Ziel der
Logischen Untersuchungen."[22]

Husserl hat in der VI. LU die Wesensschau bzw. diese ideierende oder generali-
sierende Abstraktion als eine besondere Form der kategorialen Anschauung ausführ-
lich analysiert. Hier können wir beispielsweise die Wesensschau des Allgemeinen
„Rot", wie Husserl an zahlreichen Stellen formuliert, erklären.

Nach dem zuerst in den *Logischen Untersuchungen* bestimmten
Fundierungsverhältnis[23] sind alle kategorialen Anschauungen einschließlich
der ideierenden Abstraktion oder Wesensschau in der schlichten Anschauung
einseitig fundiert, das heißt, die kategoriale Anschauung der idealen Gegenstände
oder des gegenständlichen Apriori muss sich auf die schlichte Anschauung
individueller Gegenstände berufen. Wenn wir das Allgemeine „Rot" erfassen
wollen, müssen wir von einer singulären Anschauung von etwas Rotem ausgehen.
Das ist der erste Schritt im „Dreischritt" der kategorialen Anschauung; er wird von
Husserl als „Gesamtwahrnehmung" bezeichnet.[24] In dieser Gesamtwahrnehmung
wird etwas Rotes (A) als Ganzes gemeint, zugleich wird ihr unselbstständiges
Moment „Rot" (a) nicht als expliziter Gegenstand mitgemeint.

Und „in der Einschränkung der Gesamtwahrnehmung zur Sonderwahrnehmung
wird nun die Partialintention auf das a nicht aus der Gesamterscheinung des *A*
herausgerissen, als ob dessen Einheit in Brüche ginge; sondern in einem *eigenen*
Akt wird das a zum eigenen Wahrnehmungsobjekt."[25] Die Sonderwahrnehmung als
zweiter Schritt der kategorialen Anschauung wird von Husserl auch „gliedernder
Akt" genannt. In dieser Sonderwahrnehmung wird das unselbstständige Moment
„Rot" (a) in etwas Rotem (A) „in explizierender Weise" gemeint. Das bedeutet aber
keinen Wechsel des gemeinten Gegenstandes, der immer etwas Rotes (A) bleibt,
sondern ein Wechsel der gemeinten Weise des unselbstständigen Momentes „Rot"
(a). Husserl sagt: „Der auf das a bezügliche Repräsentant fungiert als identisch
derselbe in doppelter Weise und indem er es tut, vollzieht sich die Deckung als
die eigentümliche Einheit der beiden repräsentativen Funktionen."[26] Es hat sich
nämlich eine „Deckungssynthesis" oder „Deckungseinheit" zwischen der expliziten
ten Intention der Sonderwahrnehmung auf das unselbstständige Moment „Rot" (a)
und der impliziten Partialintention der Gesamtwahrnehmung auf das Rot in dem
Übergang von der Gesamt- zur Sonderwahrnehmung ergeben.[27] Zugleich hat sich
im zweiten Schritt eine andere bestimmte Art von Deckungseinheit zwischen den
durchlaufenen gliedernden Akten, die auf das unselbstständige Moment „Rot" (a)
gerichtet sind, ergeben.[28]

In der kategorialen Synthesis, die der entscheidende dritte Schritt der kategorialen Anschauung ist, dient eine solche „Deckungseinheit" zwischen den durchlaufenen gliedernden Akten als Repräsentant für die kategoriale Intention.[29] Das Allgemeine „Rot" als idealer Gegenstand ist gerade „durch die Reihe der Sonderwahrnehmungen individueller Gegenstände und in der Überdeckung ihrer intentionalen Bestandteile" selbst anschaulich gegeben.[30] „Wir blicken auf das Rotmoment hin, vollziehen aber einen eigenartigen Akt, dessen Intention auf die »Idee«, auf das »Allgemeine« gerichtet ist.[31]

Weiter behauptet Husserl, dass „man an einem Typus, etwa repräsentiert durch die Idee »rot«, Ideen sehen und sich das Wesen solchen »Sehens« klarmachen lerne."[32] Das heißt, nach diesem Grundmuster oder Typus sind sowohl die Allgemeinheiten höherer Stufe (z. B. der Begriff der Farbe überhaupt) als auch das Wesen der Bewusstseinsakte, die ideierende Abstraktionen vollziehen, selbst anschaulich aufgefasst. Also sagt Husserl betont: „Wesensschauung birgt nicht mehr Schwierigkeit oder »mystische« Geheimnisse als Wahrnehmung. Wenn wir uns intuitiv zu voller Klarheit, zu voller Gegebenheit bringen »Farbe«, so ist das Gegebene ein »Wesen«, und wenn wir uns ebenso in reiner Schauung, etwa von Wahrnehmung zu Wahrnehmung blickend, zur Gegebenheit bringen, was »Wahrnehmung«, Wahrnehmung an sich selbst – dieses Identische beliebiger fließender Wahrnehmungssingularitäten – ist, so haben wir das Wesen Wahrnehmung schauend gefasst. Soweit Intuition, anschauliches Bewussthaben reicht, soweit reicht die Möglichkeit entsprechender »Ideation« [. . .] oder der »Wesensschauung«. Soweit die Intuition eine reine ist, die keine transienten Mitmeinungen befasst, soweit ist das erschaute Wesen ein adäquat Erschautes, ein absolut Gegebenes."[33]

Mit einem Wort: „Die Wesensschau als genuine Methode der Erfassung des Apriori"[34] erfasst das Apriori als „Wesenssein" und setzt in keiner Weise Dasein. Hierbei wird „die Priorität der phänomenologischen Methode vor der transzendental-logischen Kants" nach der Auffassung Thomas Seebohms nicht in Zweifel gezogen.[35] Husserls Kritik an Kant übt großen Einfluss auf die erste Phänomenologen-Generation aus, was natürlich auch für Max Scheler gilt. Obwohl Scheler auch Husserls Besinnung des Apriori kritisierte, nimmt Schelers Kritik an Kant bezüglich des Begriffs des Apriori die Einsicht Husserls in großem Ausmaß auf. Daher kann man sagen, dass Schelers Kritik an Kant die Kritik Husserls an Kant ergänzt und vertieft. Man kann wahrscheinlich sagen, dass Scheler mit Husserl in der Kritik an Kant bezüglich der Lehre des Apriori ungefähr übereinstimmt. Beispielsweise behaupten Scheler und Husserl beide, (1) ein gegenständliches Apriori als idealer Gegenstand; (2) die Wesensschau oder Ideation als genuine Methode der Erfassung des Apriori; (3) den Unterschied zwischen dem materialen Apriori und dem formalen Apriori.

Man kann feststellen, dass Schlicks Kritikpunkte an der Phänomenologie zahlreiche Missverständnisse auszeichnen, wenn sie nicht sogar im Ganzen ein Missverständnis darstellen. Wenn wir die kleineren Missdeutungen Schlicks außer Acht lassen, besteht der größte und gründlichste Trugschluss der Kritik

Schlicks an der Phänomenologie meiner Ansicht nach darin, dass er alle Probleme voreilig auf einen Satz und ihren Wahrheitswert reduziert, sowohl in der Kritik an der Wesensschau oder Ideation als auch in der Kritik an dem materialen Apriori. Daher vermag seine Kritik das der Phänomenologie eigene Problem nicht zu treffen, so dass man ihr sehr einfach widersprechen kann. So hat z. B. hat Scheler den Unterschied zwischen dem Intuitionismus und der phänomenologischen Anschauung bereits klar hervorgehoben. (Vgl. XI, S. 23 ff.) Es heißt bei ihm deutlich: „Doch ist diese bei Bergson wenig klare Lehre von der Intuition nicht mit der streng und eng begrenzten »Wesensschau« der Phänomenologie zu verwechseln". (III, S. 327, Anm. 1) Auch meint nicht nur das materiale Apriori in der Phänomenologie die Sätze a priori, sondern vor allem das anschauliche Apriori, das Schlick tatsächlich übersehen hat. Wie bereits bemerkt, beruht die Phänomenologie auf dem Prinzip der *Selbstgegebenheit* oder der *absoluten Evidenz*, das nach Schlick bei den Phänomenologen „viel mehr psychologistisch" als bei Kant ist.[36] Aus diesem Grund kann er jedoch die revolutionäre Bedeutung der Wesensschau und des materialen Apriori sowie der Phänomenologie selbst nicht anerkennen. Vielleicht kann man sagen, dass es Schlick eigentlich nicht besonders im Sinn lag, das Apriori anschaulich zu erfassen. Husserl hat im Voraus danach gefragt: „Wie könnten wir ihn überzeugen, unter der Voraussetzung, dass er keinen anderen Sinn hätte?"[37]

Wahrscheinlich sollten wir Schlick nicht als bloßen Opponenten betrachten, sondern als Spiegel. In diesem Spiegel kann die Phänomenologie über sich selbst möglicherweise besser und tiefer nachdenken. Wir werden einige Ergebnisse dieser Reflexion zu erklären versuchen.

SYNTHETISCHES APRIORI UND DIE FUNKTIONALISIERUNG DER WESENSEINSICHT

Am 25. Dezember 1929 fragt Schlick in einem Gespräch mit Wittgenstein: „Was kann man einem Philosophen erwidern, der meint, dass die Aussagen der Phänomenologie synthetische Urteile a priori sind?" Diese Unterhaltung wurde unter dem Titel „Anti- Husserl" protokolliert.[38] In der Tat hat Wittgenstein in eben diesem Gespräch sowohl Kant und Husserl als auch Schlick selbst kritisiert. Wir interessieren uns hier vor allem für seine Kritik an Schlick. Er behauptet: „In der Phänomenologie handelt es sich immer um die Möglichkeit, d. h. um den Sinn, nicht um Wahrheit und Falschheit."[39] Das heißt, für Wittgenstein gibt es im Gegensatz zu Schlick eine dritte Möglichkeit: Es geht in der Phänomenologie nicht um „sinnlose Scheinsätze", sondern um „den Sinn". Natürlich heißt das nicht, dass Wittgenstein die Wesensschau und das materiale Apriori bzw. Phänomenologie im Ganzen völlig übernehmen kann. Für ihn bedeutet diese dritte Möglichkeit nichts anderes als „Syntax",[40] die in der Phänomenologie als „apriorisches Wesen und apriorische Wesensstruktur (oder materiales Apriori)" angesehen wird, und die Apriorität des Wesens und der Wesenszusammenhang bedeuten nichts anderes als die Möglichkeit der Schlussfolgerung aufgrund des Gesetzes der Sprache. Dennoch gibt diese sogenannte Syntax a priori uns doch einen Anhaltspunkt, über die Lehre des phänomenologischen Apriori weiter und anders nachzudenken.

Hier werden wir wieder auf den „Großvater (mütterlicherseits)" der Phänomenologie, Bolzano, zurückgreifen. Husserl hat bereits in §11–12 der III. LU einen wichtigen Unterschied zwischen dem synthetischen und dem analytischen Apriori markiert. Der Grund, weshalb dieser Unterschied wichtig ist, besteht darin, dass er einerseits mit dem Unterschied zwischen dem materialen und dem formalen Apriori Husserls gleichgesetzt wird und die Grundlage für die Lehre der materialen und der formalen Ontologie in Husserls Phänomenologie schafft, und dass dieser Unterschied andererseits in der analytischen Philosophie diskutiert wird, wie hier bei Schlick und Wittgenstein. Es ist vor allem festzustellen, dass die echte Quelle des materialen Apriori Husserls die Lehre des synthetischen Apriori bei Bolzano ist, wie J. Benoist eindringlich hervorhebt.[41]

Für Bolzano bedeutet das Apriori im Gegensatz zu Kant vor allem „begriffliches Apriori" und er hat über das Verhältnis des synthetischen Apriori zu verschiedenen Begriffen nachgedacht. So gibt es bei ihm z. B. ein Apriori der Farbe, usw.[42] Daher bedeutet das Apriori bei Husserl vor allem anschauliches Apriori, das in der Wesensschau anschaulich selbst gegeben werden kann. In der Gesamtheit des Apriori unterscheidet Husserl die *„sachhaltigen Begriffe"* oder das materiale Apriori von den *„bloß formalen Begriffen"* oder dem formalen Apriori: „Begriffe wie *Etwas* oder *Eins, Gegenstand, Beschaffenheit, Beziehung, Verknüpfung, Mehrheit, Anzahl, Ordnung, Ordnungszahl, Ganzes, Teil, Größe* usw. haben einen grundverschiedenen Charakter gegenüber Begriffen wie *Haus, Baum, Farbe, Ton, Raum, Empfindung, Gefühl* usw., welche ihrerseits Sachhaltiges zum Ausdruck bringen. Während jene sich um die leere Idee des Etwas oder Gegenstands überhaupt gruppieren und mit ihm durch die formalen ontologischen Axiome verknüpft sind, ordnen sich die letzteren um verschiedene oberste sachhaltige Gattungen (*materiale Kategorien*), in denen *materiale Ontologien* wurzeln."[43]

Zugleich besteht ein *Satz an sich* nach Bolzano aus den *Vorstellungen an sich*, die als Teile des Satzes an sich angesehen werden. Zur Unterscheidung der analytischen Sätze von den synthetischen Sätzen ist zu überlegen, ob und wie weit der Wahrheitswert eines bestimmten Satzes mit der **Veränderung** seiner Vorstellungteile einen Kompromiss schließen kann.[44] Obwohl Husserl insofern Bolzano nicht ganz folgt, hat er diese Idee der „Veränderung" von Bolzano übernommen und eine Lehre der **„Ersetzung"** oder **„Formalisierung"** in Bezug auf das Verhältnis zwischen analytisch-apriorischen Sätzen und synthetisch-apriorischen Sätzen entwickelt. Husserl formuliert: „In einem analytischen Satze muss es möglich sein, jede sachhaltige Materie, bei voller Erhaltung der logischen Form des Satzes, durch die leere Form *etwas* zu ersetzen und jede Daseinssetzung durch Übergang in die entsprechende Urteilsform »unbedingter Allgemeinheit« oder Gesetzlichkeit auszuschalten."[45] Ganz im Unterschied zu Kant bedeutet ein synthetischer Satz a priori bei Husserl einen solchen Satz, der „sachhaltige Begriffe in einer Weise einschließt, die eine Formalisierung dieser Begriffe *salva veritate* nicht zulässt".[46]

In diesem Verständnis geht J. Benoist weiter. Er behauptet, dass das phänomenologische Apriori einen zweifachen Charakter hat, d. h. ein anschauliches Apriori und grammatikalisches Apriori.[47] Nach ihm handelt es sich in der IV. LU gerade um das grammatikalische Apriori.[48] Benoist geht so radikal vor, dass er zuletzt behauptet, die Begrenzung des grammatikalischen Apriori bestimme die Begrenzung der Anschauung selbst und die Form unserer Welt sei nichts anderes als die Form unserer Sprache.[49] Man kann mit Recht fragen, ob er hier noch Husserl oder die Phänomenologie interpretiert, oder ob er sich nicht vielmehr in die Lehre des späten Wittgenstein verläuft.[50]

Um die Radikalisierung Benoists abzulehnen, werden wir uns nun mit der Lehre der *Funktionalisierung der Wesenseinsicht* bei Scheler beschäftigen. Diese wichtige Eigenschaft aller Wesenserkenntnis gehört nach Scheler zu den noch „am wenigsten durchschauten" Eigenschaften. Die sogenannte Funktionalisierung der Wesenseinsicht bedeutet: „*Die Wesenserkenntnis funktionalisiert sich zu einem Gesetz der bloßen »Anwendung« des auf die zufälligen Tatsachen gerichteten Verstandes, der die zufällige Tatsachewelt »nach« Wesenszusammenhängen »bestimmt« auffasst, zerlegt, anschaut, beurteilt.*" (V, S. 198) Deswegen ist alles subjektive Apriori bzw. die Form a priori im transzendentalen Sinne Kants „nichts Ursprüngliches, sondern ein Gewordenes". (Vgl. V. S. 208; IX, S. 204)

Die Lehre von der Funktionalisierung der Wesenseinsicht erklärt einerseits vor allem das Verhältnis zwischen dem materialen Apriori und dem formalen Apriori, d. i. dass „alle Funktionsgesetze auf ursprüngliche Gegenstands-Erfahrung, aber auf *Wesens*erfahrung resp. Wesensschau zurückgehen". Durch die Funktionalisierung der Wesenseinsicht wird das ursprüngliche materiale Apriori zu einem subjektiven formalen Apriori: „Gedachtes wird »Form« des Denkens, Geliebtes wird »Form« und Art des Liebens." (Vgl. V, S. 198, 208) In diesem Sinne hat W. Henckmann die Funktionalisierung der Wesenseinsicht auch als „Schematisierung" bezeichnet, „wonach allerdings nur ein Wandel vom materialen zum formalen Apriori möglich ist".[51]

Andererseits wird uns das Folgende durch die Funktionalisierung der Wesenseinsicht verständlich: „Ein *Werden und Wachsen* der Vernunft *selbst*, d. h. ihres Besitztums an apriorischen Auswahl- und Funktionsgesetzen." In der Tat gibt es für Scheler im Gegensatz zu Kant keine „schlechthin ursprüngliche[n], schlechthin unveränderliche[n] und unvermehr- wie unverminderbare[n] Funktionsgesetze". (Vgl. V, S. 198) Er hat bereits überzeugend die „kantische Identitäts- und Konstanzlehre der menschlichen Vernunft" abgelehnt. (Vgl. V, S. 200; II, 20) Im Gegensatz dazu behauptet Scheler „ein Vernunftwerden durch Funktionalisierung von Wesensanschauung, und zwar ein so geartetes, das über den formalsten Gehalt dieser Wesensanschauungen hinaus innerhalb der verschiedenen großen *Gruppen* der gegliederten Menschheit zu verschiedenen Vernunftgestaltungen geführt hat; das ferner zu wahrem Wachstum und (wahrer Abnahme) der höheren und höchsten Geisteskräfte des Menschen führen kann und tatsächlich geführt hat". (V, S. 201 f.)[52]

Nun ist es für Scheler sehr deutlich, dass einerseits das grammatikalische Apriori (oder wahrscheinlich auch die Syntax a priori bei Wittgenstein) ursprünglich als materiales Apriori in der Wesensschau anschaulich selbst gegeben und dann durch die Funktionalisierung der Wesenseinsicht zum formalen Apriori wird; dass andererseits alles grammatikalische Apriori nicht schlechthin Ursprüngliches oder Unveränderliches, sondern ein Gewordenes ist.

Nun können wir auch feststellen, dass die Frage „Gibt es ein materiales Apriori?" eigentlich zwei Fragen verschiedener Stufen in sich schließen kann. Nämlich erstens: „Gibt es sowohl ein anschauliches Apriori als auch ein grammatikalisches Apriori?", und zweitens: „Welches Verhältnis gibt es zwischen dem anschaulichen Apriori und dem grammatikalischen Apriori?" Während man durch das Prinzip der Selbstgegebenheit oder der absoluten Evidenz auf die erste Frage antworten kann, können wir aufgrund der Lehre der Funktionalisierung der Wesenseinsicht auf die zweite Frage antworten. Kürzer gesagt, gibt es für die Phänomenologen im Gegensatz zu der Identifizierung des Formalen mit dem Apriori bei Kant ein materiales Apriori, das in der Wesensschau selbst gegeben wird. Durch die Lehre der Funktionalisierung der Wesenseinsicht kann der Gegensatz „a priori- a posteriori" (als absolut) von dem Gegensatz „formal- material" (als relativ) vollständig unterschieden werden.

Zugleich ist es festzustellen, dass das grammatikalische Apriori auf das anschauliche Apriori fundiert ist. In diesem Sinne kann man sagen, dass die Form unserer Welt nicht die Form unserer Sprache ist, sondern die Form unserer Vernunft bzw. unserer *werdenden und wachsenden* Vernunft. Nach dieser Funktionalisierung der Wesenseinsicht scheint es gerade möglich, dass die Struktur des Denkens und die Struktur der Sprache identisch sind.

Department of Philosophy, Sun Yat-sen University, Xingang Road No. 135, 510275 Guangzhou, China
e-mail: renzhizhangcn@gmail.com; renzhizhang@hotmail.com

NOTES

[1] W. Henckmann, *Max Scheler* (München: C. H. Beck Verlag, 1998), S. 78.
[2] Vgl. Schlick, „Gibt es ein materiales Apriori?", in *Gesammelte Aufsätze 1926–1936*, Wien 1938, hrsg. Schlick. (Nachdruck: Hildesheim 1969), S. 20–30, hier S. 29. (Zuerst erschienen in: *Wissenschaftlicher Jahresbricht der Philosophischen Gesellschaft an der Uni. zu Wien für das Vereinsjahr 1930/31*)
[3] Vgl. Ernst Tugendhat, „Phenomenology and linguistic analysis", in *Edmund Husserl. Critical assessments of leading philosophers*. Vol. IV, ed. R. Bernet, Donn Welton and Gina Zavota. (London & New York: Routledge, 2005), pp. 49–70, hier p. 49.
[4] In seinem 1918 veröffentlichten philosophischen Hauptwerk *Allgemeine Erkenntnislehre* übt Schlick Kritik an Husserl. 1921 publiziert Husserl die zweite Auflage des zweiten Teils des zweiten Bandes der *Logischen Untersuchungen* und hebt im Vorwort des Buches zu einem Gegenschlag an: „Ich muss noch ausdrücklich bemerken, dass es sich bei M. Schlick nicht bloß um irrelevante Entgleisungen handelt, sondern um sinnverkehrende Unterschiebungen, auf die seine ganze Kritik aufgebaut ist." (Hua XIX/2, B2 VIf.) Im Jahre 1925 veröffentlichte Schlick die zweite Auflage des Buches *Allgemeine Erkenntnislehre*. Er ging auf Husserls Replik ein und hat alle irgendwie entbehrlichen polemischen Ausführungen aus dem Buch entfernt. In seinem 1930/31 veröffentlichten Aufsatz „Gibt es ein materiales Apriori?" kritisiert er jedoch Husserls Phänomenologie erneut, natürlich auch oder hauptsächlich in Form der Phänomenologie Max Schelers.

Zum Vergleich Schlicks mit der Phänomenologie vgl. z. B. M. M. Van De Pitte, „Schlick's critique of phenomenological propositions", *Philosophy and Phenomenological Research* 45 (2) (1984):195–225; Jim Shelton, „Schlick and Husserl on the foundations of phenomenology", *Philosophy and Phenomenological Research* 48 (3) (1988):557–561; Philip Blosser, „The a priori in phenomenology and the legacy of logical empiricism", *Philosophy Today* 34 (3) (1990):195–205.

[5]   Vgl. M. Schlick, „Erleben, Erkennen, Metaphysik" (1926), in *Wiener Kreis*, hrsg. M. Stöltzner und T. Uebel. (Hamburg: Meiner, 2006), S. 169–186, hier, S. 171 f.

[6]   M. Schlick, *Allgemeine Erkenntnislehre*. (Berlin: Springer,[1]1918, [2]1925), A 68 f. (Wir werden mit „A" die erste Auflage des 1918 veröffentlichten Buches bezeichnen und mit „B" die zweite, 1925 publizierte Auflage.) Bei Schlick ist „Kennen" mit „Erleben" oder „Erlebnis" sinnverwandt.

[7]   M. Schlick, „Positivismus und Realismus" (1932), in *Wiener Kreis*, hrsg. M. Stöltzner und T. Uebel. (a. a. O.), S. 222.

[8]   Schlick, *Allgemeine Erkenntnislehre*, A120/B 127.

[9]   Vgl. Julius Kraft, *Von Husserl zu Heidegger. Kritik der phänomenologischen Philosophie*. (Frankfurt am Main, [2]1957), S. 108.

[10] Vgl. Schlick, „Gibt es ein materiales Apriori?", a. a. O., S. 22.

[11] Ebd. S. 25.

[12] Vgl. Schlick, *Allgemeine Erkenntnislehre*, B 69 ff.

[13] Schlick, „Gibt es ein materiales Apriori?", a. a. O., S. 30.

[14] Vgl. Hua VII, S. 235.

[15] Vgl. Richard T. Murphy, *Hume and Husserl. Towards radical subjectivism, Phaenomenologica 79*. (The Hague/Boston/London: Martinus Nijhoff, 1980), p. 24. „*Husserl will not even deign to adopt the Kantian notion of the »a priori«. Rather, he will interpret Hume's notion of the a priori in order to overcome the radical concreteness and subjectivism of Hume's skepticism and establish philosophy as the rigorous foundational science.*"

[16] Vgl. Thomas Seebohm, *Die Bedingungen der Möglichkeit der Transzendental-Philosophie. Edmund Husserls Transzendental-Phänomenologischer Ansatz, Dargestellt im Anschluss an seine Kant-Kritik*. (Bonn: H. Bouvier u. CO. Verlag 1962), S. 19.

[17] Hua XIX/1, A 124/B₁ 124 f.

[18] Vgl. Hua XIX/1, A 121/B₁ 122.

[19] Vgl. Hua XIX/2, A 634/B₂ 162.

[20] Heidegger behauptet später, dass die Konsequenz der Entdeckung der kategorialen Anschauung, insbesondere der Ideation, darin liege, „dass dadurch die philosophische Forschung instand gesetzt wurde, das Apriori schärfer zu fassen und die Charakteristik des Sinnes seines Seins vorzubereiten." (Heidegger, *Prolegomena zur Geschichte des Zeitbegriffs. Marburger Vorlesung Sommersemester 1925*, GA 20, Frankfurt am Main: Vittorio Klostermann GmbH [3]1994, S. 98) In der Tat hat Heidegger die „Intentionalität", die „kategoriale Anschauung" und den „ursprünglichen Sinn des Apriori" als die drei fundamentalen Entdeckungen der Phänomenologie bezeichnet. Wir werden die Intentionalität bei Husserl und Scheler in den Abschnitten 3.2.2 und 3.3.2 behandeln.

[21] Vgl. Hua XIX/2, VI. LU, §47 und §52.

[22] Dieter Lohmar, *Erfahrung und kategoriales Denken. Hume, Kant und Husserl über vorprädikative Erfahrung und prädikative Erkenntnis, Phaenomenologica 147*. (Dordrecht/Boston/London: Kluwer Academic, 1998), S. 183.

[23] Husserl bestimmt in der III. LU formal den Begriff des Fundierungsverhältnisses und unterscheidet zwischen der wechselseitigen und der einseitigen Fundierung. Während in der III. LU die wechselseitige Fundierung wichtiger ist, zieht Husserl in der VI. LU den Begriff der einseitigen Fundierung vor. (Vgl. Thomas Nenon, „Two Models of Foundation in the Logical Investigations", in *Husserl in Contemporary Context. Prospects and Projects for Phenomenology*, ed. B. C. Hopkins. (Dordrecht/Boston/London: Kluwer Academic, 1997), pp. 97–114.

[24] Vgl. Hua XIX/2, §48; Vgl. auch D. Lohmar, *Erfahrung und kategoriales Denken*, a. a. O., S. 169 ff.

[25] Hua XIX/2, A 626/B₂ 154.

[26] Ebd.

[27] Vgl. Hua XIX/2, A 592 ff./B$_2$ 120 ff., A 508 ff./B$_2$ 36 ff.; Vgl. auch D. Lohmar, *Erfahrung und kategoriales Denken*, a. a. O., S. 171.

[28] Diese zwei verschiedenen Arten der Deckungseinheit dienen in der Tat als Repräsentant für die zwei verschiedenen kategorialen Anschauungen: d. i. den *synthetischen* Akt und den *abstraktiven* Akt der kategorialen Anschauung. Der Repräsentant für die kategoriale Anschauung ist nicht mit dem sinnlichen Repräsentanten der Gegenstände der Gesamtwahrnehmung oder der Sonderwahrnehmung identisch. Die Deckungseinheit beruft sich nicht auf die gleichen reellen Bestände, sondern ist nicht-sinnlich.

[29] Hier kann man das Schema von Auffassung und Auffassungsinhalt, wie in der sinnlichen Anschauung, wiederfinden. Husserl selbst kritisiert dieses Schema (vgl. z. B. Hua X, S. 7, Anm. 1), und R. Sokolowski vertritt die Ansicht, dass Husserl es in der genetischen Phänomenologie fallen lässt. (Vgl. R. Sokolowski, *The Formation of Husserl's Concept of Constitution. Phaenomenologica 18.* The Hague: Martinus Nijhoff, 1970, pp. 177 ff.) Dennoch findet sich dieses Schema nach Meinung D. Lohmars an vielen entscheidenden Stellen sowohl in den LU als auch in *Erfahrung und Urteil* wieder. „For acts constituting intentional objects and categorial objects it is not defective, but unavoidable." (D. Lohmar, „Husserl's Concept of Categorical Intuition", in *Edmund Husserl. Critical Assessments of Leading Philosophers.* Vol. III, ed. R. Bernet, Donn Welton and Gina Zavota. London & New York: Routledge, 2005b, pp. 61–83, hier p. 70.) – Was die Erfüllung der kategorialen Anschauung betrifft, gibt es verschiedene Ansichten. Nach Tugendhat gibt der „aktuelle Vollzug der kategorialen Synthesis" der kategorialen Intention Erfüllung. (Vgl. E. Tugendhat, *Der Wahrheitsbegriff bei Husserl und Heidegger.* Berlin: Walter de Gruyter & Co., [2]1970, S. 111–129) Wir stimmen hier mit D. Lohmar darin überein, dass die Deckungseinheit als nicht-sinnlicher Repräsentant der kategorialen Intention Erfüllung gibt. (Vgl. D. Lohmar, *Erfahrung und kategoriales Denken*, a. a. O., S. 172)

[30] Vgl. D. Lohmar, *Erfahrung und kategoriales Denken*, a. a. O., S. 185.

[31] Hua XIX/1, A 221/B$_1$ 223; Vgl. auch Hua II, S. 56 f. Der Unterschied zwischen der individuellen Anschauung von etwas Rotem und der Wesensschau auf das Allgemeine Rot bedeutet hier nicht einen einfachen Wechsel des Blicks und Interesses oder die Änderung der Apperzeption.

[32] Hua XVIII, B XV.

[33] Hua XXV, S. 32 f.

[34] Hua IX, S. 72. Später hat Husserl unter den Titeln „eidetische Reduktion" oder „eidetische Variation" die hier besprochene „Ideation" oder „ideierende Abstraktion" oder „Wesensschau" weiter als eine besondere Form der kategorialen Anschauung kritisch bedacht. (vgl. z. B. Hua IX, S. 72–87; Husserl, *Erfahrung und Urteil*, §86–88) – Gemäß des Hauptthemas unserer Untersuchung kann es hier freilich nicht darum gehen, die Lehre der Wesensschau und der „eidetischen Reduktion" bei Husserl ausreichend zu klären. Vgl. zu diesem Thema Liangkang Ni, *Seinsglaube in der Phänomenologie Edmund Husserls, Phaenomenologica 153.* Dordrecht/Boston/London: Kluwer Academic, 1999, S. 155–187; D. Lohmar, „Die phänomenologische Methode der Wesensschau und ihre Präzisierung als eidetische Variation", in *Phänomenologische Forschungen* 2005a, S. 65–91; Burt C. Hopkins, „Phenomenological Cognition of the A Priori: Husserl's Method of »Seeing Essences« (Wesenserschauung)", in *Husserl in Contemporary Context. Prospects and Projects for Phenomenology* 1997, pp. 151–178.

[35] Vgl. Thomas Seebohm, *Die Bedingungen der Möglichkeit der Transzendental-Philosophie. Edmund Husserls Transzendental-Phänomenologischer Ansatz, Dargestellt im Anschluss an seine Kant-Kritik*, a. a. O., S. 19.

[36] Vgl. Schlick, „Gibt es ein materiales Apriori?", a. a. O., S. 22.

[37] Hua II, S. 61.

[38] Vgl. L. Wittgenstein, „Wittgenstein und der Wiener Kreis", in *Wittgenstein Schriften, Bd. 3*, Hrsg. von Friedrich Waismann. (Frankfurt am Main: Suhrkamp, 1967), S. 67 f.

[39] Ebd. S. 63.

[40] Ebd. S. 66. Wittgenstein formulierte: „Wenn jemand nie aus seinem Zimmer herauskommt, so weiß er doch, dass der Raum weitergeht, d. h., dass die Möglichkeit besteht, aus dem Zimmer herauszukommen (und wenn es auch diamantene Wände hätte). Das also ist keine Erfahrung. Es ist in der Syntax des Raumes gelegen, a priori."

[41] Vgl. zu dieser wichtigen Entdeckung J. Benoist, *L'a priori conceptuel: Bolzano, Husserl, Schlick.* a. a. O., S. 98 ff., 138 ff.

[42] Vgl. Bolzano, *Wissenschaftslehre*, § 72.

[43] Hua XIX/1, A 246/B$_1$ 252.

[44] Vgl. Bolzano, *Wissenschaftslehre*, § 148.

[45] Hua XIX/1, B$_1$ 255; Vgl. Hua XIX/1, A 247: „Die *Formalisierung* besteht darin, dass in dem vorgegebenen analytischen Satze alle sachhaltigen Bestimmungen durch Unbestimmte *ersetzt* und diese dann als unbeschränkte Variable gefasst werden." (Herv. W. Z.)

[46] Hua XIX/1, A 248/B$_1$ 256. Husserl insistierte stets auf dem Unterschied zwischen dem materialen (sachhaltigen oder synthetischen) Apriori und dem formalen Apriori in diesem Sinne. Vgl. Hua XVII, S. 26; Hua XI, S. 33 f.; Hua XXIV, S. 240; usw. Vgl. auch Elisabeth Ströker, *Husserls transzendentale Phänomenologie*, Frankfurt am Main 1987, S. 176 ff.

[47] Vgl. J. Benoist, *L'a priori conceptuel: Bolzano, Husserl, Schlick*. a. a. O., S. 106 ff., 114.

[48] Vgl. J. Benoist, „Grammatik und Intentionalität (IV. Logische Untersuchung)", in *Edmund Husserl, Logische Untersuchungen, Klassiker Auslegen, Bd. 35*, hrsg. Verena Mayer. (Berlin: Akademie-Verlag, 2008), S. 123–138 und J. Benoist, „The Question of Grammar in Logical Investigations, With Special Reference to Brentano, Marty, Bolzano and Later Developments in Logic", in *Phenomenology World-Wide*, ed. Anna-Teresa Tymieniecka. (a. a. O.), pp. 94–97. Vgl. auch Hua XIX/1, IV. LU, § 10.

[49] Vgl. J. Benoist, *L'a priori conceptuel: Bolzano, Husserl, Schlick*. a. a. O., S. 134, 178.

[50] Vgl. Claudio Majolino, „Book Review: *Jocelyn Benoist, L'a priori conceptuel. Bolzano, Husserl, Schlick* (Paris: Vrin 1999)", in *Husserl Studies 18*: pp. 223–232, 2002, hier p. 230.

[51] Vgl. Wolfhart Henckmann, „Schelers Lehre vom Apriori", a. a. O., S. 138 f.

[52] In diesem Sinne betont Scheler weiter, „dass die großen menschlichen Kulturen und Erkenntniszusammenhänge – schon auf dem Niveau des apriorischen Wissens – gegenseitig *unvertretbar und unersetzlich* sind". (V, S. 202) Man könnte auf dieser Basis eine Lehre des kulturellen Apriori entwickeln, vgl. VIII, S. 24 ff.; vgl. auch Anthony J. Steinbock, „Personal Givenness and Cultural a prioris", in *Time, Space, and Culture*, eds. David Carr and Chan-Fai Cheung. (Dordrecht: Kluwer Academic, 2004), pp. 159–176.

LITERATURVERZEICHNIS

Benoist, J. 1999. *L'a priori conceptuel: Bolzano, Husserl, Schlick*. Paris: Librairie Philosophique J. Vrin.

Benoist, J. 2003. The question of grammar in logical investigations, with special reference to Brentano, Marty, Bolzano and later developments in logic. In *Phenomenology world-wide*, ed. Anna-Teresa Tymieniecka, 94–97. Dordrecht: Kluwer.

Benoist, J. 2008. Grammatik und Intentionalität (IV. Logische Untersuchung). In *Edmund Husserl, Logische Untersuchungen, Klassiker Auslegen, Bd. 35*, hrsg. Verena Mayer, 123–138. Berlin: Akademie-Verlag.

Blosser, Ph. 1990. The a priori in phenomenology and the legacy of logical empiricism. *Philosophy Today* 34(3) :195–205.

Bolzano, B., *Wissenschaftslehre: Versuch einer ausführlichen und grösstetheils neuen Darstellung der Logik, mit steter Rücksicht auf deren bisherige Bearbeiter*, Sulzbach 1837. Die neue kritische Ausgabe in: *Bernard Bolzano-Gesamtausgabe, Bd. I, 11–14*, Hrsg. von Jan Berg, Stuttgart: 1985–2000.

Dummett, M. 1993. *Origins of analytic philosophy*. London: Duckworth.

Dummett, M. 2001. Preface. In Husserl, *The shorter logical investigations* (trans: Findlay, J. N.), edited and abridged with a new Introduction by Dermot Moran. London and New York: Routledge.

Heidegger, M. [3]1994. *Prolegomena zur Geschichte des Zeitbegriffs. Marburger Vorlesung Sommersemester 1925*, GA 20. Frankfurt am Main: Vittorio Klostermann GmbH.

Henckmann, W. 1987. Schelers Lehre vom Apriori. In *Gewißheit und Gewissen: Festschrift für Franz Wiedmann zum 60. Geburtstag*, hrsg. W. von Baumgartner, 117–140. Würzburg.

Henckmann, W. 1998. *Max Scheler*. München: C. H. Beck Verlag.

Hopkins, B. C. ed. 1997. Phenomenological cognition of the a priori: Husserl's method of »seeing essences« (Wesenserschauung). In *Husserl in contemporary context. Prospects and projects for phenomenology*, 151–178. Dordrecht/Boston/London: Springer.

Husserl, E. *Husserliana – Gesammelte Werke*, Den Haag, Dordrecht/Boston/London 1950 ff. Die Seitenzahlen beziehen sich auf die in Hua XVIII, Hua XIX/1, Hua XIX/2 (*Logische Untersuchungen*) wiedergegebene Paginierung der ersten (A) und zweiten (B) Auflage. Andere Schriften von Husserl werden nach der Husserliana-Ausgabe zitiert als „Hua Band-Nr., Seite". Kluwer Academic und dann Springer.

Husserl, E. [6]1985. *Erfahrung und Urteil. Untersuchung zur Genealogie der Logik*, redigiert und hrsg. von L. Landgrebe. Hamburg: Meiner.

Kraft, J. [2]1957. *Von Husserl zu Heidegger. Kritik der phänomenologischen Philosophie*. Frankfurt am Main: Verlag "Öffentl.Leben".

Lohmar, D. 1998. *Erfahrung und kategoriales Denken. Hume, Kant und Husserl über vorprädikative Erfahrung und prädikative Erkenntnis, Phaenomenologica 147*, Dordrecht/Bosten/London: Kluwer Academic.

Lohmar, D. 2005a. Die phänomenologische Methode der Wesensschau und ihre Präzisierung als eidetische Variation. In *Phänomenologische Forschungen 2005*, 65–91. Hamburg: Meiner.

Lohmar, D. 2005b. Husserl's concept of categorical intuition. In *Edmund Husserl. Critical Assessments of Leading Philosophers*. Vol. III, ed. R. Bernet, Donn Welton and Gina Zavota, 61–83. London & New York: Routledge.

Majolino, C. 2002. Book Review: *Jocelyn Benoist, L'a priori conceptuel. Bolzano, Husserl, Schlick* (Paris: Vrin 1999). In *Husserl Studies 18*, 223–232. Dordrecht/Boston/London: Springer.

Murphy, R. T. 1980. *Hume and Husserl. Towards radical subjectivism, Phaenomenologica 79*, The Hague/Boston/London: Martinus Nijhoff.

Nenon, Th. 1997. Two models of foundation in the logical investigations. In *Husserl in contemporary context. Prospects and projects for phenomenology*, ed. B. C. Hopkins. Dordrecht/Bosten/London: Kluwer Academic.

NI, L.-K. 1999. *Seinsglaube in der Phänomenologie Edmund Husserls, Phaenomenologica 153*. Dordrecht/Bosten/London: Kluwer Academic.

Scheler, M., *Max Scheler Gesammelte Werke*, Zuerst im Francke-Verlag, Bern/München erschienen, ab 1986 im Bouvier-Verlag, Bonn. Hrsg. von Maria Scheler & Manfred S. Frings. Die Schriften von Scheler werden im laufenden Text nach der *Max Scheler Gesammelte Werke* zitiert als „Band-Nr., Seite".

II (1980): *Der Formalismus in der Ethik und die materiale Wertethik.*
III (1972): *Vom Umsturz der Werte.*
V (1968): *Vom Ewigen im Menschen.*
VIII (1980): *Die Wissensformen und die Gesellschaft.*
IX (1976): *Späte Schriften.*
X (1986): *Schriften aus dem Nachlaß, Bd. 1: Zur Ethik und Erkenntnislehre.*
XI (1979): *Schriften aus dem Nachlaß, Bd. 2: Erkenntnislehre und Metaphysik.*

Schlick, M. [1]1918, [2]1925. *Allgemeine Erkenntnislehre*. Berlin: Springer.

Schlick, M. 1969. Gibt es ein materiales Apriori? In *Gesammelte Aufsätze 1926–1936*, Wien 1938, hrsg. M. Schlick, 20–30. Nachdruck: Hildesheim.

Schlick, M. 2006. Erleben, Erkennen, Metaphysik (1926). In *Wiener Kreis*, hrsg. M. Stöltzner und T. Uebel, 169–186. Hamburg: Meiner.

Schlick, M. 2006. Positivismus und Realismus (1932). In *Wiener Kreis*, hrsg. M. Stöltzner und T. Uebel. Hamburg: Meiner.

Seebohm, Th. 1962. *Die Bedingungen der Möglichkeit der Transzendental-Philosophie. Edmund Husserls Transzendental-Phänomenologischer Ansatz, Dargestellt im Anschluss an seine Kant-Kritik.* Bonn: H. Bouvier u. CO. Verlag.

Shelton, J. 1988. Schlick and Husserl on the foundations of phenomenology. *Philosophy and Phenomenological Research* 48 (3):557–561.

Sokolowski, R. 1970. *The formation of Husserl's concept of constitution. Phaenomenologica 18*. The Hague: Martinus Nijhoff.

Steinbock, A. J. 2004. Personal givenness and cultural a prioris. In *Time, space, and culture*, eds. David Carr and Chan-Fai Cheung, 159–176., Dordrecht: Kluwer Academic.

Ströker, E. 1987. *Husserls transzendentale Phänomenologie*. Frankfurt am Main: Vittorio Klostermann.
Tugendhat, E. ²1970. *Der Wahrheitsbegriff bei Husserl und Heidegger*. Berlin: Walter de Gruyter & Co.
Tugendhat, E. 2005. Phenomenology and linguistic analysis. In *Edmund Husserl. Critical assessments of leading philosophers*. Vol. IV, eds. R. Bernet, Donn Welton, and Gina Zavota, 49–70. London & New York: Routledge.
Van De Pitte, M. M. 1984. Schlick's critique of phenomenological propositions. *Philosophy and Phenomenological Research* 45 (2):195–225.
Wittgenstein, L. 1967. Wittgenstein und der Wiener Kreis. In *Wittgenstein Schriften, Bd. 3*, hrsg. von Friedrich Waismann. Frankfurt am Main: Suhrkamp.

SECTION III
LOGOS AND EDUCATION

RIMMA KURENKOVA, EVGENY PLEKHANOV, AND ELENA
ROGACHEVA

THE IDEA OF PAIDEA IN THE CONTEXT
OF ONTOPOESIS OF LIFE

ABSTRACT

The paper deals with the analysis of the notion of paidea in a wide context starting from antiquity till post-modern century. It is stressed that the actualization
of Greek-Roman ideal of universal education (παιδεία, humanitas universalis) by
means of phenomenological discourse lies in the tradition founded by E. Husserl
and linked with spiritual heritage of Antiquity. Within the concept of ontopoesis of
A.T. Tymieniecka, one can easily see an attempt to actualize the whole specter of
intuitive meanings of antique notion φύσις. Thanks to it paidea will find a new and
deeper interpretation.

Being the heritage of ancient thought of Greece *paidea* remains one of those
universal things of culture that in post-modern century keeps in itself the idea of
unity of the individual and society, of general and special, of objective-necessary
and subjective-valuable. Education is one of the most important values in the life
of a human being. It is a good thing not only in the sense that it gives a chance for
a person to get professional knowledge and skills, to be involved in the process of
acculturation and reach a high social status. The main task of education is to develop
a personality. An individual should be given a chance to get "a human image".
So, real possibilities provided by the system of education for every human being
to become a unique creature may be regarded as the main criterion of humanistic
approach. Humanistic measure of education is in the degree of educational ideal
limit by the inner nature of a human being. The problem is to what degree and
under what social conditions he\she is able to demonstrate internal principle of
free and unlimited obtaining of integral structure of individual spiritual life. So,
the Greek idea of *paidea* is born. It aims at restoration of the unity between an
individual and society, tradition and contemporary time, subjective valuable and
objective-necessary, individual and universal.

In *paidea* we donot deal with absorbing of one opposition by the other but with the
link of those oppositions on the basis of the third and much higher element. Being a
humanistic universal thing different in various types and forms of Western-European
education, *paidea* is kept as a cultural paradigm. Let us remember K. Jaspers'
remark about the role of Antiquity. He thought that Antiquity has provided a factual
basis to who we could be in the West as human beings.

Our position, that we try to make arguments to is the following: *humanistic
essence of paidea internally presupposes the formation of such a spiritual position
of a human being that comes from a universal link of a human being with the world*

141

A.-T. Tymieniecka (ed.), Analecta Husserliana CX, 141–146.
DOI 10.1007/978-94-007-1691-9_12, © Springer Science+Business Media B.V. 2011

of all being alive. But in order to be a projective idea of "education in general" there is a need for reconstruction of its semantic structure that as contemporary studies show, has acquired clear and finished contours only in the epoch of high Greek classics.[1] Understanding of paidea as a spiritual space where cognition in its sense becomes a sort of paradigm (a norm or a pattern) of internal life of an individual who systematically ignoring public opinion and any authority practices the acts of critical and logical thinking is being formed on the basis of the principle of ethic rationalism, that can be vividly seen in Socratic method. A thought is based on the assumption that a desire forms the basis of every action. It passes an internal sense to every action. The desire is "to make a name for oneself" and have it repeated for generations. All the highest level are "those whose procreancy is in the spirit rather than of the flesh – and they are not unknown, Socrates – conceive and bear the things of spirit. And what are they? You ask. Wisdom and all her sister virtues; it is the office of every poet to beget them, and of every artist whom we may call creative" (Simposium, 209 a).[2] So education is cognition of what is a real virtue – the way to virtue life. But the experience, known already to the Greeks show that it is possible to teach any practical skills or arts ($\tau\acute{\varepsilon}\chi\nu\eta$), but not to moral behavior. No parent can transfer his own life experience to his/her children. The same happens to a teacher who can not teach his pupil a virtue. Techne for the ancient Greeks meant craft, skill, art; it is knowledge of poiesis, involving knowing how to create what the craftsperson desires. By contrast, *theoria*, from which theory is derived, means speculation, contemplation or "a spectator above". Theory assumes an attitude of detachment and distance from everyday life and practice. The form of knowledge associated with theory was episteme, which meant certain knowledge of perfectly clear, immutable, and time-less truths. *Episteme* opposes *techne* because *techne* is knowledge of how to do things in this vague, changeable world. The Greeks put *theoria* and *episteme* at the top of the hierarchy of knowledge. *Poesis* and *techne* were at the bottom.

The way to practical morality ($\varphi\rho\acute{o}\nu\eta\sigma\iota\varsigma$) is not morality itself. If virtue ($\alpha\rho\varepsilon\tau\alpha\acute{\iota}$) was only "true knowledge" of kindness, it would never be active educative tool. In reality it contains some spontaneous and unconscious element that though cannot give us clear vision of the reasons of our actions, but still drives us to virtue and happiness ($\varepsilon\acute{\upsilon}\delta\alpha\iota\mu\nu\acute{\iota}\alpha$). The cognition of virtue itself that Socrates thought to be the basis of human virtues is only the conscious embodiment of that striving rooted in the deepness of the soul, where cognition and its results make one whole.

Plato's theory of learning is of importance to recall too. Learning is a process of growth and change. Some learning, such as learning through self-initiated inquiry, caused Plato special problems in the dialogue called *Meno*. There he set out the Meno paradox: It is impossible to learn anything through inquiry because either you already know, so there is no need to inquire, or you have no knowledge what-soever and therefore would never recognize it. This paradox results from either/or thinking. It doesnot allow for coming to know. Plato's solution looks metaphysical and epistemological. His theory of recollection presumed that before birth everyone caught a brief glimpse of what he called the immutable and eternal Forms. For him learning meant recollecting forms. Plato believed that theoretical wisdom(*theoria*) is knowledge about these metaphysical verities that he called Forms. The Forms are abstract and indubitable supernatural entities, existing outside space and time and therefore unchangeable. For Plato, everyday things located in space and time, hence subject to change to the vicissitudes of change and fortune, are but contingent and imperfect copes of the perfect Forms of true reality. As Jim Garrison says: "For Plato all knowledge is of the entirely abstract, immutable, indubitable, and eternally fixed Forms. All the rest is just opinions about things of the empirical world of space and time copied from the Forms. Plato placed a supreme harmonizing principle – the Good- above the Forms. By harmoniously structuring the Forms, 'the Good' not only guarantees that reality is rational, it also assures that reality is an aesthetic and moral order. For Plato, indubitable knowledge of the Forms (and above all 'the Good') is the source of timeless wisdom".[3]

Plato's idea of *eros* as a *daimon* is a valuable one. The desire for a better world drives a person from where he/she is now to where he/she ought to be. Eros is a mediating diamond existing midway between being and not being. It defies the law of noncontradiction and is a principle of genesis, birth, and becoming. Becoming and development are intermediate between being and not being. According to the myth, eros is the son of Poros and Penia. Poros means "plenty", "way", "method", "craft", or "skill". The myth associated the minor Olimpian Deity Poros with the virtues of practical ability. Penia was unattractive, poor and homeless. When Aphrodite, goddess of beauty, was born, the Olympian Gods feasted. Drunk from too much nectar, Poros falls asleep in the garden of Seus. In a scheme to overcome her poverty, Penia contrived to lie down beside Poros, and together they conceive *eros* (Symposium, 203b–c). Eros conception occurred in the excesses of intoxication, a kind of madness. So, conception and birth require the mediation of passionate desire. So Eros helps unite opposites, it's a powerful and paradoxical passion that mediates a multitude of opposites and brings people together. In Plato's theory we see that *eros* is not the subject of love, but a desire. Usually a person desires what he is deprived of. In metaphysical sense eros is a striving of a human being to the unity, wholeness, that is becoming an ideal seen as absolutely perfect and full of virtue. Eros is a deep need for spiritual self perfection oriented to idea. So, love for good and bliss is a sort of striving to real accomplishment of a human being's nature, hence to education in the initial sense of this word.

The origin of eros of Gods means that it should serve to good and perfection of the subject of love. Love to another person is grasped by Plato as the need to develop one's own Self, that can be only along with "you". Thanks to it, the forces belonging to every of the two parts unite and start acting. Eros is a symbol of spiritual link between the individuals and brings *paidea* into the space of human communication, where good is a norm where real friendship and love can be accomplished. Then it becomes clear that if we cannot teach virtue, but you can transfer it only by means of upbringing. The spirit of loving person is forcefull and desires to be embodied in another person. Mutual love bonds people in their striving to beautiful and eternal. It opens for the educator a way to mimesis, that allows to form in the pupil the desire for perfection. So, *paidea,* born with help of *eros,* turns into $\alpha\rho\varepsilon\tau\alpha\acute{\iota}$.

In Greece the idea of education was firstly grasped and embodied to an extent that it is applied by everyone who understands it. All great elevations of a human being took place in the West thanks to closeness and demarcation with Antiquity. Through a variety of expressions, the metaphysic of Platonic supernaturalism exercises an immense influence on Western thought.

Wherever it was forgotten barbarity appeared on the scene. Torn off its ground should sway as it loses its support. Our state would be the same in case we lose our links with Antiquity. It is our soil though it constantly changes and only then and without autonomous power of education – the past of its people.[4]

The actualization of Greek-Roman ideal of universal education (παιδεία, humanitas universalis) by means of phenomenological discourse lies in the tradition founded by E. Husserl and linked with spiritual heritage of Antiquity. The founder of historic-cultural phenomenology saw in ellinist world the sources of life intention, the horizon for constituting "spiritual image of Europe." Theoretical statement born within Greek philosophy meant radical emancipation of human consciousness from the power of utilitarian necessities of everyday being. New sense structures emboding the ideas of universal have become life meaningfull for a human being, who owing to it has become a completely new individual.

Paidea means περιαγωγή όλης τής ψυχής- a guidance to reform of the human being in his/her very essence. The role of Ariadna's thread showing to a human being the way for spiritual renovation should be given to philosophy that is the means of "such an orientation in the truth that determines the being of the truth as an idea itself" (Heidegger M.)

A historical drama of Paidea, according to E. Husserl, is in the fact, that idea of universal development of a human being was grasped from philosophy by the gaining strength science that in 17th century lost its connection with the universe of pre-predicative senses and set the task for radical reform of new European's life world. The subordination of Paidea to the activistic claims of the reason caused the transformation of the idea of "education". The formation (Formierung) starts dominating over the initial meaning of "development in accordance with a pattern, an idea" (Vorbild).

In Antiquity the pattern and the key idea for human being development was nature. The initial meaning of φύσις is organic growth. The investigations of Heidel W.A. made it possible to reveal the following meanings of this notion: 1.Birth, emergence (γένετις); 2. Internal force (δύναμις), providing the course of the process; 3. Initial state (άρχή, from which everything emerges and where everything returns; 4. Personified creative force, which is present and acts everywhere (Φύσις); 5.Individual or general constitution of a separate human being, human society or living creatures; 6. Universal characteristic of space, being presented analogically with a living creature (φύσις τοῦ πάντου); 7. Invisible force, determining internal form or the structure of this creature; 8.The spiritual or emotional of a human being, his natural "etos", that can be seen on top of all this in natural insufficiency and rudeness, that may be overcome by upbringing, teaching and exercise. Both basic and complimentary meanings of φύσις are closely connected with the idea of the animation of nature, its fullness of internal activity and life.

Deep connection existing in antique perception between φύσις и παιδεία was lost by both E.Husserl and M Heidegger. It blocked the classic leaders of phenomenological philosophy to reveal the presence of creative human being development in universal context of life. Within the concept of ontopoesis of A.T. Tymieniecka, one can easily see an attempt to actualize the whole specter of intuitive meanings of antique notion φύσις. Thanks to it paidea will find a new and deeper interpretation.

Ontopoesis is being accomplished as progressive individualization of the forms of life, emergent activity of which can form the system of world contexts supporting each other. At the highest stage of evolution vital constitutes transform into social and cultural contexts that gives possibilities to human individuals for free self-determination. It does not abolish, but on the contrary stresses the necessity of personal development in much deeper layers of world context. That is why transformation of a human being in its essence (paidea) should be understood as cooperation of a human being in emergent unfolding of life process.

Erudition is a characteristic of human state that emerges thanks to meaningful life intention. Socio-cultural context, created on the basis of intellectual, moral and aesthetic sense believing, forms the world of meanings endured and interpreted by people in their mutual every-day life. Intersubjective structure of the world of everyday life was interpreted by E.Husserl and A. Shutz as the basic, pre-predicative reality. That is why education from the socio- phenomenological point is limited to the process of sedimentation of social knowledge in the form of individual experience.

If stick to the concept of ontopoesis, then theoretical, moral and aesthetic maxims may be seen in the life-world as well as practical(natural) maxim. Thanks to it, a human being can place her/himself "within the unity of all alive" as a conscious, responsible and creative creature. Universal erudition means involvement of all the complex of those tasks that are set by the necessity of keeping life on our planet into horizon of contemporary human being experience.

Department of Musical Education, The Vladimir State University for Humanities, Vladimir, Russia; Department of Socio-Humanities, Vladimir Branch of Russian Academy of State Service, Vladimir, Russia; Vladimir State University, Stroiteley Avenue, 11 Vladimir 600024, Vladimir, Russia
e-mail: eplekhanov@rambler.ru
e-mail: erogacheva@hotmail.com

NOTES

[1] The most profound work dealing with the evolution of Greek paidea is the study of German specialist in phololology Verner. Jeger. (Paidea. Die Formirung des griechischen Menshen). Paidea. Vospitanije antichnogo greka. (Paidea. Upbring of Antique Greek. Translated from German into Russian by M.N.Botvinnik.-M., 1997. p. 151.

² Jim Garrison, Dewey and Eros. Wisdom and Desire in the Art of Teaching. Published by Teacher's College Press, Columbia University, New York, 1997, p. 10.
³ Jim Garrison, Dewey and Eros. Wisdom and Desire in the Art of Teaching. Published by Teacher's College Press, Columbia University, New York, 1997, pp. 5–6.
⁴ K. Jaspers. Spiritual situation of time. The Sense and Purpose of History. M, 1991.–C. 358–359.

ELENA ROGACHEVA

INTERNATIONAL DIMENTION OF JOHN DEWEY'S PEDAGOGY: LESSONS FOR TOMORROW

ABSTRACT

The chapter deals with international reputation of John Dewey's pedagogy in different cultural contexts in the 20th century. The actuality of the ideas of this outstanding American philosopher and educator is evident as his model of society- and child-oriented school based on the idea of communication and cooperation still attracts many educational theoreticians and practitioners. The chapter shows how Dewey's educational ideas were digested in many cultural contexts. So, the chapter contributes to the problem of educational transfer. Dewey's appeal to develop reflective capacities of teachers and to overcome dogmatic thinking is still vital in Russia. Any school reform depends on the teacher's competence. Innovative search in education in many countries is progressing only thanks to innovative teachers. For Dewey who thought of school as a co-society of researchers the basic elements of educational paradigm were the school, the child and the society. The conditions for making these three elements meaningful were "democracy", "growth" and "experience". Learning by doing has become very important as well as the creation of educative atmosphere by means of museum pedagogy and art and music education. Developing a real citizen of a democratic society was also Dewey's dream. Dewey could implement his educational program in his Laboratory school at the University of Chicago (1896–1904) that became a pioneer of laboratory school movement and stimulated innovative search in European countries, including Russia, in Eastern (Japan, China, Turkey) and Latin American ones (Mexico, Chili, Cuba, Brazil, Argentina). Of particular interest is the part in the chapter that describes the perception of Dewey's pedagogy in Russia where reputation of Dewey was changing from "the best philosopher of contemporary school" (Stanislav Shatzky in 1920s) to "the enemy of all progressive mankind" (in Stalin time, late 1930s). The materials of the chapter crush the existing ideological myth of Dewey created in Soviet Russia.

For more then 25 years educational writings of J. Dewey served as my intellectual background as I attempted to reconstruct the pragmatic paradigm in education, to question the "identity" of this paradigm and to trace its influence on the development of educational theory and practice in Russia and other parts of the world. According to J. Lovinger: "Scientists are similar to lovers: they find tokens of their beloved everywhere." While studying the process of reception of Dewey in Turkey, Japan, and Latin America I realized the fact that only in a cross-cultural dialogue one could grasp the resemblances and differences of innovative educational

A.-T. Tymieniecka (ed.), Analecta Husserliana CX, 147–169.
DOI 10.1007/978-94-007-1691-9_13, © Springer Science+Business Media B.V. 2011

developments caused by new pedagogy of activity. More to it, I came to understanding that the notion "influence" lost, to a great extent, its explanatory power as a tool in intellectual history of ideas. In such complex issues like educational transfer we deal with the process of reception the ideas within a specific cultural context and they interact with existing traditions, ideas, and practices. So, the specific context is of decisive influence on the way in which these ideas and practices are taken up, digested, translated, transformed and eventually made into something new.

As Quentin Skinner points out, there are three conditions that must be met in order to conclude that the appearance of a given set of ideas in a text may be explained by their appearance in the text of an earlier writer. First of all there must be a genuine similarity. Further, it must be the case that the ideas in the later text could not be found in the work of any other writer but the one said to have influence. And, finally the probability of the similarity being random should be very low. So, I agree with the positions of some other Deweyan scholars[1] that only taking into consideration a specific cultural context it becomes possible to explain why, for example, despite the manifest influence of Dewey on the thought of prominent educationalists in many countries this did no result in any tangible influence on educational practice, or why while Dewey's ideas were not only well-known but appear to have been integrated into existing traditions, they were other factors, unrelated to the quality or significance of Dewey's ideas, that exerted a decisive influence on the eventual course of events. The Dutch case, as well as the other ones, brings a lot of arguments against the validity of the notion of "influence" in our analysis.

Today when world integration makes the science cross the national boarders, comparative research in education is becoming of great importance and come to the focus of scientific discussions. Educators try to find out the facts and processes of cultural interlinks, though they understand how important it is to avoid superficial analogies, to deepen logical arguments in comparing different phenomena. The problems of educational transfer, perception and acceptance of the other have become of paramount significance. Educational legacy of American scholar John Dewey is considered to be the subject of interest not by chance. He was not only the leader of North American educational reform that took place around the end of the 19th century and the beginning of the 20th century, but one of the key figures in what was called "new education", "progressive education" or "reform pedagogy". The man of the 20th century, John Dewey has made great impact on the development of world pedagogy. As N.Yulina points out, he could be called the philosopher of modernism, as "he tried to grasp the dynamics of modernization, civilization and culture in the 20th century, the strings, pushing the countries towards democracy and humanism, and at the same time to understand what forces block it. He believed in human wisdom, in science and scientific methods, in active social and moral role of philosophy in society, in open systems of thought and, he rejected dogmatism and authoritarianism, being confident of humanistic capacities of liberal democracy, and what is more important, in enlightment as the main lever of democracy".[2]

Dewey's educational paradigm was an alternative to existing traditional authoritarian one that was criticized by educators in many countries. Future teachers should realize that it was the ideas in his writings – his instrumental form of pragmatism – that made his influence so great. In Dewey's case we face an extraordinary versatility. W. R. McKenzi gave a summary of what Dewey was working at in the 1890s: «...Philosophy, psychology, philosophy of education, educational psychology, psychology of selected educational subjects, child study, elementary education, secondary education, Laboratory schools, pedagogy as a university discipline and others...»[3]

Dewey wrote about his own development: "Upon the whole, the forces that have influenced me have come from persons and situations more than from books – not that I have not, I hope, learned a great deal from philosophical writings, but that what I have learned from them has been technical in comparison with what I have been forced to think upon and about because of some experience, in which I found myself entangled".[4]

The image of Dewey is multi – faced – a serious social and political leader, a reformer in education, a philosopher, a master of polemics and at the same time – a beloving father, a good family man, a generous friend and a patron of talented students, ready to join them in the strikes for democracy. Today all these characteristics cannot match the existing myth of Dewey in Russia as "weapon-carrier of American reaction", that was created during Stalin regime and "cold war" period. The President of American Psychological Association, the President of American Philosophical Association, the President of American Association of University Professors, "the Teacher of teachers" – John Dewey was defamed in Russia in 1930s after he got involved in Mexico in the International Commission of Inquiry into the charges against Leon Trotsky at the Moscow trial and a statement of the commission's findings had been published under the title "Not Guilty". Dewey's relations with Stalin were badly spoiled and he became the opponent of the "Genius and the Teacher of the Peoples".

My intention as a researcher and a lecturer in Philosophy and History of Education for many years was to shed light on the educational phenomenon of Dewey and give a chance for Russian teachers to grasp his real contribution to the development of educational theory and practice and to show the international character of his pedagogy. At the beginning of the last century Russian educator Stanislav Shatsky mentioned that future teachers should study Dewey's works very thoroughly. Dewey attracts the reader by his great ability to reflect over his own educational experience, over the vast number of well-analyzed facts.

His educational philosophy, grown out of his experiment at his school, seemed to be inspiring for many teachers because it met the demands of the changing society. In Special Collection of Regenstein Library at the University of Chicago one can see 143 items chronologically listed beginning from Dewey's letter of February 15th, 1894, optimistically viewing the opportunities in prospect at the University of Chicago. The archives give a chance to grasp the devotion of the scientist to his experiment and his reflection over it. Dewey's letter to president W.Harper about his friend G. H. Mead, who was his true and cooperative colleague at Lab School (dated April, 10th, 1894), his "Plan of Organization of the University Primary School as well as the records of his Lab School, nice pictures of it in different periods of its existence and even the letter of Dewey of June 16th, 1904, after he angrily had resigned from the University and was about to leave for Columbia and New York – all the documents show Dewey's commitment to the idea that the concerns of education are really worthy of the most serious scholarship that university can provide".[5]

Dewey's school aimed at educating a flexible, creative, thinking and cooperative pupil and not a passive person. He wanted school to be a social institution representing life as vital to the child as that carried on at home. Very relevant for contemporary teachers are today Dewey's ideas of education as a process of living each day and not a preparation for future living.

Many of Dewey's followers assumed that a subject-oriented curriculum in his experimental school was replaced with a new program consisting mainly of projects. Some of them – W. Kilpatrick and Y. Meriam – were sure that teaching «accidentally» they were exactly following Dewey. On the contrary, I agree with L. Tanner that Dewey's curriculum was not synonymous with projects. The curriculum had two dimensions: the child's side (activities) and the teacher's side (logically organized bodies of subject matter: chemistry, physics, biology, mathematics, history, language, literature, music and physical training).[6]

By studying Dewey's educational experiment teachers realize that to refer to a school subject mainly as a set of facts and principles, mastered through effort rather than interest, means to ignore child psychology. Relevant for contemporary students is Dewey's idea that something done should be something inherently significant, and of such a nature that the pupil appreciates for himself/herself that it is important enough to take a vital interest in it. «Learning by doing» was the principle proclaimed by Dewey as a reform of the methods of instruction.

Dewey was far ahead of his time when he pointed out that a school subject is just a form of communication and artistic expression, and should not be referred to as something existing for its own sake. Dewey's idea of an educational process based on communication, his insistence that children understand a subject best if they experience it as a form of communication is very appealing to contemporary teachers too.

Being misunderstood by their interpreters some ideas of Dewey after implementation into practice brought some negative results. The overemphasis of the process aspect of teaching/learning and the under-emphasis of the content itself in some American schools during the early 20th century led W. Bagley – the leader of essentialist movement – to criticize Dewey's theory. But nobody would deny that Dewey's ideas encouraged the thought, self-activity and creativity of the learners.

Thanks to progressive experiments of F. Parker, J. Dewey, W. Kilpatrick and others the American school was turned from one of «passive listening» to the «school of activity», as P. Blonsky (an outstanding Russian educator and psychologist) remarked.

In his many years of working first at the University of Chicago and then at Teacher's College of Columbia University Dewey tried to do his best to improve teachers' professional training. His Chicago summer-sessions for in-service teachers brought him popularity and the title of «the teacher of teachers». His «laboratory» approach to organization of the practical aspect of teacher education was a new idea and differed much from a traditional approach (an «apprenticeship» model). Dewey tried to find the correlation between theory and practice. He wanted to put a teacher in the position of a researcher and thanks to him a lot of interesting techniques were introduced to teacher training. Different case – studies of educational problems of the classroom stimulated a trainee teacher to develop his reflective skills and to realize the problems of the concrete school students. More to it, the students saw how theory could be applied to practical task. Many findings of the American educator have significance today.

Now from a historical perspective we can precisely value the novelty of Dewey's experiment and its shortcomings. There is a great difference between an idea and its implementation into practice, as the fate of the idea is in the hands of those who implement it. So there is a difference between Dewey and Deweyan. His idea of freedom was sometimes taken for anarchy, his statement that a teacher should not be a «mentor», but a guide, an adviser and an organizer of a child's various activities was also misunderstood by many teachers as a very easy task. Instead of grasping the more complicated role, teachers got rid of all of their responsibilities, just the opposite of what Dewey had meant.

The task for Dewey's philosophy of education was to comprehend and gather together the varied details of the world and of life into a single inclusive whole – to attain as unified, consistent, and complete an outlook upon experience as is possible on the macro, meso and micro societal levels. Dewey tried to overcome the gap between educational theory and practice. A lot could be learnt from what Dewey said and practiced.

According to D. Sidorsky, "John Dewey was the most influential figure in American philosophical thought in the first half of the 20th century. His influence was both broad in scope and deep in impact. . . The impact of Dewey's ideas upon American philosophical and social thought was so great that it must be considered a major phenomenon of American cultural history of the 20th century"[7]

John Dewey's influence on philosophical thought and educational reform was not limited to America. Before 1950 "Dewey in Europe" was somewhat of a trademark as Jurgen Oelkers stated in 2000.[8] John Dewey was firmly linked to European "progressive education" and was read and received all over the continent. In 1946 the then director of the International Bureau of Education in Geneva, Robert Drottens, hailed Dewey as the person who had had the greatest influence on contemporary education worldwide.[9] In 1961, the president of Hong Kong's New Asia College, Ou Tsui-Chen, commented: "John Dewey was one of the most important philosophers of education of this century; few educational theorists have equaled his widespread influence, which was not limited to his own society, but was felt throughout the world"[10] In Soviet Russia in 1920s Stanislav Shatsky called Dewey "the best philosopher of contemporary school". On occasion of Dewey's ninetieth birthday (October 20, 1949) W.Brickman discussed Dewey's reputation as an educator in foreign countries and pointed to some examples of Dewey's influence on educational thought and practice abroad. Mentioning that ". . . a more accessible measure of Dewey's relationship to his contemporaries in foreign countries is his reputation as an educator", he determined it ". . .by translations of books and articles, professional reviews, discussions of ideas in professional and other publications, and references to theory and practice in miscellaneous sources".[11]

We should stress that the attitude towards John Dewey throughout the 20th century dynamically changed. As Oelkers remarked about European perception of Dewey, it was relatively easy to link Dewey with the philosophy of Henri Bergson in Geneva, but impossible to bring about pragmatism and Dewey against neo-Kantianism in Germany before and after 1914. Critical theory up to Habermas showed no real interest in Dewey, at least not in his conceptions of democracy and education, although the social theories have very much in common.

After Robert Westbrook's magistral study "John Dewey and American Democracy" (1991), research and literature on Dewey has exploded. In education alone ten to twenty doctoral dissertations, books, articles or collections appear every year. "Understanding John Dewey", written by Campbell in 1995 has become of central interest to Anglo-Saxon philosophy and history of education in the nineties. After 1989 with the fall of socialist education interest in Dewey has increased in Europe too. For European education Dewey is no classic in the sense of "essential truths", his theory of education is a challenge to do better.[12]

Dewey's influence on educational thought and practice was felt on six continents and was brought about in three ways: (1) Dewey's visits to foreign countries, most notably his visits to Japan, China, Turkey, and the U.S.S.R.; (2) translations of Dewey's books and other writings into at least thirty-five languages; and (3) the thousands of students from other lands who studied with Dewey and his colleagues at Teacher College, Columbia University and other American Universities and colleges where Dewey's philosophy was taught, and then returned home to become leaders in their countries' ministries and universities.[13]

In the older European countries at the beginning of the 20th century there was little tendency to look to America for new ideas in the realm of thought. W. Kilpatrick wrote that in such old European countries like Germany and England John Dewey's ideas have been interpreted rather narrowly, mostly in connection with the place of industries in elementary education, and George Kershensteiner's "Arbeitsschule" dealt with Dewey's Critique of a child's activity.[14]

It is important to mark that John Dewey's philosophy of education was being formed under a great influence of German philosophers. The influence of European philosophical and educational thought on the formation of the first generation of academics in the United States was enormous, as many researchers point it.[15] John Dewey's educational philosophy was greatly influenced by Hegel and Herbart. It was with "naturalized Hegelianism", and "social behaviorism" in which distinctive social categories such as communication and participation played a pivotal role that philosopher and psychologist John Dewey entered the American educational stage. This stage itself was thoroughly influenced by three strains of European thought: (1) William Torrey Harris – United States commissioner of education from 1899 to 1906 and the leading American Educator in the last quarter of the 19th century had brought Hegelianism into the American schools: (2) Herbatianism, introduced in the United States by Charles De Garmo, Charles McMurry and Frank McMurry who tried to implement in America what they had learnt at Jena in teacher training schools in the country; (3) The ideas of Friedrich Froebel, introduced through the kindergarten, which was introduced by disciples of Froebel in the 1850s, first in private and from 1873 onward also in the public system.[16]

Being extential to the European Educational tradition John Dewey had close links to it and even visited Europe for some times. His first visit took place in January 1895 and he spent a year there with his wife and three young children. In 1904 Dewey visited Europe again and it was another 20 years before Dewey went to Europe to survey Turkey's educational system and to recommend ways for its improvement. In 1926 Dewey saw Paris, Madrid, Vienna, visiting museums together with art collector Albert Barnes. In 1928 he went to London, Berlin and again to Paris and afterwards visited Leningrad and Moscow to see schools in and around both cities as a member of a group of 25 American educators who were there by invitation of the Soviet commissioner of education. In April and May of 1929 Dewey was in Edinburgh, to give the prestigious Gifford Lectures. In 1930, he was in Paris to accept an honorary degree from the University of Paris and for a few weeks in Vienna. He came to Europe for the last time revisiting Paris and Vienna in 1933.[17]

Though Dewey did not receive his education in Europe, his intellectual background was closely connected with European thinking. Dewey positioned himself on a theoretical level in between the Herbartians and the Hegelians criticizing the Hegelians for their failure to connect the subject matter of the curriculum to the interests and the activities of the child and the Herbatians –the representatives of the child-study movement for their failure to connect the interests and the activities of the child to the subject matter of the curriculum.[18]

So, we can explain similarities between European and North American educational reform of the eve of 19th–20th centuries partly by the shared intellectual background of Dewey and European reformers. Though even in the recent analysis of Education and the Struggle for Democracy, Carr and Hartnett (1996) conclude that, if to take the context of Britain, "John Dewey is doubtedly the most influential educational philosopher of the 20th century".[19]

At the same time when English political leaders want to find somebody to be blamed for all the faults at their schools they speak of Dewey's "undeniable" and "disastrous" influence on English education as presented in the report of John Major and his education secretary while adopting National Curriculum in 1991. How can we explain such opposite positions in Britain?

At the beginning of the 20th century, in 1929, for example Thompson wrote that "In Great Britain, except Scotland... I have been repeatedly struck by the absence of references to Dewey's ideas and sometimes by complete ignorance of them, although the same views in other dress are often mentioned in their practical aspect".

Though Scotland has a separate educational system from England and Wales, there was "the comparative lack of penetration shown by Dewey's doctrines before the 1960s."[20]

In the first half of the 20th century Dewey was in teacher training courses on the lists of prescribed reading, thanks largely to J.J.Findlay, professor of the University of Manchester, who published a collection of Dewey's essays in 1906 under the title "The School and the Child", and did much to introduce Dewey to an academic educationalists in Britain. But his ideas were not widely assimilated into practice or theory. Herbart's was still the favored theory at the start of the century.

A modest movement towards greater recognition of Dewey was felt in the 1930s in England. Three reports of the Consultative committee show this (1926 – no explicit references on Dewey, but the one in 1931 on the Primary school marks the beginning of acceptance of Dewey's ideas by the educational establishment. Though a passing reference to Dewey by name, its recommendations clearly have close affinities to Dewey's thinking, contrasting "traditional education" with "the real business of life". But everything connected with such innovations as Kilpatrick's Project method, Parkhurst's Dalton Plan – all of which were influenced by Dewey, gets a somewhat guarded endorsement: "It would be unnecessary and pedantic to attempt to throw the whole of the teaching of the primary school into the project form. . .". Some of this may be a protest against too enthusiastic adoption of Dewey's views in English primary schools: in general, however, it is a characteristic English reaction. The Committee rejects Dewey's philosophy, and they reject his principles as principles, but they are quite prepared to accept and commend his methods where they serve their own principles. And these principles remain the traditional ones. In the report of 1933 on Infant and Nursery Schools – a whole page is allotted to Dewey's ideas with the conclusion that "Dewey's works. . . have played an important part in the evolution of modern ideas on infant education in this country".[21]

In England only pedagogical ideas of Dewey had some impact but not his epistemological, social or political ones up to 1940s and only to extent that the notion was endorsed by some as a worthwhile principle, or at least as an aspiration, and generally accepted more by academics and reformers than by teachers.

Scotland (1969) links this "comparative lack of penetration" to the intellectual climate in his country at this time: "Project method and problem teaching and activity methods were lectured on in Scottish universities and colleges, much discussed in professional assemblies, but little practiced at schools. In a country with a strong tradition of Platonic idealism, Dewey's pragmatic attitude could hardly expect to be welcome, nor could a doctrine, which stressed the need for the learner to do the work appeal in a system where the . . . teacher was the king. . ., where stern discipline was considered to build character".[22] Dewey's emphasis on social context of education was in opposition to the individualistic philosophy of Nunn which was widely favored from 1920 until at least 1940. Long-established, rigid structures in British society, in which social class divisions were endemic, subject-oriented curriculum was the obstacle for Dewey's model of school.

Only in 1960s Britain saw a marked warning of deference to authority. Old ways were to be questioned and traditional practices challenged. In primary education in 1960s there were significant changes. Plowden Report (1967) advocated a strikingly progressive approach to education and Deweyesque nature is vivid: "At heart of the educational process lies the child. No advances in policy, no acquisition of new equipment have their desired effect unless they are in harmony with the nature of the child, unless they are fundamentally accepted to him" (Plowden, p. 7). It reminds us of Dewey's change, "not unlike that introduced by Copernicus when the astronomical center shifted from the earth to the sun", and ". . . the child becomes the sun about which the appliances of education revolve; he is the center about which they are organized".[23]

From 1979 till now with he advent of a right-wing Conservative government in London, much political pressure was exerted to bring British primary education back to more traditional ways, but still despite the criticism, and the introduction of a national subject-based curriculum, the appeal of child-centered thinking continuous to influence practice in Britain. The opponent of child-centered education, Anthony O'Hear, professor of philosophy at Bradford University has criticized in 1991 what he sees as Dewey's "disparagement of didactism". But he really overestimates Dewey's influence as he writes that "Deweyesque practice is contemporary practice in many of our schools, particularly in the maintained sector, where it is all but universal at primary and junior level; and Deweyesque theory is contemporary theory in the educational establishment of our country".[24] The researcher Bretony thinks that this unprofessional judgment though very influential because of the post of the author is an overstatement that made possible the statement of "undeniable" and "disastrousus" influence of Dewey on English school to appear. The notion Deweyesque is rather vague. Sometimes in England they saw Dewey responsible for all the progressive education implementations and this is not the right way as Dewey criticized progressive education methods very much. But what is true is that his educational philosophy has found its proper place in educational discourse in Britain.

As for France, Dewey was first recorded there in 1883 after an anonymous review of a philosophical text by Dewey that had appeared in the April 1882. After this no notice was taken of Dewey in France for several years until the journal "L'Education", edited from 1909 onwards by Georges Bertie, director of Ecole Roches, listed Dewey in its editorial as a leading contributor. From this period until 1960s, the reception of Dewey was restricted to the pedagogical element of his heritage. It is important to mention that Dewey's ideas didn't penetrate deeply in French educational system, though they were popular in academic discussions. Only in 1901 the church was separated from the state in France. It was not easy for Dewey's active pedagogy to be accepted in the tradition where the center was on a teacher. In the period before the World War 1 there appeared first translations of John Dewey. The critical reception on Dewey in France may be explained by the conflict in this country between new education and traditional school. The experiments in French schools, proclaimed like Dewey's experiment sometimes didn't correspond to the original idea. In 1965 there appeared in France the book entitled "John Dewey's Pedagogy", written by Gerald Dalledalle, with the introduction by Maurice Debesse. Debesse came to conclusion that though Dewey was considered in France to be a very important author within New Education, the French didn't know him very well. Only Gerald Dalledalle tried to pay serious attention to Dewey's works. He systematically studied John Dewey's works and wrote many books on pragmatism and his founders.[25] Delladalle wrote that John Dewey's educational ideas were rather influential in France but it is very hard to trace this influence on different French educators. Dalledalle himself confirmed that American philosophy in general and Dewey's ideas in particular served his intellectual background. He accepted

John Dewey's idea of cooperative work and considered the Ecolle de Roches to be experimental sides for John Dewey's principles. "New classes", introduced in three French schools (Sevres, Montgeron, Pontoise) reflected the ideas of American reformer. This scientist stressed very important roles played by Claparede, Ferrier and Decroly in empirical reception of Dewey's ideas in France. In 1975 Dalledalle published the translation of John Dewey's "Democracy and Education". In his introduction he pointed to five central aspects of John Dewey's pedagogy: spontaneous and intellectual activity concentrated on the interests of a child, the sociality of whom should be shaped at school, reflecting the structure of the existing society, in case if the structure is based on the principle of continuity.[26] This author thought that John Dewey was falsely blamed by all the mistakes of American school system. He thought that many teachers tried to copy the ideas of Dewey's school without understanding of his experimental method. John Dewey's idea of constant reconstruction of experience demanded to take into consideration the changing conditions of life, it was incompatible with "orthodoxy of undeflected passage along a single path of salvation".[27]

In the Netherlands there was also some interest to John Dewey's progressive ideas. According to the report of Dutch researchers G. Biesta and S. Miedema (1988), in period of 1908–1988 43 writings on Dewey have been published (42 in Dutch and one in English). The opinions expressed in the writings on Dewey's influence were different. Some totally rejected his ideas, some thought that parts of Dewey's work could be used and other parts, especially his philosophy of life, should be rejected. Dewey's anti-fundamentalism, both in his theory of knowledge and in his ethics, and consequently in his educational ideas have definitely formed an important stumbling block for educators in the Netherlands, especially for those who adhered to biblical conceptions. At the end of the 19th century and the beginning of the 20th century private schools, were founded in the Netherlands, providing education based on educational principles, like those voiced by Maria Montessori, Helen Parkhurst (John Dewey's pupil), Peter Peterson, Rudolph Steiner. Dutch researcher N.L. Dodde wrote, that at the beginning of the last century "...the school system should be more conveniently arranged and more accessible for pupils. The educational institute should pay more attention to differences in interest, experience and development of its pupils and the education should, besides intellectual education, also offer space for more practical training".[28]

The most obvious proof of Dewey's influence on education seems to be the existence of a "Deweyan" educational practice. In the case with the Dutch educator Jan Ligthart (1859–1916). A principal of an elementary school in the "Tullinghstraat", we sec much in tune with Dewey; examples include bringing daily life in its totality into school and bringing about the active participation of the child. Ligthart was opposed to verbalism and stressed learning by doing. When his school was visited by Ellen Kay (1905), A. Zelenko (1910), and Eduard Clapared (1912), those familiar with both Dewey and Ligthart often concluded that there were striking and surprising similarities between them. Ligthart was aware of his similarities with Dewey, but al the same time stressed that it was not the result of the Dewey's influence but the coincidence of ideas – thinking in the same direction, as they say. The "encounter" between Dewey and Lighthart clearly reveals that the existence of a strong similarity between two sets of ideas and/or practices is not enough to conclude that the one has influenced the other. In Lighthart's case the first condition of Skinner's methodology is met, but the others are not. Skinner's point allows us to speak of influence if we can trace a direct, exclusive and unidirectional connection between one set of ideas and another. So, we see that the perspective of "influence" is hardly adequate to bring Dewey's contribution into vision. Another Dutch educator-G. Wielenga, professor of Free University of Amsterdam, a Dutch Reformed institution for higher education (founded in 1880) played a great role in bringing Dewey's ideas to the Netherlands.

Primarily engaged with Christian elementary and secondary education, in a series of lectures that were published from 1946 on wards, Wielenga had expressed a very positive interest in Dewey's work on psychological, didactical and more general educational questions related to the issue of learning how to think and attempted to legitimize the adoption of Dewey's psychological and educational ideas. At the same time he rejected Dewey's view on religion and the religious, his "humanistic" philosophy of life.[29] Wielenga tried to find a place for Dewey's ideas about the process of education within his own. Another Dutch educator Van der Velde was also very positive about Deweyan ideas. He was associate professor at the center for educational studies of the City University of Amsterdam and taught courses in the philosophy and history of education. In 1968 he published a book "Child, School, Society" together with Van Gelder. Contrary to prevailing interpretations of Dewey's conception of education as being a 100 per cent social theory of education, Van der Velde argued that Dewey was concerned both with the individual and with society, and, more specifically, with the interaction between the two. Dewey's position came close to that of the most renowned educationalist in post-war Dutch academic education, M.J. Langeveld, who contributed to a theory of education along phenomenological-hermeneutical lines, starting from the "common ground" of the phenomenon of education, and not from first (denominational) principles. The writings of Van der Velde and his colleagues and the earlier work of Wielenga had made Dewey's ideas available to the larger educational community. But at this juncture in time dramatic changes in the context took place. Educationalists in special education and curriculum studies took inspiration from the findings of German and Anglo-American empirical studies. The fighting flared up between those in favor of a value free, objective empirical paradigm for educational science, and the adherents of the phenomenological-hermeneutical approach along the lines of Langeveld and the "paradigm wars" took up most of the time of Dutch educationalists for well over a decade. The Dutch researchers S.Miedema and G.Biesta consider it to be the reason for holding the Dutch educators back from actively pursuing the Deweyan approach to education and schooling.

In the field of Dutch kindergarten Dewey's ideas were recepted. The key figure in this case was C. Philippi-Siewetz van Reesema who first wrote about Dewey in her extensive study on American educational "pioneers" and the way in which they had developed their educational philosophy and their school-systems. Philippi became a member on Montessori Dutch Association Board in 1917 and attended

Montessori's course in London and started criticizing Montessori for her dogmatic and strict use of educational tools and the "so-called" sensitive periods. She praised in her Dewey's contribution to the education of young children – his experimental, observational and experiential approach, his contention that nursery and infant school should not be separate but ought to be part of a comprehensive school system, and his genetic psychology which she perceived as being an implicit critique of formal learning (Frobel, Herbart) and the formal approach to educational tools (Montessori). In her book on the world of infant and infant education she made use of Deweyan ideas. Her students – for ex. W.Nijkamp also sustained Philippi's positive reception of Dewey's ideas. A. Stoll – another influential figure in Dutch education also paid positive attention to Dewey in her handbook for students at Christian infant teacher college.

So, Dewey's ideas had a real impact on Dutch infant education in kindergarten classrooms, but as infant education was considered to be the domain of women who were not seen as belonging to the academic circles and as the field was seen as "preparation for real education", most of the work was ignored.

In the early years of the century before the World War I, the ideas and practical suggestions of Dewey also became known in Australia largely through the interpretations of educational writers in England. In Australia this was the beginning, for Australia, of what has been called the "New Education". The second period, twenty years from the end of World War I to the beginning of World War II, had a much richer experience of progressive education. Hcrbatianism, which was seriously criticized by J. Dewey, had by then become the orthodox conservatism of educational thought and practice and was challenged by the Dewey of Democracy and Education, by the Project Method, and other new forms of instruction.

Dewey's influence may be observed in Turkey where his involvement was evoked by an invitation of the Turkish Government under the presidency of Mustapha Kemal, named Ataturk, to survey the Turkish educational system and organization and to make recommendations for its improvements. Dewey's investigations resulted in his Report and Recommendation upon Turkish Education.[30]

Dewey came to Turkey when it was changing from a Muslim theocracy into a secular state. In 1923, the Turkish government was proclaimed, State and society were secularized, all citizens got equal rights, but at the same time American educator marked that Turkish nationalism was propagated against anti-Turkish nationalism (mostly Armenian and Greek).[31] The Turkish government thought John Dewey's philosophy of education "to fit the democratic aims of Turkish educational reform movement".[32] In Dewey's report the main end to be secured by the Turkish educational system was "the development of Turkey as a vital, free, independent and lay republic in full membership in the circle of civilized states".[33] American reformer suggested that Turkish schools should: (1) "form proper political habits and ideas, (2) foster the various forms of economic and commercial skill and ability; and (3) develop the traits and dispositions of character, intellectual and moral, which fit men and women for self-government, economic self-support and industrial progress; namely, initiative and inventiveness, in dependence of judgment, ability to think scientifically and to cooperate for common purposes socially."[34] Dewey wanted to educate the mass of Turkish citizens "for intellectual participation in the political, economic and cultural growth of the country";[35] he didn't limit this aim to certain leaders. The American scholar stressed the importance of the existence of different types of schools – vocational and agricultural in addition with existing schools with only academic training. He saw in private schools an experiment station for public schools. He recommended foreign schools in Turkey (mostly French and American) because they embodied a variety of typical methods of school administration and instruction from which mainstream Turkish educators could profit. He also stressed the need for better salaries for teachers as an indication of the recognition of the society and government of the teacher's status. Dewey wanted to introduce in Turkish teacher education modern and progressive pedagogical ideas, he also suggested that teachers had to be send abroad to experience other systems and solutions. Traveling specially trained supervising inspectors and libraries were seen like good means of improving Turkish schools.[36] The American educator thought it important for Turkish government to sponsor the translation of foreign books and particularly that "those, dealing with practical methods and equipment in progressive schools" should be "widely circulated" and "carefully studied by teachers".[37] It is important to mention that while Dewey was in Turkey the schools were not in operation. He relied on impressions and information given him about the structure and climate of Turkish schools. Maxwell – Hyslops asserted that: "The aims and nature of the organization of education in Turkey today offer proof of the extent to which [Dewey's] recommendations were followed".[38]

Dewey's report had a great impact on a Turkish educational practice. His ideas on teacher training, teacher payment and differentiation between teacher training schools and training of inspectors nearly completely was set into practice. But the policy of prohibition and strict control, regarding the foreign schools didn't change.[39] It is a pity that some of Dewey's views were interpreted rather narrowly in Turkey, that led to positivistic, technological and product-oriented patterns of action. Theocratic culture and the family structure of the country blocked democratic reform in Dewey's sense. The case of Turkey is a good example of the use of progressive ideas in the modernization of the State. Though we can clearly see the misinterpretation of Dewey's educational ideas by Turkish official government that destabilized pluralism in educational system, contrary to his recommendations.

In Latin America he seems to be also famous at the beginning of the century. In Chili (1908), Cuba (1925), Mexico (1929) and Argentina (1939) the first translations of his famous books gave a chance for the educators in those countries to get to know his philosophy of education. In Brazil the educational heritage of Dewey was known thanks to Lourenzo Filho. He even gave the title to his own book "Dewey and World Educational Reform". In 1930 a famous book of Dewey "Democracy and Education" and in 1933 his famous text "How we think" appeared in translation.[40]

Though his ideas were not too influential in Latin America as socio-cultural situation in such countries like Mexico, for example, differed greatly from that one in North America, his action pedagogy was even officially adopted there in 1923 and played some role in the modernization of society. Dewey visited Mexico two times. In this country two main of his ideas – observation and experience as the means of individual efficiency and cooperative work were seen as the means to strengthen the spirit of fraternity and to provide future new social order.[41] According to M.Vaughan, progressive reform associated with Deweyan philosophy of education could not become a wide – spread movement as the situation of dependent capitalism in economy and lack of resources blocked it. It was just an experiment.[42]

The political context in Russia during the last century influenced the process of John Dewey's pedagogy digesting. Analyzing the process of Dewey's reception in Russia one can identify four distinct periods:

(1) The pre-revolutionary period (the first two decades);
(2) The 1920s – the period of his most popularity:
(3) The 1930s: the period of the de-Deweyization of Soviet education;
(4) The late 1980s–1990s when, as a part of the movement of "the pedagogy of cooperation", Dewey's ideas became the focus of attention in Russia again.

At the beginning of the century Dewey's idea of a child-oriented school penetrated Russia with the publication of his hook "School and Society". This was translated into Russian in 1907 and had a great impact on many talented educators of the time, such as N. Krupskaya, A. Lunacharsky, P. Blonsky, A. Pinkevich, and S. Shatsky. Before the revolution. In setting his "Settlement" program Shatsky and. his colleague A. Zelenko, and L. Shleger were greatly inspired by Dewey's new philosophy of education, his democratic model of the school, and his idea of the organization of the child's vital activities. The "Settlement" was the first club for children in Russia in the working men's quarter of Moscow, at Maryina Rosha. A. Zelenko was connected with the University "Settlement" in New York City. When he came back to Russia, he told Shatsky about the Hull-House.

The Hull-House as a community center for all of Chicago, organized by Jane Addams, was for Dewey, associated with it, a sort of a social center. It turned out to become "a cultural center. A social service school, a university, and a church".[43]

Shatsky was very inspired by the American experience and tried to operate along non-political lines and in the neutral fields of children's clubs, recreation, and health. A group of children was made to concentrate on agricultural work and manage its own affairs. Shatsky tried in his experiment to discover regularities in the way groups of children behave; he did his best to find ways and means to help the young generation master progressive and cultural norms. While experimenting, Shatsky met with constant opposition and embarrassment from the Tsarist regime and his experiment was soon halted. His wife Valentina Shatskaya taught aesthetics at school and made a program for the society "Child's Work and Leisure," which was in tune with Dewey's ideas.

In 1911, Shatsky organized a summer colony called "Bodraya Zhyzn" in Kaluzhskaya region. He considered the most important task of school to be the organization of children's vital activities. Later, after visiting Shatsky's colony as a member of an American delegation, Dewey wrote in 1929 in "Impressions of Soviet Russia", that his school was based "on a combination of Tolstoy's version of Rousseau's doctrine of freedom and the idea of the educational value of productive work derived from American sources."[44]

Shatsky tried lo implement many of Dewey's ideas in his practical work in the colony. For Shatsky, education meant "organization of children's life" and he tried to act in conformity with nature and did not ignore the influence of environment. Shatsky thought that the main task for a teacher was to create facilities for a child to display his/her "forces and abilities" in order to give vent to all natural instincts. Inspired by Dewey, he tried to implement Dewey's principles and practice of democracy into school life and administration, and showed increased human interest in current social affairs. But Shatsky went further than Dewey's adaptation to society idea and tried to change the environment by means of the school.

In 1922–1933 Dewey's theory and practice greatly influenced existing Soviet educational practice. J. Dewey visited Russia in 1928 as a member of an American delegation, and saw tremendous changes in the relationships of teachers and pupils in Soviet schools. Dewey's concept of a teacher as a guide, and organizer of various activities was taken by Shatsky and other Soviet progressive educators as a main principle in their experimental educational practice. While in Russia Dewey was impressed by the phenomenal achievements of the Soviet school system, which were due lo the deep and constant attention which Soviet society paid to the upbringing of the younger generation. Although he found much political propaganda at schools, Dewey noted the enthusiasm of remarkable Russian men and women, students and teachers, who were ardently convinced of the necessity place of education with a social aim and cooperative methods in securing the purposes of the revolution. After his visit Dewey wrote a series of articles very sympathetic in tone lo the USSR, which led to his being described as a "Bolshevik" and a "red" in the conservative press.

It is not by chance that Dewey gave such high evaluation of the school of 1920s. Many specialists consider this period lo be the brightest period of Soviet education, as it was a period of a dialogue in educational science and innovative search m education. The "Era of Krupskaya," as this period is sometimes called, may be characterized by the fact that many talented people such as P. Blonsky, A. Kalashnikov, S. Shatsky, A. Pinkevich and others worked with N. Krupskaya – at that time the Deputy Chairperson of the People's Commisariat of Education (headed by A. Lunacharsky) on school programs, plans, and textbooks. According to P. Blonsky, under the guidance of Krupskaya "all kinds of public dialogue took place, as did public criticism of various pedagogical positions and undertakings".[45]

During Dewey's visit lo Russia he met Krupskaya and had fruitful discussions with her on the problem of the labor school. Krupskaya knew the works of Dewey well and in her book Narodnaje Obrazavanije i Demokratija (Popular Education and Democracy) she analyzed the theory and practice of education from a historical perspective. Dewey's school of activity appealed to Krupskaya, as she also thought that schoolwork should be inseparably connected with science and culture. The Soviet educator B. Komarovsky published in the 1920s two books devoted lo the analysis of J. Dewey's ideas. Komarovsky called Dewey a prominent researcher in the fields of logics and epistemology, pedagogy and psychology, ethics and social philosophy.[46] M. Bernstein named Dewey as the best American educator and "the best of the best

Americans". A. Lunacharsky gave Dewey the title of "one of the greatest educators of our century".[47]

The innovation movement in Soviet education at that lime reflected the American influence. M. Pistrak, a member of the State Academic Council, confessed in the pedagogical discussion of 1928 that the Russians adapted the Dalton System from Western Europe and America and tried lo apply it, but not very successfully. Russian pedagogues as P. P. Blonsky, S. T. Shatsky, and A. Zelenko and other Soviet educators in the 1920s tried to learn about experiences in American high and secondary schools (M. S. Bernstein, G. F. Svadkovsky), some visited the Unired States and thought that it was the main educational laboratory at that time.[48]

Soviet educators actively applied the testing and project methods. But in the late 1930s Stalin's directives and "the iron curtain" blocked close cooperation between Soviet and American educators. Any signs of the American way of life were to be condemned and abolished. In 1932. The Dalton System in Soviet Russia was abolished by a special statement of VKP (B). The official reasons for this were the low role of the pedagogue and the disregard for the individual capacities of pupils. But the real aims of Stalin's policy were to make Soviet school a part of a command-administrative system, and make the pupil a small screw in the state machine. The fear caused by the statement prohibiting the Dalton System and other American methods lasted for a long lime and is even nowadays a blocking factor today to educational reform in Russian high and secondary schools. The complex programs that were elaborated by the members of the scientific-pedagogical section of the State Academic Council may be considered to be an example to combine Marxist principles with progressive educational ideas. The subject matter in the program was organized in three columns: nature, labor, and society. All the teaching was based on "integral instruction" through themes and not on regular discipline. The programs were to be filled with regional materials, corresponding to the vital needs of the environment in which the child lived.

The influence of Dewey's ideas is clearly observed in the complex programs. The Soviet educators were looking for a new school that could focus its attention on the children, their interests, and their inclination for action. The educators thought that the programs should reflect the growing complexity of children's lives and their personalities. The new programs were aimed at getting children acquainted with something essential for their present life and future. The implementation of the ideas embodied in complex programs proved to be not so good in practice as it seemed in theory. First, the programs were applied universally, to all schools in the Soviet Union. This was problematic for such a vast country where schools differed greatly in material resources and facilities. Second, the teachers were not prepared for the creative implementation of the ideas. Sometimes the task of linking the program with the local needs of the school and its surroundings led to frivolous things; some pupils devoted much time to such complex themes as "The Duck", "The Birch Tree", and so on. These links seemed to be very artificial. The authors of the new programs did their best to improve the complex programs until the 1930s. Al that time all the school experiments were ended by the authoritarian regime of Stalin. In hindsight, we can sec that the ideas that informed the complex programs were not accompanied by the necessary means of implementation, trying to give new content to their schools, but having no forms in which to stack and organize it. Sometimes complex programs were simply ignored by the teaching staff, or simplified lo such an extent that the essence of the complex method was completely lost.

The project method, originated by Dewey's pupil W. Kilpalrick, was also introduced in the Russian school of the 1920s. Being a modification of Dewey's problem method, it was adapted to the Soviet system with the aim of realizing the principle of education in the collective – the main principle of Soviet school of that time. Soviet educators made an attempt to compile new textbooks for schools practicing the project method. Soon the method was established to such an extent that it led to the neglect of scientific knowledge, reading, writing, and arithmetic skills. These extremes were most characteristic in the educational practice of the Lefts (V.N. Shulgin, A.V. Shapiro and V.M. Pozncr).

It is remarkable that Dewey's ideas were adopted both in pre- and post revolutionary Russia. The Revolution marked a decisive change in the outlook of Russian educators with regard to the role of the school in the transformation of society. Dewey's ideas happened to be fit first, because they stressed on the continuity between school and society, on the intrinsic relationship between learning and work, and on the cooperative attitude.[49]

The ban of pedology in 1936 and its liquidation as a humanistic discipline by the Resolution of the Central Committee of the Party paralyzed the development of all sciences dealing with childhood and stopped a very serious experiment in education. Slalin's command-administrative system was strengthening step by step. It is worth mentioning that later in his autobiography Dewey wrote that the reports that came to him after the high-pressure five year plan was put into effect of the increasing reglamentation of the schools and of their use as tools for limited ends were a great disappointment to him. The process of de-Dcweyization in Russia started with the elimination of encyclopedia articles on Dewey in the period of the late I930s-1950s and also with the criticism of progressive experimentation in the schools. During the Cold War Dewey was labeled in Russia as "the wicked enemy of all the freedom loving peoples on our earth" and in the 1950s all the articles and hooks written about him belittled his educational contribution and stressed his misguided social and political orientation. His pragmatic philosophy was criticized, too. The publications of Soviet researchers on Dewey in the 1960s and 1970s were in the same lone. Only in the late 1980s and early 1990s was there a shift in the perception of John Dewey's philosophy of education in Russia, that brought a sort of revival of interest to the ideas of American reformer, this time in connection with the category of experience, active learning, dialogue-oriented pedagogy, cooperative and interactive methods of teaching and idea of inter-subjectivity. Trying to find a democratic model of school Russian Educators turned to historical legacy of progressive educators in Russia and abroad. Dewey's ideas serve as an instrument in the change of society.

The experience of Japan in perception of Dewey's educational ideas is of particular interest as it helps to see how Japanese tradition tried to meet and interpret innovative western ideas. John Dewey's reputation as the recognized leader of the pragmatic movement in philosophy and pedagogy came to oriental countries like Japan and China in the beginning of the 20th century. When Dewey settled in New York at Columbia's University in 1905, he was already rather famous. In January 1918 as Dewey and his wife were about to sale for a vacation to the "Orient", he received an invitation to deliver some lectures in Japan. At the end of the 19th century Japan was very open to western innovations. In creating Japanese educational system, the Japanese had full confidence in foreign educators and counselors. As for Dewey's pedagogy, it became known in Japan even at the end of the 19th century. A famous book by Sudzi Ivasa was written under the influence of J. Dewey's philosophy of education and became a manual for teachers of Japan. Among supporters of westernization movement in Japanese school there were different positions. Some of them supported Herbert and didn't accept pragmatism, though many others positively accepted many key elements of Dewey's philosophy of education. It is remarkable that Americans first drew attention to Dewey after in one of the journals there was a paper of Japanese author Motoro Yujiro in 1887. The paper was devoted to psychology and the author was the first of Japanese pioneers – a Christian protestant, studying American philosophy. He evidently heard of Dewey at the lectures of professor Stanley Hall at the University of John Hopkins. Later Motora became the Head of Japanese Association of Child Study founded in 1902 after coming back home Motora became the professor of the University in Tokyo and Tokyo High Normal School. It was Motora who let it possible for Japanese to know one more representative of pragmatism – W. James. Motora wrote some papers about him and was the editor of the first translation of W. James' "Principles of Psychology".[50]

One more Japanese scientist Nikaima Rikiso has made his contribution to Dewey's reception in Japan by discussing Dewey's work "The Outlines of Critical Theory of Ethics", that first appeared in Japanese translation in 1900. Next year Japan saw Dewey's book "School and Society" and in 1905, 1923, 1935 and 1950 – four more of his main translated in Japanese books.

Japan is a country of traditions. When Dewey came to Japan during his two-and-a-half-month's visit he delivered a series of eight lectures at the National Imperial University in Tokyo. These lectures were organized around a general theme dealing "with the problem of reconstructing moral and social thinking and he benefits to be derived from a democratic way of life".[51]

Dewey thought that the lectures would give him a chance to express his ideas for world peace. Since Dewey's visit in 1919 Dewey's influence on Japanese educational thought seems to have been continuous and reached its peak, in all probability, during the "Americanization" of Japanese education following the World War II.[52]

The name of Dewey is often mentioned in the lectures and papers of Japanese educator Naruse. He admitted that Dewey's educational idea appealed to him greatly. In 1912 Naruse visited Dewey in New York and Dewey got his chance to pay him a visit later in 1918 when he had a lecture at Imperial Tokyo University on "New Tendencies in Philosophy, Religion and Education".[53]

At the beginning of the 20th century many young Japanese students who studied in U.S. took interest in Dewey's ideas in Japanese educational thought and wrote that Naruse took some elements of Dewey's didactics in his school, but he was not a very good specialist in Dewey's philosophy. One of the serious researchers of Dewey's pragmatism was Tanaka Odo (1867–1932). He listened to Dewey's lectures at the University of Chicago in 1889 after the graduation from the University of Chicago Tanaka taught at High Industrial School in Tokyo and then at Waseda University. He did not share all the positions with Dewey on societal problems and was greatly influenced by Hegel. Tanaka was an idealist in the case of social progress but reproached his Japanese colleagues for "Philosophy in armchair" in tune with Dewey and criticized them for "being isolated from a real world in an iron tower". In this book "Off the library to the street" (1911) Tanaka asked the scientists to leave their study-rooms and to study a real social world. It was Tanaka who did his best for Waseda University to become the center of pragmatism. A famous "Waseda group" consisted of Sugimoro Kojiri, Hoashi Rijichiro and Tanaka. Hoashi called himself "The pupil of Dewey's pupil".

Though in Kobajashi's view Dewey's brief lecture tour in 1919 did not have a significant impact but his ideas as transmitted through his writings in the years following did influence Japanese thought. The popularity of Dewey in the postwar period was amazing as Japan was an Asian country long known for its authoritarian tradition in education. Still Japanese kept to look for "Western technology" but tried to adhere to "Eastern morals".

Between the two world wars of the last century dedicated Dewey's scholars who had studied in Northern America tried their students with democratic ideas of American reformer. When in 1927 William Heard Kilpatrick visited Japan and lectured on his version of the project method, which had been inspired by Dewey's ideas, his lectures reached a very wide audience through various media, including radio. The Dalton Plan and the project method became very popular with Japanese education at that time. A number of schools had been founded following the pattern of progressive schools that had been started in the United States.[54]

Kobajashi was writing his study on Dewey in 1964 and he marked that two years earlier, a Japanese journalist had stated "no one can deny Dewey's great influence on educational thought in Japan in the last eighteen years. It exceeds that of any other educational thinker".[55] The Japanese Bibliography of Education for 1945–1957 contained 176 entries under the heading "Studies of Educational Thinkers". Almost half – eighty-one – dealt with John Dewey; the U.S.S.R. educator Makarenko was second in frequency with only eighteen entries. Nagano's General Introduction to Dewey's Philosophy, published in 1946, was in its sixteenth printing by 1948.

Kobayashi cites other evidence of popularity of Dewey in post – World War II Japan brisk book sales, 21 translations of Dewey's works, papers presented on Dewey at meetings of educational research associations from 1946 – every year at least one, but in 1951 – 8 papers on Dewey. Many students at the Universities did their master's papers on Dewey. The popularity of Dewey was so high that in 1959 in the year of his 100th Anniversary of birth there appeared a catalogue on Dewey's studies in Japan. In the country of festivals there was the festival of Dewey on Shikoku Island. On June 1, in 1953 the University of Hokkaido organized "The Night of John Dewey". In 1957 the Japanese Society of John Dewey was organized and by 1962 it united 130 educators and philosophers.[56]

Many Japanese educators perceived the "New Education" of the Occupation years following World War II as continuing the "New Education Movement" that had existed in Japan between 1912 and 1926 (which had led to Dewey's influence at the time) and that had been curtailed by the rising militarism and the war.[57] Though we can make a strong case for John Dewey's influencing Japanese educational thought it is not easy to determine Dewey's impact on school practices even those Japanese educators, who viewed the educational reforms promulgated by the U.S. Occupation as "based on Deweyan principles, differ among themselves on the extent to which Deweyan ideas have penetrated classroom activities. Furthermore, the Deweyan approach being more an attitude rather than a set procedure of teaching is difficult to observe directly and to judge objectively".[58]

According to the French historian L.Fevre, "the only lesson of history is precisely that it offers no lessons". This is true, but at the same time the historical material can be very effective in solving contemporary problems not by giving ready answers, but by searching for unused ways and conditions of successful implementation into practice of this or that idea.

Dewey's philosophy of education and his experimental practice had to pass national filters. The "Russian Dewey", the "English Dewey", the "Turkish Dewey" or the "Japanese Dewey" were just cultural interpretations of Dewey's ideas and practices. In any country – Britain, the Netherlands, Russia, Turkey, Latin America or Japan, the cultural canvass every time would correct the model sample digesting and interpreting it. In some countries they were used as a means in modernization of the state, in some – to stimulate educational discourse in school reform. In any case, it is hard to deny that Dewey's reputation as a world famous pedagogue and thinker was observed in many countries. In 2009 October, 20 all progressive educational community has celebrated 150th anniversary of John Dewey's birth. Luckily in Russia he is now also considered to be one of the outstanding philosophers and educators of the last century. The myth created in Stalin time about Dewey as "the enemy of all progressive mankind" was crushed in 1990s by the efforts of our researchers and now intending teachers read about him not ideological staff but objective truth. Publication of the translated books of American scholar in the last two decades gave the possibility for Russian readers to see the texts of Dewey itself instead of many stereotypic interpretations of his ideas.

But pragmatic pedagogy should be considered to be rather a way "to think about" education than a way "to do" education. This is not to suggest that Dewey's ideas are by definition impractical. It is only meant to draw attention to the fact that pragmatic pedagogy is not a sharply defined educational program that can easily be put into practice in a variety of different settings.[59] Dewey's ideas on developing reflective capacities of a teacher, his stress on competence of the teacher and necessity of profound psychological training, his laboratory approach to organizing intending teachers' practice and his democratic model of school are still actual. His theory of civic education and idea of school based on the idea of communication and cooperation, creation of the school scientific community remain relevant today.

It was said that the 21st century would be the one of the interpretation of the text. Looking at Dewey's pedagogy as a sort of a text, a phenomenon interpreted by the consciousness of different cultures we did not try to measure the degree of its influence it had marked but aimed at observing various results of its perception or rejection in different countries. Dewey's reputation as a foreign educator cannot be limited to the countries, mentioned in this chapter. Of great interest could be also the experience of China, Israel, Germany and other countries in reception of John Dewey's active pedagogy. But even the reaction of the above given ones widens our vision of the educational reform of the eve of 19th–20th century implementation.

Vladimir State University, Stroiteley Avenue, 11 Vladimir 600024, Vladimir, Russia
e-mail: erogacheva@hotmail.com

NOTES

[1] Biestat, Gert and Siebren Miedema, Context and Interaction. How to Assess Dewey's Influence on Educational Reform in Europe? in Dewey and European Education. General Problems and Case Studies, Eds. by Jurgen Oelkers and Heinz Rhyn, repr. From Studies in Philosophy and Education 19:21–37, 2000, pp. 33–34; Bretony, K.J., "Undeniable" and "Disastrous Influence"? Dewey and English Education (1895–1939), Oxford Review, Vol. 23, 1997, 428–429.

[2] Julina, N. Filosifija D. Djui I postmodernistskij pragmatism R. Rorti // Filosofskij pragmatism Richarda Rorti i rossijskij kontekst. M.: Tradizija, 1997.cc.172–173.

[3] McKenzi, Introduction Towards Unity of Thought and Action, in John Dewey Early Works, New York, 1975, pp. IX–XYI.

[4] Adams G.P. and W.P. Montague, editors, Contemporary American Philosophy, Vol. 2, New York: Macmillan, 1930, p. 22.

[5] Special Collection, University of Chicago, Regenstein Library, Laboratory Schools. Records. 1891–1986. Elementary School, 1898–1934 Vol. 1.1898, box. 1. Folder1; Vol. II (1899–1900), box.2.Folder1; Vol. iii (1900–1901), box. 3.Folder1: pp. 1–639: July 10 The Plan of Organization of the University School. University Presidential Papers.Box30.Folder 23.

[6] Tanner, Laurel N. Dewey's Laboratory School: Lessons for Today; Foreword by Philip W. Jackson, New York: Teachers College Press, 1997, 200p.

[7] Sidorsky D. and John Dewey, The Essential Writings, New York: Harpe & Row, 1977, p. vii.

[8] Dewey and European Education, in General Problems and Case Studies, Eds. by Jurgen Oelkers and Hein Rhyn, reprinted from Studies in Philosophy of Education, Vol. 19, Dordrecht: Kluwer, November 1–2, 2000, intr.

[9] Brickman, W.W., John Dewey's Impressions of Soviet Russia and the Revolutionary World: Mexico-China-Turkey, 1929, New York: Teacher's College Press, 1964, p. 1.

[10] Tsui-Chen, Ou, A Re-evaluation of the Educational Theory of John Dewey, Educational Forum, Vol. 25, March, 1961, 277.

[11] Brickman, W.W., John Dewey's Foreign Reputation as an Educator, School and Society, Vol. 70, October 22 1949, p. 258.

[12] Ryan, A., John Dewey and the High Tide of American Liberalism, New York/London: W.W. Norton, 1995.

[13] Harry Passow, A., John Dewey's Influence on Education around the World, Columbia University Record, 1982, 83, p. 402.

[14] Kilpatrick, W., Dewey's Influence on Education, in The Philosophy of John Dewey, Eds. by P.A. Shlipp and L.E. Hahn, Illinois: LaSalle, 1989, p. 471.

[15] See in: Wilson, D.J., Science, Community, and the Transformation of American Philosophy, 1860–1930, Chicago: University of Chicago Press, 1990; Stowe, W.W., A History of American Philosophy, 2d edn, New York: Columbia University Press, 1963.

168 ELENA ROGACHEVA

[16] See in Gert J.J. Biesta and Siebren Miedema Dewey in Europe: A Case Study on the International Dimentions of the Turn –of-the Century Educational Reform, American Journal of Education, Vol. 105, No.1, November 1996, 8.

[17] Dykhuizen, G., The Life and Mind of John Dewey, Carbondale: Southern Illinois University Press, 1973, pp. 79, 115, 224–225, 222–223, 235, 230–240, 271.

[18] Dewey, John, Interest in Relation to Training of the Will (1899), in John Dewey: The Early Works, 1882–1898, Vol. 5, Ed. by Ann Boydston, Carbondale: Southern Illinois University Press, 1972, pp. 115, 146.

[19] Carr, W. and A. Hartner, Education and the Struggle for Democracy, Open University Press, 1996, p. 54.

[20] Scotland, J., The History of Scottish Education, Vol. 2, University of London Press, 1969, p. 262.

[21] Darling, J. and John Nisbet, Dewey in Britain, pp. 40–44.

[22] Scotland, J., The History of Scottish Education, Vol. 2, University of London Press, 1969, pp. 262–263.

[23] Dewey, John, School and Society, p. 51.

[24] O'Hear, A., Father of Child-Centeredness: John Dewey and the Ideology of Modern Education, London: Center for Policy Studies, 1991, p. 27.

[25] Gerald Delledalle. Histoire de la philosophie americaine (Paris, 1954), Ecrits sur, le signe (Paris, 1979), Charles S.Pierce, phenomenologue et semioticien (Amsterdam, 1987) and La Philosophie americane (Paris, 1987).

[26] Schneider, H., John Dewey in France, Studies in Philosophy of Education, Vol. 19, Dordrecht: Kluwer, 2000, pp. 76–77.

[27] Ibid., 77.

[28] Dodde, N.L., The Netherlands// International Handbook on History of Education// edited by Kadrya Salimova & N.L. Dodde// International Academy of Self-perfection// "Orbita-M" 2000// p. 317.

[29] Biesta, G. and S. Miedema, Context and Interaction. How to Access Dewey's Influence on Educational Reform in Europe, p. 28.

[30] Dewey, John, Report and Recommendation upon Turkish Education (1939), in John Dewey: The Middle Works, 1899–1924, Vol. 15, ed. by Ann Boydston, Carbondale: Southern Illinois University Press, 1983, pp. 273–297.

[31] Dewey, John, Foreign Schools in Turkey (1924), in John Dewey: The Middle Works, 1899–1924, Vol. 15, Ed. by Jo Ann Boydston, Carbondale: Southern Illinois University Press, 1983, pp. 144–149.

[32] Dykhuizen, G., The Life and Mind of John Dewey, Carbondale: Southern Illinois University Press, 1973, p. 224.

[33] Dewey, John, Foreign Schools in Turkey (1924), in John Dewey: The Middle Works, 1899–1924, Vol. 15, Ed. by Jo Ann Boydston, Carbondale: Southern Illinois University Press, 1983, p. 275.

[34] Ibid., p. 275.

[35] Dewey, John, Foreign Schools in Turkey (1924), in John Dewey: The Middle Works, 1899–1924, Vol. 15, Ed. by Jo Ann Boydston, Carbondale: Southern Illinois University Press, 1983, p. 275.

[36] Biesta, Gert J.J. and Siebren Miedema, Dewey in Europe: A Case Study on the International Dimensions of the Turn-of-the-Century Educational Reform, American Journal of Education, Vol. 105, No. 1, November 1996, p. 11.

[37] Brickman, John Dewey's Impressions of Soviet Russia and the Revolutionary World, p. 14.

[38] Ibid.

[39] Büyükdüvenci, S., John Dewey's Impact on Turkish Education, in The New Scholarship on Dewey, Ed. by Jim Garrison, Dordrecht: Kluwer, 1995, pp. 228–230.

[40] W.W. Brickman, p. 261.

[41] BSEP2, nos. 5, 6 (1923–1924): 294–295.

[42] Vaughan, M.K., Action Pedagogy in Mexico in the 1920s in State, Education and Social Class in Mexico, 1880–1928, DeKalb: Northen Illinois University Press, 1982, p. 171.

[43] Addams, J., Twenty Years at Hull-House, New York: A signet Classic, 1981, XI.

[44] Dewey, Impressions of Soviet Russia and the Revolutionary World: Mexico-China-Turkey, New York: New Republican, 1929, p. 64.

[45] Blonski, P., Moi Vospominanija.M., 1971, c.174.

[46] Komarovski, B.B., Pedagogika Djui (Dewey's Pedagogy) Ch. 1., Filosofskije predposilki, Baku, 1926, c.8.

[47] Lunacharski, A.V., O Vospitanii I obrazovanii, On Upbringing and Education, M., 1976, c.470.

[48] Bernstein, M.S., Po Pedagogicheskoi Amerike? Acress Educational America, M., 1930, c.4.

[49] Dewey, John, Impressions of Soviet Russia (1928), in John Dewey: The later works, 1925–1953, Vol. 3, Ed. by Jo An. Boydston, Carbondale: Southern Illinois University press, 1984, p. 235.

[50] Kobayashi, Victor N., John Dewey in Japanese Educational Thought, University of Michigan Press, 1962.

[51] Howlett, C.F., Troubled Philosopher: John Dewey and the Struggle for World Peace, Port Washington, NY: Kennikat Press, 1977, p. 45.

[52] Passow, A. Harry, John Dewey's Influence on Education Around the World, Columbia University, Teacher's College Record, 1982, 83, p. 402.

[53] Collected lectures of Professor Naruse, Tokyo, 1940. Vol. 6. p. 2. in John Dewey in Japanese Educational Thought, by Victor Kobayashi, University of Michigan Press.

[54] Passow, A. Harry, John Dewey's Influence on EDUCATION Around the World, Columbia University, Teacher's College Record, 1982, 83, p. 413.

[55] Collected lectures of Professor Naruse, Tokyo, 1940. Vol. 6. p. 2. in John Dewey in Japanese Educational Thought, by Victor Kobayashi, University of Michigan Press, p. 1.

[56] Kendo, Nishitani, The History and Present Status of the John Dewey Society, Tokyo, 1962, September 12, p. 2.

[57] Kobayashi, John Dewey in Japanese Educational Thought, University of Michigan Press, 1962, p. 7.

[58] Ibid., p. 8.

[59] Jackson, P.W., Introduction, in The School and Society. The Child and the Curriculum, Ed. by J. Dewey, An Expanded edition with a New Introduction by P.W. Jackson, Chicago & London: The University of Chicago Press, 1990, pp. xxxiii–xxxiv.

MARA STAFECKA

THINKING CONDITIONED BY LANGUAGE AND TRADITION

ABSTRACT

Tradition is embedded in language. Language assures continuity of tradition. When the strength of words to express meaning weakens, there is a rebellious linguistic eruption. Writers, poets, and thinkers experiment and play with words and their meaning to restore the expressive power of language. It isn't someone's capricious willfulness and selfishness that brings new layers of language into existence. It is our human existence grasping and reflecting itself in the word. Thinking and understanding depend on language, on the flexible balance between expressive and conceptual powers of language, between metaphoric and conceptualizing forces of language.

In our global world, more and more we encounter situations when people move from one culture to another, switch languages and countries, and run into different sources of information, different ways of describing and interpreting things. People are born into one tradition. Then they move and acquire something else. Can we expect instant enrichment from the doubling or tripling of cultural backgrounds? Blindfolded by excitement, we forget how challenging it is to live in a state of constant and unending translation.

Understanding "what means what" becomes more and more difficult and requires a well-trained faculty of judgment to correct contextual preconditions that influence our thinking. In the old days, cultural upbringing was more straightforward, with sources of information concentrated in universities and libraries, but even then conflicts and contradictions arose when meaning was grasped within opposing systems of ideas and beliefs. In today's world, sources of information do not have a historically subordinate hierarchy extending into the depth of previous centuries. Today information floats horizontally. Contradictions and latent conflicts are permanently woven into the fabric of our daily lives, but it is immensely difficult to identify them because in the cultural background hidden a priori preconceptions are determining understanding.

Existing surroundings in which we are immersed pre-program our perception of this world. We arrive into an established tradition and culture, and we are surrounded by concepts and notions some of which we accept as self-evident and understand spontaneously, without activating, turning on our conscious attention. Other concepts can be acquired only after many hours of laborious work. We are aware of our conscious processes. We know what subjects are involved and which theories and strategies we use to clarify and organize our thinking. We are familiar with the words we use to express our thinking and represent our understanding. But do we

A.-T. Tymieniecka (ed.), Analecta Husserliana CX, 171–180.
DOI 10.1007/978-94-007-1691-9_14, © Springer Science+Business Media B.V. 2011

always know how our thinking is conditioned historically? Do we truly know how our minds are culturally wired and how our cognitive acts are existentially preconditioned? How does the historicity of understanding affect our worldview? In his major philosophical study *Truth and Method*, Gadamer researches the historical nature of understanding. When the circular structure of understanding was discovered in the 19th century, it was perceived as an obstacle in the process of cognition and was labeled "vicious". Gadamer points out that it was Heidegger who uncovered the ontologically positive importance of the circularity of understanding. Heidegger understood that, "In the circle is hidden a positive possibility of the most primordial kind of knowing. To be sure, we genuinely take hold of this possibility only when, in our interpretation, we have understood that our first, last, and constant task is never to allow our fore-having, fore-sight, and fore-conception to be presented to us by fancies and popular conceptions, but rather to make the scientific theme secure by working out these fore-structures in terms of the things themselves" (Heidegger 1962, p. 195). We are fooling ourselves if we do not keep in mind that our approach to any object is preset by our previous experiences. This preset casts a shadow on everything that our understanding touches. We can call this preset linguisticality of understanding. Forms of language reflect historical paths of thinking.

If I am telling someone who speaks the same language as I do about my daily experiences, he most likely will fully understand my descriptions and my emotions. If I am explaining to the same person my worldview, it is possible that some words may create misunderstanding or even result in a complete rejection of my way of thinking about the world. The differences of our fore-conceptions may clash and abort any bridging effort to establish common ground. Whenever we encounter an obstacle, a barrier in the way, a contradiction, anything that obscures the clarity of our understanding, it is a possible sign that our pre-conditioned perception should be addressed and looked at. Not always can our mutual anticipation be a full match depending on differences linguistically and culturally imbued in our cognitive acts.

What helps us to stand on a common ground? What are those linguistically conditioned mechanisms that influence our thinking, increasing its sensitivity and attentiveness? Well-known linguist Guy Deutscher explains that metaphors are essential elements of our thinking process. "Metaphor is an essential tool of thought, an indispensable conceptual mechanism which allows us to think of abstract notions in terms of simpler concrete things," writes Deutscher (Deutscher 2005, p. 142). The mind's ability to use metaphors indicates its adaptability to the cognitive demands of our existence. The use of language also predicts the level of authenticity a human being may achieve in the search for its existential self. According to Deutscher, "metaphors are everywhere, not only in language, but also in our mind. Far from being a rare spark of poetic genius, the marvelous gift of a precious few, metaphor is an indispensable element in the thought-process of every one of us. . . . . We use metaphors not because of any literary leanings or artistic ambitions, but quite simply because metaphor is the chief mechanism through which we can describe and even grasp abstraction" (Deutscher 2005, p. 117). I would like to stress this idea that "a metaphor is a chief mechanism of thinking". It tells us that the words we use to express our thought matter extremely. If the words I use to say something do not touch the cords of your thinking, they are useless or could even be harmful. Overspent and overused words do not emanate a magnetic field that captures another's thinking. To transmit meaning, words have to be active, able to create forces of attraction. A human being is the meeting point between the past and the future. It lives, and this living extends a human being into the future. It leaps forward in being and comes back to itself as a reflecting thought grasping itself in language.

Gadamer directs our attention to the fact that we move in a linguistic world and our worldly experience is pre-structured by language. Most metaphors in our everyday language have penetrated the very structure of our perception and have specifically mapped out our world-view. For example, linguists acknowledge that images of "more is up" and "less is down" are very common in our use of language. "The conceptual metaphor 'more is up' has taken over much more than just language," acknowledges Deutscher, "and has become so deeply entrenched in our minds that it even influences how we plot graphs and design control panels" (Deutscher 2005, p. 123). We accept this as our reality without questioning or focusing our thinking on it. This is the fore-conception that makes up the first part of a hermeneutical circle. This is our a priori understanding that we cast over the object of our thinking. Undercurrents that determine the thinking being's rootedness in history, culture and tradition are responsible for forming pre-conceptions that a priori bond our perception.

Deutscher underlines that over-familiarity inevitably weakens the force of the meaning. He also stresses that "the strength of meaning of a particular word depends on its distinctiveness, so the more often we hear the word, and in less discriminating contexts, the less powerful the impression it makes. When certain intensifiers are used more and more often, it is only natural that an inflationary process will ensue, resulting in attrition of meaning" (Deutscher 2005, p. 97). Sensing this problem led to the assertion that language is failing to deliver and disclose meaning. In the beginning of the 20th century, the critique of Enlightenment ideas focused on the inability of language to provide impeccable conceptual tools needed to understand human beings and their world. Language failed to envelop the experience of thinking in words because worn-out and overused conceptual notions weakened the expressive capacity of language. Thinking became aware of the limits of the existing language and engaged in linguistic creativity to expand its expressiveness.

French linguist Roland Barthes also examined different situations where linguistic forms carry meaning and asked what makes them more or less expressive and mentally engaging. In his article, "The Rustle of Language," Barthes described how to detect meaning and how language conducts meaning's appearances and disappearances. He imagined himself being the ancient Greek who was described by Hegel. According to Hegel, this ancient Greek passionately and uninterruptedly interrogated the rustle of branches, springs, and winds, which are the shudder of nature. The ancient Greek was a genuine element of nature. "And I", wrote Barthes, "it is the shudder of meaning I interrogate, listening to the rustle of language, that language which for me, modern man, is my Nature" (Barthes 1986, p. 79).

Barthes compared meaning and music. To have an ear for meaning is similar to having an ear for music, for sound. Understanding is a pleasurable moment similar to experiencing something sensually delightful. "And language – can language rustle?" asks Barthes, "Speech remains, it seems, condemned to stammering; writing, to silence and to the distinction of signs: in any case, there always remains *too much meaning* for language to fulfill delectation appropriate to its substance. But what is impossible is not inconceivable: the rustle of language forms a utopia. Which utopia? That of a music of meaning; in its utopic state, language would be enlarged, I should even say denatured to the point of forming a vast auditory fabric in which the semantic apparatus would be made unreal; the phonetic, metric, vocal signifier would be deployed in all its sumptuosity, without a sign ever becoming detached from it (ever naturalizing this pure layer of delectation), but also – and this is what is difficult – without meaning being brutally dismissed, dogmatically foreclosed, in short castrated. Rustling, entrusted to the signifier by an unprecedented movement unknown to our rational discourses, language would not thereby abandon a horizon of meaning: meaning, undivided, impenetrable, unnamable, would however be posited in the distance like a mirage, making the vocal exercise into a double landscape, furnished with a 'background'; but instead of the music of the phonemes being the 'background' of our messages (as happens in our poetry), meaning would now be the vanishing point of delectation" (Barthes 1986, pp. 77–78). As an example, where meaning was not just conceptually but also sensually incorporated in the text, Barthes cited Michelet, whose history excurses were impassioned not because their author was overly emotional or hotheadedly judged about historic facts, but because he did not arrest language at the fact (Barthes 1986, pp. 197–198).

What happens to all those powerful, vibrant, and potent words that come into language, take part in a cognitive break-through, perform their mission and then fade away on the sidelines for the next generation of human beings. Deutscher explains that historically, "The simplest and sturdiest of words are swept along, one after another, and carried towards abstract meanings. As these words drift downstream, they are bleached of their original vitality and turn into pale lifeless terms for abstract concepts – the substance from which the structure of language is formed. And when at last the river sinks into the sea, these spent metaphors are deposited, layer after layer, and so the structure of language grows, as a reef of dead metaphors" (Deutscher 2005, p. 118).

Thinking is greatly influenced by thinker's native tongue. Language plays an exceptional role in conditioning thinking and worldview. Words reflect the ways people think and behave and in that sense they are the keys that help to explain and understand their culture. It is difficult to match in translation the meanings of words in different languages because languages have culturally specific notions or categories. They are the cognitive tools that contain the experientially unique experience of every particular society. Those tools influence extensively the manifestation of thought of members of society. In every culture there are some words with a very distinctive, deeply embedded meaning that permeates all layers of that functioning society. They are the "key words" that define the cultural undercurrents of every nation.

Not always are linguistic experiments and rebellions understood and appreciated. Most likely, society will perceive changes as an unnecessary interruption and inertia will prevail at first. Unfortunately, even the creative and talented minds greet tearing down parts of tradition with suspicion. At the beginning of the 20th century, Russian poet Iosip Mandelshtam incriminated his fellow Russian writers and poets of straying away from the pristine Hellenistic tradition of use of words. He described Andrey Beliy as being sickly phenomenon in the history of the Russian language because he mercilessly and brutally chased words following only the temperament of his speculative thinking. Choking in his ornate verbosity, Beliy would not sacrifice a single nuance, a single facet of his capricious thought (Mandelshtam 1987, p. 59). He would destroy existing bridges between words and meanings because, according to Mandelshtam, he was too lazy to explore how to cross them. As a result of this instantaneous firework, when we read Beliy, indicated Mandelshtam, we have to deal with a pile of debris, a dismal picture of destruction. At the same time, Mandelshtam accepted even more radical experimentation with language, seeing in it a race towards the future. He viewed the pace of language as different than that of life. It is impossible to adjust language mechanically. Sometimes language leaves everyday life far behind. According to Mandelshtam, this is the case of Russian poet Velimir Khlebnikov, who busied himself with words like a mole, digging tunnels into the future, and moved language years ahead (Mandelshtam 1987, p. 60). However, Mandelshtam did not elaborate on why Khlebnikov experimented with language wishing to fortify its emotional and, thus, conceptual power.

Both, Beliy and Khlebnikod, played with language to attain the most powerful effect and to push their readers into chopped-up perception routines, to ambush them, to overwhelm them, to snowball them with words they could not anticipate receiving. Linguistic rebellions are visible and impressive, with words cannonballed at each other to tear them apart and prepare for something different. At the same time, side by side with the linguistic fireworks, minuscule change occurs daily and it takes a while until it can be noticed. Hans-Georg Gadamer indicated that the "life of language consists in the constant playing further of the game that we began when we first learned to speak. A new word usage comes into play and, equally unnoticed and unintended, the old words die. This is the ongoing game in which the being-with-others of men occurs" (Gadamer 1977, p. 56).

When we are articulating our thought, the first word brings along the next one and this is repeated many times. When we choose the word, it has to break itself out of the preconditioned context and emerge with unintended consequences and associations. Whenever we use a word, its appearance adds something to an existing thesaurus. When human beings communicate, the purpose of this communication is not to exchange well-defined facts, but to transmit meaningful content, to reveal something. If the disclosure of meaning is happening, the human being adds a new facet to his existence.

When we approach language as a research subject, we also encounter a very specific problem. Gadamer points out that "all thinking about language is already once again drawn back into language. We can only think in a language, and just this residing of our thinking in a language is the profound enigma that language presents to thought" (Gadamer 1977, p. 62). We can agree with Doede that, "For Gadamer, humans dwell in a world that is linguistically saturated; language is the historical-cultural *a priori* that makes possible the human way of being in the world. From this perspective, the thinking most expressive of human being is essentially dependent upon language. And language, as these thinkers conceive of it, is both the product of social relations *and the producer of social beings* – self-reflexive beings whose identities are socially forged through mutual linguistic expressivity" (Doede 2003–2004, p. 7). Gadamer underlines that "we are always already biased in our thinking and knowing by our linguistic interpretation of the world. To grow into this linguistic interpretation means to grow up in the world. To this extent, language is the real mark of our finitude. It is always out beyond us. The consciousness of the individual is not the standard by which the being of language can be measured" (Gadamer 1977, p. 64). Language is the mark of our finitude and at the same time a dialogue that opens up infinity for thinking and meaning. Gadamer also speaks of the self-forgetfulness of language. It is not natural for a subject to be aware of the structural components of language while using it. "The structure, grammar, syntax of a language – all those factors which linguistic science makes thematic – are not at all conscious to living speaking" (Gadamer 1977, p. 64). The more natural is the use of language; the less conscious we are about it.

Gadamer has elaborated about the role of translator. And it is true not only in translating from other languages but also in every case when someone is interpreting a literary text or anything else. Hermeneutics embraces the universality of translation. Let us listen to Gadamer when he states that a translator "cannot simply convert what is said out of the foreign language into his own without himself becoming again the one saying it" (Gadamer 1977, p. 67). Understanding is both an interpretation and translation. Every interpretation is already a move towards overcoming naiveté, the initial assumption that perception gives us understanding of the world as it is. According to Husserl, at the natural standpoint we just passively perform our observations of presented reality following the preset rules of cognitive inquiries. Husserl uses the term "phenomenological reduction" to explain that thinking being has to overcome psychological and empirical actuality to claim understanding.

When we are glancing towards an object in our view, we are apprehending it based on an intuitive perception of its nature. Our previous sense-experience has built a certain fundamental framework, which lets us intuit with self-evidence. It is very important to think about the remarkable words of Gadamer,"It is not really ourselves who understand: it is always a past that allows us to say, 'I have understood' " (Gadamer 1977, p. 58).

We are immersed in the tradition and cannot understand ourselves as a separate entity. And it is also true that we cannot understand tradition without understanding our being in it. According to Gadamer, "The operation of the understanding requires that the unconscious elements involved in the original act of knowledge be brought to consciousness. Thus romantic hermeneutics was based on one of the fundamental concepts of Kantian aesthetics, namely, the concept of the genius who, like nature itself, creates exemplary work 'unconsciously' – without consciously applying rules or merely imitating models" (Gadamer 1977, p. 45). Thus, interpretation is needed to understand the results of those unconscious acts and to make them available for the broader society. Romantic hermeneutics also contributed to the understanding of the historicity of tradition. Gadamer makes clear that "romanticism began with the deep conviction of a total strangeness of the tradition (as the reverse side of the totally different character of the present), and this conviction became the basic methodological presupposition of its hermeneutical procedure. Precisely in this way hermeneutics became a universal, methodical attitude: it presupposed the foreignness of the content that is to be understood and thus made its task the overcoming of this foreignness by gaining understanding" (Gadamer 1977, p. 47).

What belongs to that circle of understanding? What are the steps towards true knowing? At the beginning a person projects understanding, inherited from the tradition and cultural surroundings. This subconscious, unreflected knowing a priori belongs to the past and is forwarded by the tradition. Gadamer writes, "a person who is trying to understand a text is always projecting. He projects a meaning for the text as a whole as soon as some initial meaning emerges in the text. Again, the initial meaning emerges only because he is reading the text with particular expectations in regard to a certain meaning. Working out this fore-projection, which is constantly revised in terms of what emerges as he penetrates into the meaning, is understanding what is there" (Gadamer 2006, p. 269).

Gadamer's editor and translator in English notes that for "Heidegger and Gadamer alike, man not only uses language to express 'himself,' but, more basically, he listens to it and hence to the subject matter that comes to him in it. The words and concepts of a particular language reveal an initiative of being: the language of a time is not so much chosen by the persons who use it as it is their historical fate – the way being has revealed itself to and concealed itself from them as their starting point" (Gadamer 1977, p. LV). We are born in language and it is our tradition. We define and understand ourselves in words. What happens when we start feeling that language is failing to serve this need and is loosing its expressive power? Do we think that language is in crisis? Do we think we have reached its limits? Do we think that we are misusing it?

There are several assumptions about language, which we encounter in our culture. The history of philosophy reveals that mankind strives to reach the truth, thus acknowledging that the proper language of philosophizing is conceptual. From here we can see that conceptualizing power of language was always considered a unique and privileged duty that language performs in philosophy (Schmidt 2004, p. 35).

Heidegger and Gadamer focused on language's capacity to transcend conceptually the experiential space and time of a human being. They both tried to capture in words the dilemma of human cognition: the rapture between the two sides of the being -the experiential side of existence and the reflective side of concept building.

What is language capable of expressing? Where is the power of the word hidden – in its conceptual rigidity or in its expressivity, its ability to touch existential chords of being? Already Hegel brought everyday language into philosophy to sharpen thought's expressive capacity. He brought into philosophy new terminology that came from German and that was suited better to describe new experiences. Heidegger also came into philosophy with his own conceptual language. Terminological changes were aimed to expand and deepen the expressive capacity of language. For Heidegger, language performs two distinctive functions – it discloses meaning and communicates information. He allows words to resonate in his mind and become a match with existentially experiential content. What happens naturally and spontaneously during the long stretches of time in history, Heidegger is testing in the phenomenological approach to being.

While working on his own concept of language, Gadamer constantly returns to Heidegger's ideas. For Gadamer, "the role that the mystery of language plays in Heidegger's later thought is sufficient indication that his concentration on the historicity of self-understanding banished not only the concept of consciousness from its central position, but also the concept of selfhood as such. For what is more unconscious and 'selfless' than that mysterious realm of language in which we stand and which allows what is to come to expression, so that being 'is temporalized' (sich zeitigt)? But if this is valid for the mystery of language it is also valid for the concept of understanding. Understanding too cannot be grasped as a simple activity of the consciousness that understands, but is itself a mode of the event of being" (Gadamer 1977, p. 50).

Language we can perceive better as a game or as a play where the flow of words rub against things or appear from the secludedness of our unconscious knowing of the world.

It is not astounding that Gadamer eloquently elaborates about the fluidity and creativity of this process. "No one fixes the meaning of the word," Gadamer notes, "nor does the ability to speak merely mean learning the fixed meanings of words and using them correctly" (Gadamer 1977, p. 56). Whenever we discuss understanding, we cannot avoid discussing language because all thinking is confined to language either as a limit or as a possibility.

Understanding defines a human being. But understanding is not something that is consciously done or achieved; that is only a result of cognitive effort. Understanding is always an event. It is an ontological happening that changes the nature of being. Understanding is not an epistemological problem. Representatives of Enlightenment philosophy simplified the problem of understanding and interpreted it in the framework of cognitive theory. Understanding was equated with knowing, with a conscious possession of knowledge. Adorno was one the harshest critics of this legacy of Enlightenment. At the same time Adorno refused to accept Heidegger's correction of traditional rationalism. He equates Heidegger's terminology with a jargon, which carries an ideological mandate. Adorno believes that Heidegger uses words that have not been part of thought and thus are just empty shells resembling linguistic signifiers. Adorno insists that in Heidegger's philosophy, "hypocrisy thus becomes an a priori, and everyday language is spoken here and now as if it were the sacred one" (Adorno 1973, p. 12). When jargon "dresses empirical words with aura, it exaggerates general concepts and ideas of philosophy – as for instance the concept of being –so grossly that their conceptual essence, the mediation through the thinking subject, disappears completely under the varnish" (Adorno 1973, p. 12). Adorno blames Heidegger for stealing words from language and for manipulating the elements of empirical language. According to Adorno, Heidegger's "jargon" uses "disorganization as its principle of organization, the breakdown of language into words in themselves" (Adorno 1973, p. 7). Jargon confuses and manipulates, creating an illusion that a person can be in charge of his own thinking. In his foreword to the English edition of Adorno's *The Jargon of Authenticity*, Trent Schroyer emphasizes that "the jargon shares with modern advertising the ideological circularity of pretending to make present, in pure expressivity, an idealized form that is devoid of content, or, alternatively, just as the mass media can create a presence whose aura makes the spectator seem to experience a nonexistent actuality, so the jargon presents a gesture of autonomy without content" (Adorno 1973, p. XIV). Adorno states that "while the jargon overflows with the pretense of deep human emotion, it is just as standardized as the world that it officially negates; the reason for this lies partly in its mass success, partly in the fact that it posits its message automatically, through its mere nature. Thus the jargon bars the message from the experience, which is to ensoul it. The jargon has at its disposal a modest number of words, which are received as promptly as signals" (Adorno 1973, p. 6). One would wonder why Adorno extremely politicized Heidegger's position on language and emphasized its ideological ambiguity, highlighting the possibility that it can become an oppressive tool and be used against human beings.

Interestingly Foster represents the point of view that Heidegger and Adorno were attempting to deal with the same problem – crisis situation in language – feeling that language is losing its ability to communicate the truth. He stresses that "Heidegger wants to allow the moment of meaning disclosure to come to expression in language, without allowing that moment to be corrupted by the natural drift of language towards disengaged representation" (Foster 2007, p. 197). Foster also underlines Heidegger's inclination towards everyday language and its use in building terminology. Heidegger is using this strategy all over *Being and Time*, hinting that beyond the simplicity of words is the hidden depth of meaning. Heidegger focuses on a disclosive force of a word from the everyday language. According to Foster, Adorno criticizes Heidegger for his inability to properly explore and implement the idea of the aesthetic origin of cognition. The moment of artisticity in cognition appears as a blind spot linguistically and causes the illusion that it is possible to grasp meaning that cannot be expressed as conceptual content. For Adorno this means that Heidegger is giving up the critical function of thinking and is succumbing to repressive and destructive social structures. Apparently, Adorno's point is backed by Heidegger's flirtation with the Nazi regime.

Roger Foster writes, "Adorno believes that the unsayable must be brought to language in such a way that language at the same time lights up its own speechlessness before the unsayable. In other words, the task must be to bring the unsayable to language in such a way that shows it as what is unsayable. This requires a type of articulacy that is neither a refusal to speak that is, silence, nor is it the statement that subsumes something under a concept. Adorno perceived that the modernist problem required an understanding of philosophical writing as a process in which language reveals what is currently unsayable" (Foster 2007, p. 201). In the end, Adorno acknowledges that thinking may not have adequate linguistic tools to reveal its meaning, and that "unsayable" is a reality. Similarly, did not Heidegger speak about the gap between what we say and what has to be said?

180 MARA STAFECKA

Rockford, IL 61103, USA
e-mail: mstafets@yahoo.com

REFERENCES

Adorno, Theodor. 1973. *The Jargon of authenticity*. Evanston: Northwestern University Press.

Barthes, Roland. 1986. *The rustle of language*. New York: Hill & Wang.

Deutscher, Guy. 2005. *The unfolding of language*. New York: Henry Holt & Company.

Doede, R. 2003–2004. Tradition and discovery. *The Polanyi Society Periodical* XXX(1), 5–18.

Foster, Roger. 2007. Adorno and Heidegger on language and the inexpressible. *Continental Philosophical Review* 40(2):187–204.

Gadamer, Hans-Georg. 1977. *Philosophical Hermeneutics*. Berkely, LA, and London: University of California Press.

Gadamer, Hans-Georg. 2006. *Truth and method*. New York: Continuum.

Heidegger, Martin. 1962. *Being and time*. New York-Evanston: Harper & Row.

Mandelshtam, Iosip. 1987. *The word and culture*. Moscow (in Russian): Sovetskiy Pisatel.

Schmidt, Dennis. 2004. On the incalculable: Language and freedom from a hermeneutic point of view. *Research in Phenomenology* 34:31–44.

J . C . C O U C E I R O - B U E N O

H O W   T O   C O N D U C T   L I F E
(*A R E T E* A N D *P H R O N E S I S*)

A B S T R A C T

Taking Husserl's thematisation of Greek philosophy as a starting point, the aim of this paper is to explore and recover the full sense of the concept of arete (virtues) and especially that which is considered to be the "virtue of all virtues": *phronesis*. *Phronesis* is a dianoetic approach that deals with the deliberative aspects of the human condition. It is the "practical intelligence" that guides us in our actions and provides us with a sense of awareness of the world, enabling us to conduct our lives. It must also be remembered that *phronesis* is a constant presence in life's practical situations. This paper will also discuss its conceptual foundations, namely *proairesis* (the capacity for personal choice) and boulesis (deliberation). In this sense, *phronesis* is a form of knowledge that facilitates decision making and the correct governance of our lives. Indeed, it will allow us to use our acquired habits in the correct manner, as it is responsible for articulating intellectual and moral virtues. It facilitates learning in order to enable us to face the complexities of life and to do things as they should be done: *phronesis* invites us to adopt the best decision on each occasion. It is man's most reliable and immediate truth in an uncertain world. It is a means of foresight in the light of what may occur, a form of knowledge that can use the experiences of the past as a means of anticipating future events. For many centuries, *phronesis* has been a form of practical knowledge, and as such, represents a type of "emotional intelligence" that plays a key role in our experiences and situates us in a world where choice is a necessity; the choices we make at any given time shape our very nature.

I N T R O D U C T I O N

The irrepressible driving force underlying the impact of Greek philosophy based on an ideal of philosophical life, is the central theme of Husserl's well-known work *Die Krisis der europäischen Wissenschaften und die transzendentale Phänomenologie*. As the father of phenomenology reminds us, the world around us is not only made up of "empirical facts" (with all the issues entailed in the understanding of this concept), but it also involves experiences of idealised objectivities, and by extension varying values, ethics and aesthetics. Taking this as our starting point, it must be remembered that in Greek philosophy in general, and the thinking of Aristotle in particular, the key question is "How should life be conducted?", "How should I lead my life?" The answer is by cultivating *arete* (virtues), as this is the only way of living

A.-T. Tymieniecka (ed.), Analecta Husserliana CX, 181–187.
DOI 10.1007/978-94-007-1691-9_15, © Springer Science+Business Media B.V. 2011

as a human being in the true sense of the expression. Seen from this perspective, virtues are not acquired either naturally or by working against nature, but instead through a natural disposition to receive and perfect them through habit and custom. Greek ethics is essentially rooted in the nature and life of the individual seen as a whole.

Aristotle's ethics are based on virtue, unlike the moral philosophy that began with Kant and the Utilitarianists and has continued until the present day. Their sense of ethics is concerned with formal criteria or the good and evil of our acts; in other words, with our sense of duty. In contrast, Greek ethics is the shaping or education of individuals with the aim of ultimately reaching perfection. Virtues are moral habits and personal choices that human beings are required to interiorise and display socially in order to conduct their lives efficiently. These are essentially acquired through *paideia* and become second nature to the individual.

In his *Nicomachean Ethics*, Aristotle defines two major categories of virtues: ethical *arete* (strength, temperance and justice) and dianoetic virtues (wisdom and *phronesis*). The former are the so-called "moral virtues", and the latter "intellectual virtues". By shaping our character, they guide us in the way of conducting our own lives. Aristotle defines *arete* –virtue or excellence– as "a characteristic of a person that renders good, of which it is the excellence and causes it to perform its function well" (*Nicomachean Ethics*, 1106a). He goes on to clarify this by stating that: "Virtue is a state of character concerned with choice, lying in a mean, i.e. the mean relative to us, this being determined by reason and by that reason by which the man of practical wisdom would determine it" (*Nicomachean Ethics*, 1107a).

Virtue –*arete* – is therefore a habit –*exis*– that is deliberately chosen –*proairesis*: "the deliberative appetition of things within one's power" (*Eudemian Ethics*, 1226). This means that through the *arete* of *phronesis*, man is shaped in our interests by dint of repeating those acts that guide him towards good.

It must be stressed that Aristotle considers *phronesis* to be the most decisive of all *arete*. *Phronesis* is more than a simple virtue; instead it is the "governing virtue" that determines the nature of all other virtues. It is practical wisdom with an ethical capacity for determining human acts. A faculty that is concerned with making ethically desirable choices and the discovery of what is good for the individual. In other words, *arete* is the practical intelligence that enables us to choose what is fit (moral beauty) and the means best suited to achieving our proposed objective (*Nicomachean Ethics*, 1144a).

Phronesis is therefore the ability to direct our own lives, which involves deliberating on what is or is not convenient for the human being. It also helps us to reflect (the realisation of future possibilities) on man's purpose, as it has the capacity to highlight the true mission of our lives.

I

In Classical Greece, *aisthesis* (sensitive perception, beauty, aesthetics) was the overriding concept of life. This idea is directly linked to *logos* in terms of its core meaning of proportion or harmony. Indeed, this is a manifestation of life whose internal and external aspects are proportionately balanced. The Greeks referred to harmony as *arete*. *Paideia* was also governed by this concept of harmony, as art represented a permanent horizon for Greek educators. Art in the human sphere is known as a means of overcoming natural deficiencies. Its mission is therefore to create a sense of harmony between *physis* and *logos*. Aristotle frequently resorts to the metaphor of nutrition when referring to *paideia*, which he conceives as the educational interiorisation of our being within *arete*.

According to Aristotle, the essence of *paideia* lies in: a) the search for *aletheia* (seeking truth through disclosedness); and b) the shaping of the character governed by *arete* (*Nicomachean Ethics*, 1143a).

For the Greek spirit, the key lay in educating the character, enabling man to be modified by the most powerful forces that shape and govern our lives, namely music and poetry.

Educating the character implicitly aims towards the realisation of the ideal of life in a practical sense. For Aristotle, the materialised ideal is the search for *arete* (*Politics*, 1371). A quest that is developed through *phronesis*, which provides man with the capacity to choose the right form of life at each moment: for the Greeks, fulfillment in life comes with a life that is chosen (*proairesis*) and that allows self-creation.

Art (words, harmony, sound, rhythm) was not an autonomous professional activity (as it is today), but instead a means of inner education (*Bildung*) and a means of shaping our spirit – a strategy that leads to a sense of harmony within the self. The art that was created during this period must be seen as a source for the production of new entities. Rather than being "created" from nothingness, they stem from the indeterminate.

Aristotle claimed that art is the predisposition for creation accompanied by a series of rules, whilst *phronesis* is the willingness to act in accordance with those rules (*Nicomachean Ethics*, 1140b). He also distinguished between contingent objects whose principle resides in the producer and those contingent objects whose principle resides in the thing produced, "which are the things that are by nature" (*Physics*, I, 192b).

As mentioned previously, the role of *aesthesis* in Greek *paideia* is crucial. Education was seen as the passing down of *paideia* from one generation to the next. Initially, the means for this was a song. Indeed, laws were chanted long before they were first written down. And rhythm provided the means for memorising them. In truth, *nomos*, means both law and song.

The ideal of aesthetics and harmony was expressed through *kalokagathía*: beauty and goodness. Music and other similar arts were the ingredients of this "balm for the soul". The aim was to achieve beauty and goodness by extracting from musical arts the essence that fired human passions and left the spectator (observer) in complete control in order to reach *kalokagathía*. Arete was present in each part of this maximum aspiration.

The Greeks were aware of man's potential on entering the world, yet also that this potential required development in order to overcome disharmony and disorder. What was needed was a cleansing process known as *catharsis*, which played a key role in *paideia*: just as medicine healed the body, *catharsis* was responsible for man's "emotional healing". In other words, the aim was to dispel the causes that exacerbated the perturbing and unsettling emotions and passions. In the Greek world, the role of *catharsis* essentially targeted the experience of art (represented through music and tragedy).

It was within this context, and once *paideia* and *catharsis* had fulfilled their mission, *kalokagathía* (beauty and morality) would make its appearance. The Greeks' aim was to use man to create the most complete work of art through *arete*, *paideia* and the experience of art.

Humans idealistically aspire to virtue and *eudaimonia*, yet not all are successful in this quest. *Paideia* aims to fill the voids left by nature and to act as an instrument that can help man to become fulfilled and virtuous. In this sense, it has the capacity to become an artefact for life, for choosing the best possible life, together with its corresponding *praxis*.

Aristotle uses the nutrition metaphor to provide us with a clear vision of what he understands by *paideia*. In fact, it is the spiritual nutrient humans require, allowing them, through *arete* to become what they should be whilst at the same time enabling them to shun all that they should not be.

Arete consequently plays a crucial role as it is not generated by or acts against nature; instead it possesses the natural conditions necessary for reception and perfection through habit.

Arete epitomises the educational ideal of Classical Greece: practical intelligence manifested fully in *phronesis*.

Paideia therefore aims to bring about the practical realisation of the ideal of life through the development and assimilation of *arete*. The education process itself is the suitable context in which to acquire the habits for a good life. This positions us within *phronesis,* as it is this that will guide our *praxis* through practical judgement. Under such condition, education is nourished by *phronesis*, capacitating humans to appreciate and value their lives.

In *paideia*, in the education of men, intellectual considerations converge smoothly in close relationship with *praxis*. It is a question of aesthetic sensitivity, of receiving a good education through *arete*. Sensitivity that is proof of practical intelligence (*phronesis*). Wisdom cannot be taught; all that can be done is to point out the path which leads to its ultimate acquisition.

II

In the modern age, the Greek concept of virtue has gradually lost ground amongst ethical theories in favour of the idea of duty or moral law, essentially due to the influence of Kant. In his *Fundamental Principles of the Metaphysics of Morals* and *Critique of practical Reason*, Kant formalises the concept of duty, moral law and their capacity for universalisation. To put it another way, the move away from principle-based to virtue-based ethics. At all events, the concept of virtue, and by logical extension *phronesis,* falls into decline from the 18th century onwards.

The contemporary recovery of the concept of virtue was known as "the ethics of virtue" and is an attempt to provide an ideological response to Liberalism and a philosophical alternative to Utilitarianism and Kantism.

However, my aim in this paper is to recover the Greek concept of virtue in its fullest sense, and specifically that which is considered the "virtue of all virtues", namely *phronesis*. In part VI of the *Nicomachean Ethics* it is defined as a practical arrangement accompanied by true rules concerning what is good or bad for man (1140b). *Phronesis* is an *arete* associated with knowledge whose limits are those of man himself. Specifically, it is a dianoetic *arete* that assigns and points to the deliberative aspects of the human condition. It is practical intelligence capable of guiding the actions of man. It allows us to be alert and aware of the world, enabling us to conduct our lives. Furthermore, it fills the void that exists between the mechanical laws of nature and the elusive and shifting human *praxis*.

Science has always aspired to creating a totally transparent world. A world in which nothing could be any other way. Were this true, there would be no place either for the experience of art or *paideia*. If we attempt to clarify absolutely everything within our temporal and spatial sphere, then there would be no room for human initiative, which is what art and *phronesis* feed on. An interpretable world that invites us to reflect upon it is a vital space in which there is still a place for the genuinely human.

Knowledge cannot consist (or at least not exclusively) of "knowing the truth" as science would have it. Should we come face to face with the truth, we would be freed from the need to deliberate, to choose the path our lives should take (*proahiresis*), which is what makes us human. Freedom always implies choice in a life filled with complexities and shadows; it is being in a position whereby we can choose whether or not to take a certain course of action.

A man with *phronesis* has chosen the path of knowledge, and is on his way to becoming wise. As Aristotle knew, a wise man is he who knows how to take the setbacks life brings and even turn them to his advantage. A man with *phronesis* is capable of freeing himself from life's blows through an inner attitude (*catharsis*). In short, it is an attempt to be wise, to realise that wisdom depends on the self in a world which is beyond our control. *Phronesis* is attempting to obtain an insight into and to anticipate an inevitably uncertain future. It is foresight against what might happen in an attempt to safeguard ourselves from danger.

As a dianoetic virtue, *phronesis* deals with human acts in general and through them the essence of man and the man present in a world on which it must act.

In philosophical terms, Aristotle saw *phronesis* as the human quality that enables us to guide ourselves towards the act that is good and desirable for man.

Phronesis is knowledge that enables us to make choices that will guide our lives in the right direction. In this sense, it allows us to use the habits we have acquired correctly, as it articulates both our intellectual and moral virtues.

Those in possession of *phronesis* are referred to as *Phronists* – men with the capacity to deliberate in the light of any eventuality. *Phronists* are those that lean towards action and productive art. *Phronesis* is the practical disposition to exercise choice (*proairesis*), which is executed by *Phronists*. They do not merely interpret the rule or regulation, but instead are themselves the rule (like his disciple, Plato also gave precedence to man over the law).

As an extension of the concept of *Phronists*, Aristotle also invokes the *Spoudaios* (virtuous men, similar to the *Phronist*, who set an example to all). These are citizens whose actions have earned them the trust and confidence of others, and whose actions transmit a sense of certainty and reassurance. Indeed, men lacking in *spoudaios* are considered to be infirm – in other words they take decisions that are detrimental to the self.

In this sense, it must be remembered that Aristotle agreed with his master Plato (*Meno* 94a) in that virtue is not an object of education, as it is closer to wisdom and poetic inspiration than *episteme*. At this stage, it is also worth bearing in mind that for the Greek scholar, *phronesis* differs from *episteme*, in that this form of knowledge is a deliberating *arete* of the self (1140b).

The Aristotelian concept of *Phronists* would therefore be in keeping with Plato's idea of the philosopher king (*Politics*, 294a), if we do away with the idea of the world of ideas as the sole foundation. For *Phronists*, everything is susceptible to intelligibility. Furthermore, they are in optimum conditions to deal with the blows that life deals and to turn them to their advantage. *Phronists* face a complex, harsh and dangerous world in which they are forced to venture towards an uncertain future in order to safeguard themselves as far as possible from the unforeseen events that may befall them. In such an unpredictable and uncertain world, *Phronists* must make choices (*proairesis*) and deliberate (*boulesis*), as they are aware that this is the means to obtaining an insight into this future.

If man takes on an indeterminate world, his apprehension may only come from an indeterminate rule, that is represented by *phronesis*. It is from this that the concept of equity as a moment of *iuris-phronesis* appears: in order to be just, the law must be corrected through equity. *Epieikeia*, thus conceptualised, is the correct nature of law in its specificity, a means of seeking a higher justice. This idea must be based on the consideration that the law is always at a disadvantage when pitched against the complexities and elusive nature of human fatalism. In this sense, together with Aristotle, we must differentiate between what is just by virtue of the literal nature of the law and due to deliberation (*Nicomachean Ethics*. E 10). The capacity of deliberation to project itself into the future and to gain further knowledge and insight must always take precedence over written rules. Equity is not so much related to the literal sense of the law, but instead to the meaning that underlies its literality, the hidden meaning that required interpretation. All legal texts (and indeed any type of text in general) not only represent that which can be gleaned from an initial interpretation, but also a deeper meaning that requires deliberation and further interpretation. In any text, we must seek that which the writer or legislator has unwittingly included in order to bring it to the fore, making it accessible to comprehension.

III

As we have seen, *phronesis* is wisdom that enables man to cope with life, providing an insight on how it should be conducted. It should also be remembered that *phronesis* is not the knowledge of the immutable (*episteme*), but rather that of contingency. It confers a knowledge that albeit lacking in scientific character, is filled with *logos*.

We have also seen that the conceptual foundations of *phronesis* lie in the capacity to choose (*proairesis*) and deliberate (*boulesis*). Nor must we overlook the fact that it intrinsically contains the fundamental form of experience.

Phronesis is the capacity for self-advice and is not limited to personal objectives. It is knowing how to act and therefore consists of self-knowledge. It is destined for immediate application based on *praxis*.

Unlike skills, which are always specific, *phronesis* is always destined for the ends to which we live. They cannot form part of a knowledge that can be taught; we are only shown how to display them.

Phronists are those that possess *phronesis*. Individuals who are not dictated to by fury or passion, but instead who are guided along the right path to making the decisions that affect their lives. *Phronists* are aware that only those that have acquired learning (*paideia*) can adopt certain determinations.

Phronesis unquestionably involves learning to do things as they should be done and to face complex situations: it is foresight for the future, the capacity to learn lessons from the past in order to foresee what lies ahead in a world full of uncertainty.

Finally, it must be stated that had the concept of *phronesis* been the object of study in antiquity, we would not be speaking today of the concept of "emotional intelligence" as an innovative idea. Indeed, it is anything but new: for centuries.

Phronesis has been a "practical form of intelligence" and as such an inexorable form of "emotional intelligence" that plays a major part in our lives, placing us in a world where choices must be made. The choices we make at any given time shape our very being. In conclusion, *phronesis* requires preparation, habit, deliberation and the capacity to make decisions. However, once these prior requirements have been integrated, it immediately exerts a governing influence on our actions, enabling us to reach fulfilment as human beings.

University of A Coruña, A Coruña, Galicia, Spain
e-mail: juacobu@udc.es

BIBLIOGRAPHY

Aristóteles. 1959. *Ética a Nicómaco.* Madrid: IEC.
Aristóteles 1985. *Ética eudemia.* Madrid: Gredos.
Aristóteles 1995. *Física.* Madrid: Gredos.
Husserl, E. 1976. *Die Krisis der europäischen Wissenschaften un die transzendentale Phänomenologie.* Den Haag: Martines Nijhoff.
Kant, I. 2000. *Crítica de la razón práctica.* Madrid: Alianza.
Kant, I. 1996. *Fundamentación de la metafísica de las costumbres.* Madrid: Espasa/Calpe.
Plato. 1977. *Menón.* Madrid: Gredos.
Plato. 1998. *Político.* Madrid: Gredos.

SECTION IV
HUSSERL IN THE CONTEXT OF TRADITION

WITOLD PŁOTKA

THE REASON OF THE CRISIS. HUSSERL'S RE-EXAMINATION OF THE CONCEPT OF RATIONALITY

ABSTRACT

According to Edmund Husserl's diagnosis, we can speak of the crisis of sciences which generally consists in the loss of faith in reason. In the light of the Husserlian analysis, reason had become merely a technical and computational power of cognition leading humans to the control of nature, and to an oblivion simultaneously, the oblivion about human essence, i.e., governing life by reason. Conversely, Husserl sought an authentic view on reason, and, as he argued, the authentic notion of reason was present in antiquity. Precisely ancient Greeks defined reason as the foundation of the human life. The main purpose of the article is to present Husserl's immanent development of his considerations on rationality which led him from the theory of objective reason to a formulation of the theory of reflection on a practical level. By referring to the ancient ideal of reason, as it is argued in the article, Husserl re-examined cognitive model of reason and he introduced practical and communal dimensions of rationality into the mentioned model. Such a theoretical step can lead to the reinterpretation of so-called phenomenological movement in general, and of Husserl's phenomenology in particular.

INTRODUCTION

From the very beginning, phenomenology claimed to be a reformative program of philosophy in general. For this reason, in the *Logical Investigations* Edmund Husserl struggled with a psychological interpretation of logic in particular. While discussing the empirical account of logical laws, Husserl presented as an alternative to the account such crucial ideas as the theory of meaning as an ideal entity, the intentional interpretation of consciousness, and the thesis about the superiority of the objective expressions over the subjective expressions. The ideas were relevant to the methodological level of philosophical investigations, rather than to the level concerning such themes as human life, or the world. In *Ideas I*, Husserl enlarged his project by introducing the notion of constitution. The latter notion emphasized correlation between transcendental subjectivity and the world. The constitutive phenomenology claimed to be comprehended as a pure science, which is able to ask about the conditions for possibility of experience of the world. In order to examine the conditions, the phenomenologist has to bracket a worldly character

191

A.-T. Tymieniecka (ed.), Analecta Husserliana CX, 191–204.
DOI 10.1007/978-94-007-1691-9_16, © Springer Science+Business Media B.V. 2011

of being, i.e., his researches cannot be determined by presuppositions. Rather, the phenomenologist's investigations have to be free from presuppositions, as Husserl (2001b, p. 177) formulated the principle in the *Investigations*. From this perspective, however, Husserl's last published work – *The Crisis of European Sciences and Transcendental Phenomenology* – presented a specific "breakthrough," which meant the introduction of such themes as crisis, culture, and reason. Is it true that within phenomenology one is confronted by two different tendencies: on the one hand, by leading the phenomenologist towards a non-worldly being, the strict scientific one and an abstract tendency; on the other, a critical tendency. The former is associated with eidetic claims, the latter, by contrast, consists in asking about human life and his world. How, if at all, are these two tendencies non-contradictory? How, again, Husserl was able to define and examine such themes as the crisis, culture and reason in the *Crisis*, without suspending his phenomenological method?

In the article it is argued that within Husserl's phenomenology one is faced rather with a permanent development, than with the series of "breakthroughs." The notion of reason plays the crucial role in this context, because precisely the notion is a leading clue which can lead commentators to adopt the thesis about Husserl's continual re-examination of the entire phenomenology, at least in regard to the concept of rationality. Hence, inasmuch as in the *Crisis* reason is grasped as the source of the crisis of sciences (Husserl 1970, p. 9), reason is specifically understood in this context. Of course, Husserl had in mind rationality founded by a modern science. Nevertheless, a philosophical reflection on reason, grasped in its broadest sense, according to Husserl, can lead entire culture out of the crisis. Therefore, reason is the source of the crisis, but it is the only defense simultaneously. Therefore, following Maurice Natanson, "[p]henomenology is a defense of Reason" (Natanson 1973, p. 17).[1] The reflection means a critique, but the latter notion is here understood twofold. On the one hand, the critique involves answers to the following questions: What does the crisis mean? In what sense does scientific rationality base the crisis? Thus, such a critique is close to a contemporary criticism of rationality (Plotka 2009, pp. 4–5, 10). On the other, the critique has its sources in the Kantian philosophy, which asked for the conditions for possibility of experience. As soon as 1906, it was evident for Husserl that the most general purpose for entire phenomenology was equivalent to the Kantian purpose of the critique of reason; in a personal note, written on the 25th of September 1906, Husserl (1984, p. 445) noted that the critique of reason, understood as investigations of sense, essence, and methods of the rational reflection, shall be a point of departure for each philosopher. Thus, also entire phenomenology can be grasped as a permanent re-examination of the concept of rationality.

The main purpose of the article is, then, the presentation of Husserl's re-examinations of the concept of reason, at least with regard to two aforementioned senses of the critique of reason. Firstly, I will reconstruct phenomenological critique of calculative reason. The latter concept involves rationality of modern sciences. As it will become clear in the following, Husserl began this manner of criticizing rationality much earlier than in the *Crisis* published in 1936. Secondly, Husserl's account of "authentic" reason is to be sketched. As Husserl supposed, "authentic" reason can overcome calculative rationality of sciences, and it leads towards "new" rationality. It is important to note, however, that the "new" rationality will require to refer to Eugen Fink's ideas from the period of his collaboration with Husserl the reinterpretation of the method of reduction. Moreover, it is argued that this "new" rationality had its sources in the ancient Greek ideal of rational life; finally, the difference between scientific and "authentic" rationality, I will assert, consists in the introduction of practical aspects into human life. Therefore, in contrast to commentators who stress merely a theoretical character of Husserl's discussion on the crises of sciences, the article asserts that the crucial sense of his inquiry is practical at the very heart.

THE CRITIQUE OF CONVENTIONAL REASON

As phenomenology has stated clearly, the philosopher always stands opposite the world, in which it is possible to construct infinite world views. In a word, he is in the lifeworld, i.e., in the pre-scientific world of ordinary actions. Elisabeth Ströker sought to define "pre-scientific" character of the world, and she emphasized: "[t]o call a world 'pre-scientific' could naively be interpreted as if the foundations of science" (Ströker 1997, p. 305). According to Ströker, the lifeworld cannot be "given" in any way; also sciences are not able to reconstruct the world. Rather, sciences are the equivalents for certain world views. How, then, is the crisis of science possible? After all, sciences presented during the 1930s, and still present, a series of their successes. But, as James Dodd argued, the science's "very success does not preclude the possibility of crisis is a key insight of Husserl's; but it means that to talk of the crisis of science is, paradoxically, to talk of the crisis of a success" (Dodd 2004, p. 29). Therefore, one can ask again: What does it mean to speak of the crises of sciences, if the crises is parallel with its successes?

To answer the question, following Ströker (1997, p. 305), let me stress that the lifeworld is the field of practice. Although modern sciences present merely world views, they developed techniques and methods which are able to determine human practice. However, they "forgot" its genuine purpose to determine life with regard to rational principles. Instead of giving the principles and inquiring about them, they constructed its own "outer" rationality, i.e., scientific rationality. Yet, this rationality is not derived from life. Furthermore, to quote Dodd once again, modern science "no longer seems to order life, to give life the sense of itself necessary for its pursuit of itself, thus its future" (Dodd 2004, p. 140), or, to quote Husserl's *Crisis*:

[i]n our vital need ... this science has nothing to say to us. It excludes in principle precisely the question which man, given over in our unhappy times to the most portentous upheavals, finds the most burning: questions of the meaning or meaninglessness of the whole of this human existence. (Husserl 1970, p. 6)

Thus, modern sciences are in the crisis in this sense that they lost its essential connection with life and the world.[2] Nonetheless, while they were and are still successful, sciences order human life indeed, but indirect, because these successes concern at the end practice. Therefore, it should not be surprising that Husserl's main purpose was to reveal sciences presuppositions in regard to practice.

First of all, however, it is necessary to reconstruct Husserl's understanding of sciences, at least of positive sciences, with regard to the concept of rationality. As Husserl has stated clearly, the sciences are expressions of the processes of rationalization in general. In turn, rationalization is "a mental operation . . . which leads all factual descriptions through upgrading of the factuality to a pure possibility in appropriate essential establishment" (Husserl 2001c, p. 48). Obviously, by affirming the given world as the ultimate reality, the sciences are grounded on an uncritical relation to the world (Husserl 2003, p. 3); they discuss the world as if it were objective. Does it mean, then, that the special kind of mental operation which comprehends factual world as an objective and pure possibility, reflects the objective structure? Not at all, because the description of specialized fields of nature is guided not by certain essences of objects, but rather by mathematical method; only if the scientist uses this method, he is able to achieve a description of the world as the objective one (Husserl 2001a, p. 544). Husserl has proceeded to investigate scientific method by pointing at the "technicization of the method". He has stated that "[t]echnical method involves the use of the unreasonable [elements – W.P.], and namely . . . empty words and signs" (Husserl 1993, p. 35). Yet, human rationality is determined by the primacy of calculative, but "unreasonable" practice. Husserl did not provide a precise definition of "technicization;" he only emphasized that it operates with "substitutes," which are determined by methodological aims. Therefore, the "unreasonable" elements denote simply "technical" practice as mathematical. "Mathematics", according to Husserl, "is the biggest technical wonder" (Husserl 1993, p. 35).

Two aspects of scientific rationality, i.e., calculation and technicization, were analyzed elsewhere broadly (Plotka 2009, pp. 7–8). Yet, in order to understand Husserl's point in his discussion on calculative reason, let us emphasize that calculation which leads to technicization consists in the formulation of abstract "truths in themselves." By replacing the factual being and life with the universal constructions of ideal calculus, the Husserlian reflection on scientific rationality provides an observation that only by calculating ideal entities the sciences have a "kind of predication" which "infinitely surpasses the accomplishment of everyday predication" (Husserl 1970, p. 51). Moreover, the predication grasps the world which is constructed by the calculus as "given" once and for all. Nevertheless, the calculating scientist "forgets" that the proper object of his current activities is not the world, but a kind of ideal being, viz. the calculus itself. The world is the field of relative relations, rather than it is defined once and for all. As Husserl once put it, "[a]ll being is relative" (Husserl 2008, p. 5).

According to Husserl's lectures on the *Analyses Concerning Passive and Active Synthesis*, the calculative method introduces the sphere of objectivity in such a way that it allows to treat each question as settled for "everyone" (Husserl 2001a, p. 542) who practices the method. Thus, the use of the same method for all of us implies the treatment of nature as non-differentiated, i.e., as the same for "everyone." This "everyone," simply stated, is an abstract entity. It does not refer to an individual. Rather, it pointes at the abstract subject of science. The "ratio of natural sciences", as Husserl wrote, is "the ability . . . of calculating future and past relations of possible fields of the givenness of an experience" (Husserl 2008, p. 733). Therefore, calcula-tive rationality of science misinterprets its world view as the world itself. This thesis is a kind of premise of positive science.

While keeping aforementioned concept of positive science in mind, one can un-derstand Husserl's "uneasy," as he defined it, question: "What does the premise about the rationality of man who cognizes in a real and possible way mean?" (Husserl 1993, p. 30). Precisely, the question set out to challenge the positive sci-entists, casting doubt on the latent premises of science. It is alleged that science is naïve to the extent that it relies on a calculative concept of rationality. Inasmuch as the concept is "normal" for the scientist, a kind of "normality" is an "uncovered pre-supposition of the scientist" (Husserl 1993, p. 30). But, again, all presuppositions are important for the phenomenologist, who is asking the "uneasy" question and by doing so, he is heading for the formulation of the presuppositions which finally concern practice. For this reason, the proper questioned object of the question is the practice of the scientist. In this sense, Husserl suggested that we assume that cer-tain activities are rational. Hence, if one wants to describe the concept of scientific rationality, he should examine how is it practiced, rather than investigating scien-tific theories themselves. Therefore, at the beginning of the 1922–1923 lectures on *Einleitung in die Philosophie*, Husserl (2003, p. 6) emphasized that such an investi-gation might allow us to formulate a theory of rationality which is immanent to the theories constructed by the scientist. To put it clearly, as the practitioner, the scien-tist does not question the foundations of his practice. Only philosophy can do this. By contrast, the scientist just knows what he can do, and this is the reason why he does not care about the premise of rationality. He focuses on his actions or, rather, on actual operations, and he does not address the theme of rationality in his inves-tigation (Husserl 1993, p. 31). For this reason, each question about latent premises of practice is "uneasy" for him and, lastly, the proposition that scientific rationality is determined by the technical method becomes obvious. Precisely from the per-spective of calculative rationality which determines human practice, it is possible to speak of the crisis of science and rationality simultaneously.

So far, positive sciences were characterized as naïve, because they took a given world for granted; the scientist did not recognize the proper object of his activi-ties, i.e., he forgot, following Husserl, that the object is the ideal world denoted by calculus. For this reason: "[t]he natural sciences have not in a single instance un-raveled for us actual reality, the reality in which we live, move, and are" (Husserl 1965, p. 140). It is obvious that sciences concern rather a certain world view, than the world itself. As Husserl put it repeatedly, rationality of positive sciences aims at the construction of related "world views,"[3] or it founds certain "representations of the world" (Husserl 1989a, p. 189). All related "world views" parallel the cal-culative actions of the scientist. In other words, positive sciences have their own "rationality," inasmuch as they introduce certain order into the "representation of the world" which is reflected in the ideal construction of related science. According to phenomenological philosophy, then, nature is a theoretic construction of physics (Husserl 2001a, p. 543), and as such it does not equal the world in which a physi-cian lives and works (Husserl 1976, p. 390). It is important to note, however, that the world of actual practice cannot be given in any way at all. Just as Ströker (1997, p. 305) emphasized, "pre-scientific" character of the world does not mean that it is somehow pregiven in sciences; rather, the lifeworld encompasses all sciences in such a sense that it is the "ground" (*Boden*) of all practice.[4]

What is Husserl's key insight with regard to the critique of conventional reason, however, is not only the statement that the sciences determine human practice, but also that the determination involves a special kind of rationality. According to Husserl, the rationality of scientific actions became an equivalent for rationality in general. Hence, by criticizing calculative reason, Husserl (2002a, pp. 12–13) aimed *eo ipso* at "the tragedy of scientific culture," as he put it metaphorically while lecturing on nature and spirit in 1919. To phrase it differently, rationality which claimed to be comprehended as determining entire human practice limited itself to merely abstract actions. To use Husserl's notions which involve the concept of rationality, ratio of sciences did not equal ratio inner of human actions. Of course, Husserl (2002a, p. 5) agreed with Francis Bacon at least that, as they famously stated, "knowledge is power." Both philosophers attributed the power of sciences to their grounding in mathematical, as well as positive methods, which allowed to reduce the scientist's workload to merely an abstract and automatic calculus. Husserl, however, in opposition to Bacon, was conscious of twofold consequences of applying scientific method to human life. Indeed, the method "is progress," following Husserl's *Natur und Geist* lecture series from 1919, "but it is a danger as well: it saves the scientist much intellectual effort, but due to the mechanisation of method, many branches of knowledge become incomprehensible" (Husserl 2002a, p. 6). As he continued: "outer rationality, which is understood as justification based on changing conclusions, does not correspond to inner rationality, to the understanding of inner senses and aims of thoughts and to basic elements of method" (Husserl 2002a, p. 6). First of all, Husserl's words indicate that the critique of the calculative character of method is closely linked to the critique of rationality in phenomenology in general. In this context, Husserl spoke of an "inner" and "outer" rationality. While the "inner" rationality was an equivalent to the essence of genuine human rationality and thinking, forming its aim and meaning, the "outer" rationality of method reduced rationality to its own ideal constructs of justification and the "outer" mechanisms of practice. Therefore, the rationality of calculative reason transformed human rationality into a mechanism that belonged to a dogmatic science.

By replacing "inner" rationality with "outer" rationality of scientific method, the crucial consequence arises for the understanding of human being. One can ask: *Who* is the subject of science? After all, the scientist is not any individual, rather, while focusing on "hard" facts, he must suspend his subjective being, his beliefs, and his own life to proceed scientific researches. In a word, he must resign from the claims to determine his own life by science.[5] Therefore, because of the definiteness of nature, he does not use a kind of private language in opposition to the language of sciences. As Husserl put it in a note added to *Analyses Concerning Passive and Active Synthesis*: "[t]he definiteness of nature, its being-in-itself implies an intersubjective being-thus of nature that is identifiable for 'everyone' in relation to everyone – a being-thus of all that is, and according to all its things and properties" (Husserl 2001a, p. 543).

Hence, positive science replaced an individual with an abstract "everyone." Namely, "everyone" is able to use "exact" methods of science, and despite who he is, or what he believes, he is able to achieve the same truth as anyone else. As it will become clear in the following, this aspect of positive science provides a decisive difference in comparison of naïve science to authentic.

Summing up, according to Husserl's diagnosis, contemporary sciences are in the crisis. The crisis consists in the loss of the sciences' significance for human life. In order to understand this thesis, the Husserlian account of the lifeworld was reinterpreted in such a way that the lifeworld was presented as the "ground" of all practice. Inasmuch as the world involves human actions, science can determine the world, but only in an indirect way, i.e., by regulating the practice through a distinctive world view. Science offers humans a certitude and an easiness in using the mathematical method. Otherwise, it will be merely the one of the many possible world views. Indeed, positive science constructed indirectly a special kind of practice – calculative. Nonetheless, Husserl's key insight was in this context that the practice involved a kind of rationality. The latter Husserl called "outer" rationality and he indicated that it implies an abstract understanding of human life. At the end, let us emphasize that Husserl's reflection on calculative rationality is not characteristic only for the last period of his work, but, just the opposite, the question of calculative rationality became a life-long concern of Husserl right through the unfinished *Crisis*. Therefore, one is able to find the sources for Husserl's interests in rationality even in the *Investigations* (Husserl 2001b, p. 223).

TOWARDS A NEW FORMULATION OF RATIONALITY

In a personal note, written on the 25th of September 1906, Husserl (1984, p. 445) defined the main purpose of phenomenological inquiry as the critique of reason, which can reveal a sense, essence, and methods of the rational reflection. The critique defined in such a manner, based on the power of reflection, namely, on a self-referential character of reason. Just as aforementioned critique of calculative reason based on the rational power of reflection also Husserl's re-examination of calculative rationality and his way towards a new formulation of rationality had its point of departure in the question about the power of reflection. In general, this comparison demonstrates that entire Husserl's reflection on rationality is based on a fundamental ability of reflection. Indeed, it is reflection which makes possible to define and then to evaluate the mechanization and technicization of method. Husserl built this concept of rationality through examining the idea of modern sciences. Furthermore, the concept of calculative reason derived from the sciences justified the thesis about the crisis of reason, because reason was reduced to the ideal construction of calculative laws. Additionally, Husserl's thesis about the crisis had further important implication, i.e., rationality manifested itself in a factual practice. This implies, however, that the crisis concerns the fields of culture, science and philosophy itself (Buckley 1992, p. 9) simultaneously.

Although, the crisis of calculative rationality seemed to be necessary and unresolved, at the same time one has to be conscious of the possibility of non-naïve account of reason. Therefore, on Husserl's view, when one speaks of the crisis "we must not take this to mean that rationality as such is evil or that it is of only subordinate significance for mankind's existence as a whole" (Husserl 1970, p. 290). Hence, the crisis of calculative rationality did not equal the crisis of reason in general. Rather, as Philip R. Buckley argued,

> "[t]he breakdown of rationality is, for Husserl, not a sign that rationality (in its true sense, that is, philosophy) is no longer possible," but "it is a sign that the 'old' rationality is in fact no true rationality, it is a sham, and its bankruptcy has finally been exposed." The crisis makes evident for Husserl the need for the true form of rationality, for true philosophy, for transcendental phenomenology. (Buckley 1992, p. 123)

Of course, the "old" rationality equals the "outer" one and hence it indicates the concept of rationality related to modern sciences. Husserl, just the opposite, aimed at the formulation of "new" rationality, and in this context we agree with Johanna Maria Tito (1990, p. xlv), who emphasized that Husserl sought the method of investigating the essence of reason to reach a *new* reason as distinct from rationality of the calculative method.[6]

Yet, in opposition to calculative reason, rationality towards which Husserl was leading while pointing at the power of reflection has to present a real alternative for the possibility for regulating human life. In a word, it cannot be merely a theoretical critique, as the critique of calculative reason was in fact, but it must become, to paraphrase the title of one of Husserl's texts from the beginning of the 1920s, *the radical critique*. This kind of critique, has to take a regulative function of human practice to the fore. Therefore, by doing so, the critique provides a point of departure for a broader understanding of the lifeworld as the "ground" of practice. By comparison with the world view of positive science, the lifeworld claims to being comprehended as a "new" world. As Husserl put it in *Radikale Kritik*: "[a]n autonomous man will build ... this new world" (Husserl 1989a, p. 107). Thus, following Marcus Brainard (2007, pp. 17–18), one can speak of the practical impulse of Husserlian efforts. But, what kinds of practical implications does the concept of the radical critique have?

We introduced a "being-thus" as the category of naïve sciences which reduced an individual to the abstract "everyone." Namely, the "everyone" must go follow rationality immanent to science, or, to phrase it differently, the scientist follows necessary "being-thus." Furthermore, all being grasped by sciences is defined once and for all. Yet, the scientist has the entire field of his possible activities as given, defined by necessary "being-thus." For this reason, again, the scientist is determined and enslaved by calculation and technicization of scientific method; after all his activities whatever they concern are determined and defined. The phenomenologist who is re-examining calculative rationality can ask: Is this a genuine consequence of science? Was science constructed to enslave humans? Not at all! Just the opposite, the genuine intention of science, at least for Husserl, was to determine action of a free person. Hence, according to Husserl, rationality of sciences as shaped by the methodological mechanization contains in itself a fundamental contradiction. More precisely, positive science claimed that "[s]cience should make us independent ...

in all our practice and aspirations. However, as science is subordinated to the mechanisation of method, it does not make us free even theoretically" (Husserl 2002a, p. 12). By contrast, the radical critique has to make a man free, whether theoretically or practically. Here, then, the crucial practical implication of the critique arises: The rational critique involves practice by making a man free, but, again, this claim cannot be comprehended on a naïve level.

In order to grasp the non-naïve account of the claim to be free, Husserl's and Fink's view on reduction is to be analyzed briefly. Husserl presented the principle of freedom from presuppositions already in 1900–1901 in the *Investigations* (Husserl 2001b, p. 177). Also after two decades, consequently, he saw in philosophy a proper way of making the philosopher free from presuppositions (Husserl 1958, p. 479). In Husserl's view, the "presupposition" denotes "unjustified judgment" (Husserl 2002b, p. 441). It is important to note that although the purpose of phenomenology is the same in both texts, the understanding of "freedom" changed for Husserl essentially. On the one hand, we can speak of "freedom from presuppositions" if one asks for "grounds" of the presuppositions, and by doing so, he justifies the related judgment. However, from the transcendental viewpoint, this taking a certain judgment for granted, i.e., the end of asking about further presuppositions, equals naïveté. Therefore, on the other hand, "freedom from presuppositions" can be parallel rather to the understanding of presuppositions, than to the exclusion of them. To understand why critique is able to make a man free, one must take into consideration the latter understanding, and then he can speak of what Husserl (2002b, p. 303) defined as "a state where prejudices are universal" (*Universalität von Vorurteilen*). In a word, a non-naïve account of freedom denotes the situational character of reflection which, paradoxically, *always* has certain presuppositions. But, again, our *understanding*, and *not* the suspension, of the situation provides a "real autonomy" and an "absolute self-establishing" of our life (Husserl 1958, p. 506). At this point, we are confronted with the reinterpreted idea of phenomenological reduction.

What is philosophically interesting, in the context of the paradoxical account of "freedom," is Husserl's idea that the phenomenologist must grasp the state of presupposition as permanent. To understand this paradoxical structure, one can speak of what Fink defined as "the situation of reduction;" in his notes written in the period of his collaboration with Husserl, he wrote:

[p]henomenological reduction is no method which cannot be taught once for all, but inasmuch as … [philosophical – W.P.] telos is human freedom, [reduction – W.P.] is the task of philosophy. Philosophy wants to exist only for freedom. The motivation of the reduction is only the will to freedom. (Fink 2006, p. 222)

Thus, in Fink's view, freedom is the proper task of reduction, because while reducing the phenomenologist discerns himself as the subject of presuppositions, therefore, at the same time, he is conscious of them and he understands his situation as having presuppositions. In a word, only while doing reduction, he is able to see presuppositions of his activities. Obviously, before reduction the man is enslaved, because presuppositions enslave a naïve subject of actions, however, to quote Fink once again, "[a] man is enslaved essentially. And only because he is not free, he might be free" (Fink 2006, p. 222). This is precisely the context in which one shall read Fink's assertion that "[a] man exists as a paradox. He combines in himself matters, which seem to be contradictory. He understands the being in original strangeness and original confidence" (Fink 1958, p. 30). In other words, the reducing scientist lives in a natural attitude, but the very heart of this life stays still unknown, because he simply forgot about himself paradoxically. Rather, with regard to the scientific method he transformed himself from an individual into aforementioned abstract "everyone." Only if the phenomenologist started doing and redoing reduction (Cairns 1976, p. 43), he became capable to see himself as the acting person and in consequence he became free. In a word, due to reduction the phenomenologist established himself, despite calculative rationality. Here, reduction is an equivalence for the radical critique which founds authentic ratio.[7] Yet, to stress it clearly, the phenomenologist is free *only* while reducing. For this reason, he achieves authentic ratio during the radical critique, however, he returns to naïveté necessarily.[8]

The reflective power of reason made evident that the modern science deformed the idea of rationality (Mall 1972, p. 135). The decisive step towards the deformation was dualistic interpretation of reason: on the one hand, reason was grasped as merely a factual power, on the other, it was associated with objective laws of thinking. As it was presented, the naïve science negated the former and affirmed the latter at the same time; finally, the positive science replaced factual actions with abstract constructions. Conversely, the "new" rationality involves a non-dualistic understanding of reason. Almost always when Husserl spoke about the "new" rationality he referred literally, or indirectly to rationality founded by ancient Greeks,[9] because, following *The Vienna Lecture*, "[s]piritual Europe has a birthplace, ... not a geographical birthplace ..., but rather a spiritual birthplace .... It is the ancient Greek nation in the seventh and sixth centuries B.C." (Husserl 1970, p. 276). According to Husserl, precisely in ancient Greece, reason was not comprehended as a mathematical power, but rather as the authentic power of self-reflection. As Husserl once put it,

[r]ationality, in that high and genuine sense of which alone we are speaking, the primordial Greek sense which in the classical period of Greek philosophy had become an ideal, still requires, to be sure, much clarification through self-reflection. (Husserl 1989a, p. 290)

Additionally, the unified Greek view on reason as self-reflection, has one important consequence: inasmuch as one overcame the dualism of rationality, self-reflection involved rather factual actions in the world, than ideal calculus. Therefore, be referring to the Greek rational tradition, Husserl spoke to us much more than he gave as the speculative thesis on the "spiritual birthplace of Europe." Namely, he indicated that the Greek rationality proposed a concrete account of a human being, i.e., in the ethical-political sense. Hence, authentic reason apart from the abstract "everyone" as the subject of ideal actions provided a point of departure for the understanding of *animal rationale* as an individual who acts in a community. To quote James Hart's appropriate assertion: "[b]ecause it [reason – W.P.] is the power to unite and bind humans, we may say that a most decisive articulation is in the way each actualizes the latent plural dative ('us all') and anonymous 'we' and sees his or her action in terms of the good of others" (Hart 1992, p. 651). In other words, Husserl sought in the Greek rationality the supplementary of the abstract modern rationality, i.e., political and ethical involvement of subjects in the world.

To sum up, by replacing calculative rationality with authentic reason Husserl re-examined the understanding of rationality, in consequence, broadening it significantly. First of all, Husserl combined authentic reason with the power of self-reflection. Only due to the power, the critique of calculative rationality was possible, and precisely the power led the phenomenologist towards his radical autonomy. However, following Fink, the autonomy is understood as a permanent confrontation of the phenomenologist with presuppositions, rather than as a naïve denial of them. Finally, the autonomy showed its whole significance by the introduction of the practical meaning of the lifeworld and actions, because at the practical level humans are grasped as individuals, rather than as abstract subjects of calculative rationality. Moreover, while referring to the Greek sources of rationality, Husserl made it clear that practical, i.e., political and ethical,[10] aspects of reason are crucial for taking into account also intersubjective context of human actions. Only in such a way one is able to overcome solipsistic *milieu* of calculative rationality.

CONCLUSION

In conclusion, let me recall that the main purpose of the article is the presentation of Husserl's re-examination of the concept of rationality with regard to the twofold understanding of the critique of reason. On the one hand, the critique denoted the investigation, which aimed at the solution of the questions: In what sense are we able to speak of the crisis of rationality? How do sciences involve the crisis? On the other, the critique referred to the Kantian tradition in which the critique referred to the investigation of sense, essence, and methods of the rational reflection. Both understandings were connected essentially, however, inasmuch as the former understanding based on the latter, and the latter had its proper point of departure as early as in 1906, namely in Husserl's declaration that the main purpose of phenomenology is the critique of reason (Husserl 1984, p. 445), it is evident that phenomenology presents a permanent development, rather than the series of "breakthroughs." Therefore, the *Crisis* is a result of Husserl's great efforts (which he started already in the *Investigations*) to re-examine the concept of rationality.

Summing up, in regard to the first understanding of the critique, it became clear, following Søren Overgaard (2002, p. 213), that science can tell as *how* things happen as they do, however, it did not inquiry about its grounds, and for this reason it fallen into naïveté. Furthermore, the main reason of the crisis was identified as the calculative interpretation of rationality. Calculative rationality, then, led humans towards the abstract field of operations defined by mechanization and technicization of scientific method. At the end, the scientist operated within the ideal filed, but at the same time he forgot that his activities involve the field, and not the factual one; in a word, calculative rationality replaced the factual world with the ideal one, and moreover an individual with the abstract "everyone."

By contrast, Husserl's way towards a new formulation of the theory of rationality led him through inquiring about the reflection itself, rather than about the state of culture. In doing so, Husserl defined the proper aim of the reflection as the "freedom from presuppositions," however, this state of freedom is not achievable once and for all. Rather, as Husserl and Fink made it clear, the process of inquiring about the presuppositions is endless. It was argued that a permanent inquiry claimed to have its sources in ancient Greece. In ancient Greece precisely, rationality was formed not only as non-calculative, but moreover it had a broad practical sense, leading the philosopher to adopt the thesis about a communal character of investigations. The thesis made possible to transform the abstract "everyone" of calculative rationality into concrete "us" of authentic rationality. The authentic rationality which involved a practical level, let us suggest in the end, expressed the crucial sense of rigorous science. In Husserl's note from 1935, we find the following ironic question: "You still tell the same old story about Your radical rationalism, do You still believe in philosophy as a rigorous science? Have You slept through the end of the new time?" In light of our findings so far, Husserl's answer expresses practical aspect of rationality: "Oh no. I do not 'believe' or 'tell stories': I work, I build, I am responsible" (Husserl 1989a, p. 238). Therefore, as Husserl tried to show us, there is no another response to calculative reason than rational practice itself.

University of Gdańsk, Gdańsk, Poland
e-mail: witoldplotka@gmail.com

NOTES

[1] In his recent book on the crisis and reflection, James Dodd emphasized, "Husserl could perhaps be considered one of the last great philosophers of the Enlightenment, and the *Crisis* his grand defense of reason" (Dodd 2004, p. 169).

[2] Ernst Wolfgang Orth (1999, pp. 46–49) emphasized that the diagnosis of the crisis of culture and the critique of scientific reason were typical for intellectual discussions of the beginning of the twentieth century.

[3] Cf. Husserl 1989a, p. 175; Husserl 2002b, p. 321; Husserl 2008, pp. 202, 673.

[4] Husserl stressed the understanding of the lifeworld as the "ground," or the horizon of all practice repeatedly. Cf. Husserl (1970, p. 142); Husserl (1993, p. 45); Husserl (2002b, p. 394); Husserl (2008, pp. 308, 351). See also Claesges (1972, p. 88) and Park (2001, p. 109).

[5] As Ulrich Melle once put it, "[b]y concentrating on so-called objective, hard facts modern science in its positivist distortion left us without any firm guidance in making hard choices" (Melle 1998, p. 329).

[6] Ram Adhar Mall emphasized: "[p]henomenological reason does not copy the mathematical reason. Unlike the latter it does not consist in construction. It does not formalize; it does not create either. It is a reason which shows itself as a task and is clearly seen as 'lived' as such" (Mall 1973, p. 115).

[7] Inasmuch as naïve sciences which constituted calculative rationality enslaved man, according to Husserl's observation, "an authentic ratio can heal those losses" (Husserl 1989a, p. 239). In the course of examining the nature of authentic ratio, Husserl wrote about a "renewal" of humanity. It is important to note that in his articles on the renewal for *The Kaizo*, written in 1923 and 1924, a point of departure for the Husserlian critique is the observation that technique became a real practical rationality (Husserl 1989a, p. 6). In consequence, also the overcoming of the crisis can be grasped as a certain "renewal" of rationality.

8 Within the Husserlian reflection on two basic attitudes, i.e., the naïve attitude and the theoretical, or philosophical one, one shall stress essential correlation of both, rather than their contradictory character. As Husserl emphasized briefly in the *Epilogue* to the English translation of the first book of *Ideas*: "the necessary point of departure . . . is the natural-naïve attitude" (Husserl 1989b, p. 416). Nonetheless, the necessary point of departure makes possible to build the phenomenological reflection on the higher, non-naïve level. Just as any way is impossible without a point of departure, also the non-naïve phenomenology is impossible without the naïve attitude. To paraphrase Husserl's (1958, p. 478) own words, every time the phenomenology is split.

9 Cf. Natanson (1973, p. 17). About Husserl's turn towards the ancient Greek rationality, see also Ströker (1988, p. 214).

10 One of the ethical aspects introduced into the Husserlian interpretation of rationality is the concept of *responsibility* which broadens the meaning of rigorous science. Following Buckley: "Another way of defining rigour is to say that to be rigorous is to be *responsible*, to be able to justify each and every position taken, to be willing to provide the evidence for one's beliefs. This definition expresses the ethical imperative which Husserl felt regarding philosophy. To be sure, there was also an 'epistemological' imperative: philosophy is, after all, about 'knowledge.' But true knowledge for Husserl is that for which one can 'answer' (*verantworten*)" (Buckley 1992, p. 22). Cf. Kuster (1996, pp. 38–39).

REFERENCES

Brainard, Marcus. 2007. 'For a new world': On the practical impulse of Husserlian theory. *Husserl Studies* 23: 17–31.

Buckley, Philip R. 1992. *Husserl, Heidegger and the crisis of philosophical responsibility*. Dordrecht: Kluwer.

Cairns, Dorion. 1976. *Conversations with Husserl and Fink*. The Hague: Nijhoff.

Claesges, Ulrich. 1972. Zweideutigkeiten in Husserl Lebenswelt-Begriff. In *Perspektiven transzendentalphänomenologischer Forschung. Für Ludwig Landgrebe zum 70. Geburtstag von seinen Kölner Schülern*, eds. U. Claesges, and K. Held, 85–101. The Hague: Nijhoff.

Dodd, James. 2004. *Crisis and reflection. An essay on Husserl's "Crisis of the European sciences"*. Dordrecht: Kluwer.

Fink, Eugen. 1958. *Sein, Wahrheit, Welt. Vor-Fragen zum Problem des Phänomen-Begriffs*. The Hague: Nijhoff.

Fink, Eugen. 2006. *Phänomenologische Werkstatt. Teilband 1: Die Doktorarbeit und ersted Assistenzjahre bei Husserl*. Freiburg, Munich: Verlag Karl Alber.

Hart, James G. 1992. The rationality of culture and the culture of rationality: Some Husserlian proposals. *Philosophy East and West* 42(4):643–664.

Husserl, Edmund. 1958. *Erste Philosophie (1923/24). Zweiter Teil. Theorie der phänomenologischen Reduktion*. The Hague: Nijhoff.

Husserl, Edmund. 1965. *Phenomenology and the crisis of philosophy* (trans: German by Lauren, Q.). New York, NY: Harper Torchbooks.

Husserl, Edmund. 1970. *The crisis of European sciences and transcendental phenomenology. An introduction to phenomenological philosophy* (trans: German by Carr, D.). Evanston, IL: Northwestern University Press.

Husserl, Edmund. 1976. *Die Krisis der europäischen Wissenschaften und die transzendentale Phänomenologie. Eine Einleitung in die phänomenologische Philosophie*. 2nd ed. The Hague: Kluwer.

Husserl, Edmund. 1984. *Einleitung in die Logik und Erkenntnistheorie. Vorlesungen 1906/07*. Dordrecht, Boston, MA, Lancaster: Nijhoff.

Husserl, Edmund. 1989a. *Aufsätze und Vorträge (1922–1937)*. Dordrecht, Boston, MA, London: Nijhoff.

Husserl, Edmund. 1989b. *Ideas pertaining to a pure phenomenology and to a phenomenological philosophy. Second book. Studies in the phenomenology of constitution* (trans: German by Rojcewicz, R., and A. Schuwer). Dordrecht: Kluwer.

Husserl, Edmund. 1993. *Die Krisis der europäischen Wissenschaften und die transzendentale Phänomenologie. Ergänzungsband: Texte aus dem Nachlass 1934–1937*. Dordrecht: Kluwer.

Husserl, Edmund. 2001a. *Analyses concerning passive and active synthesis. Lectures on transcendental logic* (trans: German by Steinbock, A.J.). Dordrecht: Kluwer.

Husserl, Edmund. 2001b. *Logical investigations, vol. 1*. 5th ed. (trans. from German by Findlay, J.N.). London, New York, NY: Routledge.

Husserl, Edmund. 2001c. *Natur und Geist. Vorlesungen Sommersemester 1927*. Dordrecht: Kluwer.

Husserl, Edmund. 2002a. *Natur und Geist. Vorlesungen Sommersemester 1919*. Dordrecht: Kluwer.

Husserl, Edmund. 2002b. *Zur phänomenologischen Reduktion. Texte aus dem Nachlass (1926–1935)*. Dordrecht: Kluwer.

Husserl, Edmund. 2003. *Einleitung in die Philosophie. Vorlesungen 1922/23*. Dordrecht: Nijhoff.

Husserl, Edmund. 2008. *Die Lebenswelt. Auslegungen der vorgegebenen Welt und ihrer Konstitution. Texte aus dem Nachlass (1916–1937)*. Dordrecht: Springer.

Kuster, Friederike. 1996. *Wege der Verantwortung. Husserls Phänomenologie als Gang durch die Faktizität*. Dordrecht, Boston, MA, London: Kluwer.

Mall, Ram Adhar, 1972. Phenomenology of reason. In *Perspektiven transzendentalphänomenologischer Forschung. Für Ludwig Landgrebe zum 70. Geburtstag von seinen Kölner Schülern*, eds. U. Claesges, and K. Held, 129–141. The Hague: Nijhoff.

Mall, Ram Adhar. 1973. *Experience and reason. The phenomenology of Husserl and its relation to Hume's philosophy*. The Hague: Nijhoff.

Melle, Ulrich. 1998. Responsibility and the crisis of technological civilization: A Husserlian meditation on Hans Jonas. *Human Studies* 21:329–345.

Natanson, Maurice. 1973. *Edmund Husserl. Philosopher of infinite tasks*. Evanston, IL: Northwestern University Press.

Orth, Ernst Wolfgang. 1999. *Edmund Husserls "Krisis der europäischen Wissenschaften und die transzendentale Phänomenologie." Vernunft und Kultur*. Darmstadt: Wissenschaftliche Buchgesellschaft.

Overgaard, Søren. 2002. Epoché and solipsistic reduction. *Husserl Studies* 18:209–222.

Park, In-Cheol. 2001. *Die Wissenschaft von der Lebenswelt. Zur Methodik von Husserls später Phänomenologie*. Amsterdam, New York, NY: Ropodi.

Plotka, Witold. 2009. The riddle of reason: In search of Husserl's concept of rationality. *Bulletin d'analyse phénoménologique* 2:1–27.

Ströker, Elisabeth. 1988. Edmund Husserls Phänomenologie: Philosophia Perrenis in der Krise der europäischen Kultur. *Husserl Studies* 5:197–217.

Ströker, Elisabeth. 1997. Science and Lifeworld: A problem of cultural change. *Human Studies* 20: 303–314.

Tito, Johanna Maria. 1990. *Logic in the Husserlian context*. Evanston, IL: Northwestern University Press.

KOUSHIK JOARDAR

LOGOS AS SIGNIFIER: HUSSERL IN THE CONTEXT OF TRADITION

ABSTRACT

The present article has tried to establish a bond between the phenomenology of Husserl and the philosophic tradition of the West, between intentionality and the *logos*. However, the Platonic-Aristotelian tradition, in which *logos* has been taken to be *mere* reason, cosmic or human, helps little in understanding Husserl, because that actually is based on the separation between thought and its object. The real meaning of *logos* Heidegger finds in pre-Socratic philosophy of Heraclitus and Parmenides. The tradition I take into consideration is the Heideggerian interpretation of pre-Socratic philosophic heritage in which *logos* means "basic gathering". If Husserl revives antiquity, he does so by re-establishing this "basic gathering" in the name of intentionality. The concept of intentionality may be looked upon as the reunion of thinking and being.

LOGOCENTRISM

Logocentrism is detested by postmodern thinkers. According to them, it has been a prejudice of the philosophers. If one is still logocentric in our time, Foucault would call him a mere nostalgic one. However, the term is popularised as something to represent a notorious concept by Jacques Derrida. The history of metaphysics, maintains Derrida, has always been a search for the ultimate truth and this origin of all truths has been found to be in the *logos*. Or is itself the *logos*. The history this truth of all truths again "has always been . . . the debasement of writing, and its repression outside 'full' speech".[1] The belief that speech is prior to writing, which is termed as phonocentrism, is only a variation of more general theory of logocentrism. The general theory of logocentrism assigns the truth of any particular object to something outside it. However, as Derrida deals mainly with phonocentrism, "logocentrism" and "phonocentrism" became interchangeable terms. In short, logocentrism is the belief in the *logos* which provides meaning to anything from outside.

LOGOS AND THE GOSPEL

The Greek word *logos* is translated in English as "word". However, "*logos*" is used in many other senses in different contexts e.g., an utterance, principle, law, reason, an account, etc. All these senses are somehow related but "word" is one of the most

207

A.-T. Tymieniecka (ed.), Analecta Husserliana CX, 207–215.
DOI 10.1007/978-94-007-1691-9_17, © Springer Science+Business Media B.V. 2011

frequent translation for "*logos*". The association of these two may have its source in pre-Christian antiquity, but it is strengthened by the Gospel. The Gospel of John identifies Christ with *logos* or the word. When it states that "Before anything else existed, there was Christ with God", it is actually referring back to the Genesis in which God is said to be creating just by uttering words. Then God said, "Let there be light" and light appeared. The Gospel of John goes on saying that "He has always been alive and is himself God." Thus, in a sense, it is the *Logos* that is responsible for the very creation. This may remind one the concept of *Śabdabrahma* of the great Indian thinker Bhartŗhari, according to which, word is the ultimate reality. However, the Gospel also establishes at the same time the primacy of speech because the first words were uttered and not written.

LOGOS AND THE TRADITION

The concept of *logos* as an utterance is much older than the Gospel. To reiterate, the word is a Greek one and is even was present in the writings of the earliest Greek thinkers. "It is cognate with the verb *legein*, which normally means 'to say' or 'to state' ".[2] It is believed that when Heraclitus refers several times to "this *logos*" in his book, he simply means "this utterance" or "this statement". The belief is strengthened by his insistence "Listening not to me but to the account [the *logos*], it is wise to agree that all things are one"[3] In his obscure poetry, Parmenides also writes ". . .preserve the account [*logos*] when you hear it. . ."[4] We listen to speech and not to writing. One interesting point to note is that Aristotle writes, "Learning is reserved for those that in addition to memory who also have the sense of hearing. . . ."[5] *Hearing*, and not seeing. Here also speech gets priority over writing. However, let us back to the point. By transference, *logos* begins to mean an explanation and then the faculty of human beings which explains. Since then, *logos* comes to be used in the sense of "reason" also.

HEIDEGGER REINTERPRETING TRADITION

It is a common belief that Heraclitus perceived the world only as becoming and Parmenides saw a being behind becoming. Aristotle himself writes about Heraclitean view that "all perceptible things were in a permanent state of flux and that there was no knowledge of them. . . ."[6] And everybody knows that it is Heraclitus who said that one cannot step twice into the same river, although, it is possible that he did not say this exactly in the same manner. History remembers Heraclitus as a preacher of diversity, whereas, although Parmenides speaks several times of "change", he is remembered as the preacher of unity. It is generally over-looked that Heraclitus also perceived a unity behind what appears. However, Russell observes that the "One" Heraclitus perceived in all changes is the "fire" which itself is ever changing.

Heidegger rejects the traditional interpretation of Heraclitus initiated by Plato and Aristotle. The unconventional way of Heidegger's interpretation of Heraclitus and pre-Socratic Greek philosophy presents a very different, thought provoking and unique concept of *logos*. The concept of *logos* Heidegger presents is in the context of his explanation of the meaning of being in his magnificent work – *An Introduction to Metaphysics* (*Was ist Metaphysik?*).

It is difficult to grasp the meaning of being, says Heidegger. We see a building over there. It is an essent (*Seindes*). Essent is approximately what Sartre calls in-itself (*en soi*). We encounter corridors, staircases, rooms etc. in that building but where is the being of that building? We utter such expressions like "being of the building", "the building is. . ." etc. Do we really encounter being? Etymology shows that "being" is a noun but it does not belong to the class of nouns like "house", "bread", "chalk" etc. "Being" is a noun like "falling", "going", etc. Thus, "being" is a substantive formed out of an infinitive – "to be". It is clear then, "being" cannot be encountered like chalk and duster can be. Nevertheless, we see an apple falling, a boat sailing etc. Being is encountered neither in that way. Do we have to agree with Nietzsche then that "being" is an empty, vaporous term?

The emptiness of "being" does not make it meaningless to Heidegger. "[T]he word 'being' has the emptiest and therefore most comprehensive meaning".[7] For him, "being" is the most universal word. So many things are expressed with the help of the single word "is". "God is", "The earth is", "The lecture is in the auditorium", "The book is mine" and many other expressions he cited. In each case the "is" is meant differently. This extreme universality makes being indefinite in meaning and yet we understand it definitely. "Being proves to be totally indeterminate and at the same time highly determinate."[8] This would have been rejected by the traditional logic as meaningless. Heidegger embraces this contradiction. But how to explain the determinateness of "being"?

UNDERSTANDING "BEING"

One of the ways of understanding a concept is to understand it as opposed to its limitations, i.e., to that from which it is distinguished. In understanding the concept of being, Heidegger contrasts it with the concepts of becoming, appearance, thinking and ought.

There is a tendency to reduce the distinction between being and becoming to that between being and appearance and vice versa. Although connected, the two distinctions are different. Becoming is genesis, the "not yet". Whereas, being appears as the pure fullness of the permanent, completely untouched by the changing process, the unrest. Being, as opposed to appearance is understood differently, as real and authentic. Appearance is unreal and inauthentic. Thus, a strong sense of evaluation goes along with the distinction between being and appearance which is absent in the former case. "The distinction implies an evaluation – the preference is given to beings."[9]

One important thing is to be noticed that although Heidegger clarifies being by contrasting it to becoming and appearance, on a closer analysis of the concepts, he rejects any essential difference between being and becoming or being and appearance. Not only that the opposition between being and becoming is a misunderstanding, the opposition between Heraclitus and Parmenides is a misinterpretation of their works. Heraclitus "says the same as Parmenides". Appearing, on the other hand, is the very essence of being. His analysis of the German word *schein* for "appearance" provoked him to conclude that appearance means exactly the same as being. Both "being" and "appearance" indicate to a presence. In fact, the opposite of being as well as appearance is non-being, which means to withdraw from the presence. To falsify appearance as something merely imagined or subjective has been a trend set by Plato and is a deviation from the intention of the pre-Socratic philosophers.

The next formula Heidegger examined is the distinction between being and thinking. Being has often been understood as opposed to thinking but thinking differs from the other counterparts of being, viz., becoming and appearance. Thinking is the foundation of all the other distinctions. Being is placed before thinking as an object and being takes on its entire interpretation from thinking. "Thinking brings something before us, represents it."[10] But what does it mean to say that thinking represents being? Something represents some other thing if the two things are different, separate and the one acts as a sign for the other, speaks for the other. The relation between representative and the represented is contingent. Does thinking represent being in this sense? What is the original meaning of representation? How exactly thinking and being are related? For answers to these questions, Heidegger looks back to logic, the science of thinking.

HEIDEGGER'S CRITIQUE OF LOGIC

Heidegger detests Logic that is practiced in the universities and colleges. "Logic is an invention of schoolteachers and not of philosophers."[11] Traditional Logic is so formal that it has lost essent (being). Rather, it teaches us to think without reference to *physis*. Thinkers like Leibnitz, Kant and Hegel did not mean traditional logic when they spoke of Logic. What is Logic in Hegel is otherwise known as Metaphysics. The Logic that separates thinking from being has its origin in Platonic-Aristotelian philosophy. Heidegger liquidates the supremacy of logic because it does not allow asking the most fundamental metaphysical question of being and nothingness. The question regarding nothingness would be considered as contradictory by the traditional logic for violating its rule.

LOGOS REINTERPRETED

However, in order to explain the relation between being and thinking, Heidegger does take Logic into consideration but a different one altogether based on a novel interpretation of *logos*.

Logic comes from the word *logos*. *Logos* means word/discourse or speech as in "dialogue". *Logos* also means reason or thinking and Logic is a science of thinking. The transference of *logos* as word to that as thinking is already discussed in this article. However, this is not the only way of interpreting *logos*. This interpretation does not tell us why traditional (post-Socratic) Logic is a deviation from the earlier one. Heidegger finds that the fundamental meaning of *logos* stands in no direct relation to language. The original meaning of *logos* is "to gather" or "to collect". He quotes from the fragments of Heraclitus to justify his claim:

[E]verything becomes essent in accordance with this *logos*. . ... Therefore it is necessary to follow it, i.e., to adhere to togetherness in the essent; but though the *logos* is this togetherness in the essent, the many live as though each had his own understanding (opinion).[12]

But did not Heraclitus himself say that "If you have heard not me but to the *logos*. . ." etc.? Heidegger says that the hearing mentioned here has nothing to do with lobes of our ears; rather it is used in the sense of paying heed. Because men do not pay heed to the *logos* as gathering, they fail to see that "all is one" and hold different opinions. Moreover, Heidegger opines that *logos* in Heraclitus (and also in Parmenides) is rather opposed to words (or speech) and interprets the above as "do not listen to the voices (others opinions) but pay heed to the *logos*".

In what sense *logos* is gathering? Heidegger offers several senses of the term. "Gathering" does not mean a mere driving together and heaping up. Meaning of "gathering" he gives in different places of *An Introduction to Metaphysics* may be listed as following: unity, harmony, bond, relation, disclosing, making manifest, etc. If *logos* means "gathering" and "gathering" means "unity", then what is united? Wherein lies this unity? It is the unity of what? Since "being as *logos* is basic gathering",[13] unity is that in becoming, it is the harmony among appearances. As a bond, it does not let the opposites and the conflicts in the essent to fall apart. It rather ". . .by uniting the opposites maintains the full sharpness of their tension."[14] The unity or this basic gathering is not however an independent objective phenomenon. Without the apprehension of such gathering, there is no unity or harmony. It is interesting to note that one of the dictionary meanings of "gather" is "to understand". Heidegger also finds in Parmenides that "There is a reciprocal bond between apprehension and being."[15] *Logos* is both the apprehension of the unity and the unity apprehended. Heidegger takes language so much into consideration because human understanding of being is reflected in language and language is constructed in accordance with this understanding. Language relates subject to the object. However, the distinction between subject and object is rational. Heidegger's interpretation is pre-rational. According to this interpretation, language unites, gathers. *Logos* in this sense is language as well as thinking. It is "thinking" not in the sense of "reasoning" but "apprehending". The relatedness of being to apprehension makes it clear why *logos* is also "unconcealment" or "manifestation". Thus, Derrida's critique of logocentrism is meaningful only in the context of tradition in which thinking has the priority over being and the two are separated, i.e., in the context of Platonic tradition. For Heidegger, Plato is the beginning of the fall of Western civilization. He apprehends the inseparable bond between thinking and being.

H U S S E R L ' S   I N F L U E N C E   O N   H E I D E G G E R

Heidegger's interpretation of tradition and his concept of *logos* as well as of being are influenced by the philosophy of Husserl, although the former is not always expressive of the fact. Husserl once said, "Phenomenology, that is me and Heidegger." Heidegger worked with Husserl as the latter's *privatdozent* and surely had played a significant role in developing phenomenological movement. Heidegger dedicated his *Being and Time* to Husserl. However, by moving away from "Cartesian emphasis" and *epoché* of Husserl, and by putting more emphasis on "being-in-the-world", Heidegger sought to give phenomenology a new dimension. Nevertheless, "At least two of Husserl's concepts were to prove especially important for Heidegger: the concepts of evidence and categorical intuition".[16] I perceive that the concepts of "phenomena" and "intentionality" are two most important influences not only on Heidegger, but on a number of philosophers to come immediately after Husserl in the continent. Heidegger's explanation of the relation between being and thinking is a result of this influence.

H U S S E R L ' S   C O N C E P T   O F   L O G I C

Husserl was not as harsh against traditional and formal logic as Heidegger was. What Heidegger calls a matter of "schoolteachers" is for Husserl a matter of technicians of logic. He considered himself as a philosopher of logic and therefore, devoted himself to the semantic aspect of logic rather than the syntactic aspect. He was concerned with propositions, not with sentences.

Propositions are ideas expressible in language. Formal logic deals with the forms of expressions, whereas, Husserl's concern is the forms of ideas which are expressed in language. Ideas, according to Husserl, are of two kinds: subjective and objective. Subjective ideas are those of which Frege writes ". . .every idea has only one bearer, no two men have the same idea."[17] The world of subjective ideas is the inner world of the person comprised of sense impressions, imaginations, sensations, feelings, desires etc. This world is extremely personal and cannot be shared by others. If Logic is all about language expressing subjective ideas, then there would be no explanation about how communication is possible. Formal logic can only show the validity of certain reasoning, but cannot explain why such logic is valid for everyone. That is why Husserl leaves subjective ideas for psychologists. He deals with the other sort of ideas, called by him "objective ideas", which again in Frege's terminology known as "thought". Objective ideas are propositions or otherwise known as ideal meaning (*sinn*). In perceiving a tree, one certainly has private sensations of the tree but there must be some ideal meaning without the transfer of which there could be no communication about the tree. Logic, according to Husserl, is to discover the structure of this ideal meaning or "thought", i.e., the *logos*, which is the ground of all the other meanings.

PHENOMENOLOGIC

Husserl was strongly convinced that logic is objectively valid and that psychology does not and cannot address this. The real task of a philosopher is to discover the structure which makes logic valid and communication possible. To fulfil this aim, one must assume a method so that psychologism may be eliminated. The method Husserl assumes is Phenomenology which is to discover and analyse only phenomena. A phenomenon is that which is immediately given to consciousness or in other words, that what is directly evident. We must recognise phenomenon before it is interpreted. To ensure the purity of description, he again takes recourse to the mechanism of bracketing or what is more popularly known as *epoché*. *Epoché* is the technique of suspending our belief in actual existence of worldly objects which we experience. It is to bracket out all that are interests of empirical sciences. If psychological confusion is avoided, transcendental phenomenology allows us to notice that *epoché* with respect to worldly beings does not alter the fact that our manifold cogitations relating to worldly beings still bear this relation within themselves. It is revealed then that every conscious process is consciousness of something, no matter whether this "something" actually exists or not. This characteristic of aboutness of consciousness is known as intentionality which is the most universal and fundamental property of consciousness. Every directional conscious act, which is otherwise known as *noesis,* bears within itself its content or *noema.* The philosopher as a disinterested ego is to provide us with *noetic-noematic* description of this most universal structure.

Consciousness is always consciousness of something. But that is not all. "This something ... in any consciousness is there as an identical unity belonging to noetically-noematically changing modes of consciousness...."[18] Something as a content of *noetic* act becomes "something" only as a result of synthesis of various changing *noematic* modes into "one" thing. This unity also involves a simultaneous synthesis of the cogito present in every *noematic* mode into one identical ego. The synthetic act not only enables us to perceive this die as "this" die, it is also there in consciousness that is intended to non-identicals (e.g., plurality, relational complexes, etc.). More to say, a synthetic unity of cogito is not exhausted with particular subjective process. Every actuality involves its potentialities. These potentialities are marked by a horizon of the actual process itself. A horizon changes when potentialities marked by it are actualized. Thus, "Every subjective process has a process horizon."[19]

INTENTIONALITY AS GATHERING

Let us reiterate that for Heidegger, being, whether it is *seinds* or *dasein,* is to be understood not as separated from but with respect to its relatedness to its counterparts like becoming, appearance, thinking and ought. In the context of his elucidation of the relation between being and thinking, he interpreted *logos* as gathering. It is gathering in the sense of unity or bond or collection which establishes the necessary relation between being and thinking. He did not claim novelty but maintained that this was the original sense of *logos* in the pre-Socratic tradition of philosophy. I also repeat that Heidegger's interpretation of *logos* derives its sense from Husserlian phenomenology. Not only that Heidegger expressed his indebtedness clearly; neither Husserl uses and explains "*logos*" as gathering directly. But in going through Husserl's works, I perceive that the most universal structure of consciousness which Husserl discovers and in accordance to which everything becomes meaningful is nothing but what Heidegger calls "gathering". That is, I perceive intentionality as *logos*.

A. By introducing the concept of phenomenon, Husserl overcame dualisms that have created so many problems in the history of philosophy. As Sartre points out in his *Being and Nothingness* that in the phenomenology of Husserl and Heidegger, there is no being as opposed to becoming, neither there is any appearance behind which being lies. Phenomenon is "absolutely indicative of itself" and does not hide any Kantian noumenon behind it.
B. Analysis of phenomena reveals that consciousness is intentional. Intentionality rejects subject-object dualism and establishes a bond between consciousness and its content. Heidegger's rejection of the dichotomy of being and thinking is to be understood in the background of intentionality.
C. That intentionality is gathering is also evident from its essential characteristic of synthesis. The apprehension of being as content is not exhausted in a momentary act. Being is not some static existent, "being" is not a common noun. Being is an event every moment of which indicates to many other possibilities of *noetic* acts of which one will be actualised. Being thus, is gathered as a synthetic *noematic* whole, a process which never points to a dead end.
D. Synthesis of being is at the same time synthesis of cogito. As being is a synthesis of various momentary aspects (appearances) of an essent, the "I" involved in the apprehension of that very essent is a synthesis of self awareness involved in each of the said moments. The real meaning of synthesis of intentional acts is that these are acts of one consciousness.[20]

CONCLUDING REMARKS

Any effort to understand consciousness as opposed to and as separated from its objects is to objectify consciousness. Failure of such an endeavour is obvious because knowing-consciousness is never an object. Neither would it lead to a proper understanding of the object of consciousness. An object gets its sense only in relation to the consciousness. Thus, anything becomes meaningful only under the most universal structure of intentionality. The only meaning a knowing-consciousness can have for itself is that it is the meaning provider to its objects without being meaningful like any of its contents because of the necessary and unidirectional nature of intentionality. The presence of this signifier in any meaning, any sense and any interpretation is unavoidable.

*Department of Philosophy, University of North Bengal, Darjeeling, West Bengal
734013, India*
e-mail: koushik_jrdr@yahoo.co.in

NOTES

[1] Jacques Derrida, *Of Grammatology*, trans. Gayatri C. Spivak (Delhi: MLBD, 2002), p. 3.
[2] Introduction to Jonathan Barnes *Early Greek Philosophy* (London: Penguin Books, 2001), p. xxiii.
[3] Ibid., p. 50.
[4] Ibid., p. 80.
[5] Aristotle, *Metaphysics*, trans. Hugh Lawson-Tancred (London: Penguin Books, 2004), p. 4.
[6] Ibid., Alpha 6, p. 23.
[7] Martin Heidegger, *An Introduction to Metaphysics* (Delhi: MLBD, 2005), p. 75.
[8] Ibid., p. 78.
[9] Ibid., p. 98.
[10] Ibid., p. 118.
[11] Ibid., p. 121.
[12] Ibid., p. 127.
[13] Ibid., p. 133.
[14] Ibid., p. 134.
[15] Ibid., p. 167.
[16] Richard Polt, *Heidegger: An Introduction* (London: Routledge, 2003), p. 14.
[17] "The Thought" by Frege in *Phenomenology and Existentialism*, Edited by R. Solomon (New Delhi: Harper and Row, 1972), p. 101.
[18] Edmund Husserl, *Cartesian Meditations*, trans. Dorion Cairns (The Hague: Martinus Nijhoff, 1977), p. 41.
[19] Ibid., p. 44.
[20] Ibid., p. 41–42.

SUSI FERRARELLO

THE AXIOLOGY OF ONTOPOIESIS
AND ITS RATIONALITY

ABSTRACT

Ontopoiesis is a fascinating concept introduced by Anna-Teresa Tymieniecka in the phenomenological context to express a rich range of meanings. In this chapter I would like to draw its rational bounds. Especially, what I aim to show is the relationship between *ontopoiesis*, *logos* and *antiquity*. En effect, I would like to sketch in the balance between creative energies of *ontopoiesis* and the layers of reality on which they are applied. From a phenomenological point of view, namely from an Husserlian one, Tyminiecka shows a path by which the phenomenologist marks out the meaning of this concept. Here, I am going to be focused particularly on Husserl's definition of spiritual and creative life. Effectively, in the sixth chapter of *Husserliana* XXXVII, he talks about the spiritual life by the hermeneutic instrument of *dynamis*, pointing up its rational and irrational aspects. In fact, it is not clear if Husserl conceives this kind of *dynamis* as rational at all. Therefore what I want to examine here is whether *ontopoiesis* might be enlivened by irrational sentiments like ancient *ate* or *menos* or if it is an Apollinean energy that inspires our souls.

INTRODUCTION

Ontopoiesis is a polyhedral term purposely introduced by Anna-Teresa Tymieniecka within the phenomenological context to express a rich range of meanings.[1] It has been applied by scholars on many fields of research, mainly belonging to the domain of spiritual life. Its meaning goes from a philosophical or a pedagogical to a religious one and it denotes always a very creative power of our spirit. In this chapter I would like to draw the rational bounds of this term. What I aim to show is the relationship between *ontopoiesis*, *logos* and *antiquity*.

From an etymological point of view, *ontopoiesis* means the creation (*poieō*) of being (*on*). In Greek, we have different verbs to indicate the action of doing something, for example *prattō*, *ergazomai*, *draō*. Every verb stands for a specific kind of acting: *draō* for performing, *ergaozomai* for working, *prattō* for taking care of. On the contrary *poieō* means a creative original deed. It points to a free and absolute creation of being from the origin. It depicts the action nearly in a biblical denotation. Therefore, the questioning of its meaning, above all in reference with the *logos*, is very challenging. In fact, how can an act of creation be *logikos*? How might this power give origin to different explicable layers of reality?

A.-T. Tymieniecka (ed.), Analecta Husserliana CX, 217–225.
DOI 10.1007/978-94-007-1691-9_18, © Springer Science+Business Media B.V. 2011

As Dodds wrote in *The Greeks and the Irrational*, the spiritual power of creation was regarded by Greeks as a complex concept. In Homer for example *menos* is a form of psychological interjection performed by Zeus in order to infuse power in the soul (or *thumos*) of injured hero.[2] In *Odyssea*, *ate* is another kind of sway hurled by Erinys to accomplish a *moira*. Both represent a sort of blindness through which human archetypes (performed by Homeric characters) mold their life. Next to these irrational powers, Greek spiritual pantheon took into consideration also *Apollinean* energies, as Nitezsche called them, which explained the human *kosmos* and *nomos* in a harmonic and rational way.

Therefore in this work, I would like to sketch in the balance between these creative energies and the layers of reality from a phenomenological point of view, namely from an Husserlian one. In the sixth chapter of *Husserliana* XXXVII, Husserl grapples the issue of spiritual life employing the hermeneutic instrument of motivation to figure out the rational and irrational aspects of life.

D Y N A M I S O F " O N T O P O I E S I S "

In a work entitled *Impetus and Equipoise* Tyminiecka tackles the issue of ontopoiesis from its point of origin. She sketches out it as a sort of energy which belongs to the human life and brings forth to a stream of different energies. It is nor an intellective nor a blind whole of all the subjective energies which can be freed along with a subjective perception.[3] As she wrote: "The balance of impetus and equipoise is the innermost law of logos, its First Principle. This is the first principle of becoming and beingness, the first principle of ontopoiesis of life. When we ponder the progress of life, from its initial outburst through its unfolding, we see a tremendous impetus sending infinite streamlets through life's arteries – an impetus that once in motion reinvigorates itself at each step and diversifies its proficiencies in ever new radiation".[4] Ontopoiesis and life seems to be posed on the same stage: ontopoiesis is a sort of infinite life. Indeed, it represents the beginning power of human development and life constitutes its actual result. The *logos* represents the link between the two. Indeed, it is that law which rules the living power of ontopoiesis. In every act of life there is "an ontopoietic principle that functions as an axis for the coordination of the preconscious, vitally significant life carrying operations that, although they remain at the circuits of the pre-intellectual, carry on the mute order of nature that will then nourish and launch the intentional system. This principle may be called equally (a) the point of reference for the distribution of the sense initiating operation at horizontal level (. . .) (b) the entelechially charged indicator of direction for the genetic construction of individual beingness and (c) (. . .) the measuring rod for the constructive attainment of types of complexity".[5] Thus, the ontopoietic principle works at the pre-intellectual stage and it gives sense in a sort of horizontal way. It is the initial burst of signification by which life takes its rational power and it is linked up to the logos because of its aims. As the word entelechia means, every act is directed to one goal. By *entelechia* we intend a sort of equipollence between one action and one aim. Etymologically this term comes from Greek, namely from the numeral adjective *en* (one thing), and the substantive *telos*, in English goal. Thereof the rationality of ontopoiesis lays in the scope of its ends.

Besides, interpreting Husserlian analysis of *Erfahrung und Urteil* and *Crisis*, Tyminiecka describes the historical perspective in which the impetus of *ontopoiesis* operates.[6] "It is Husserl's last work, the *Crisis of Culture*, that has sharpened the phenomenological stand on all these matters. Husserl (. . .) brought his vast inquiry into the human powers, human rationality, human cultural/scientific/technological unfolding to the culminating concept of the human life word. (. . .) Husserl sees the great difficulty into which the present world has got (. . .) with a loss of a firm point of reference for human individual existence given the 'artificiality' (. . .) of scientific cultural development. (. . .) Husserl searches for the sedimentations of rational cultural foundations (. . .) really aims at reaching the border between human constitutive activity (. . .) and rudimentary, elementary, pre-human Nature".[7] From this point of view, for Husserl *ontopoiesis* is a rational power that has to be recovered in the different layers of a human and historical present of crisis. Everyone can grasp it in order to overcome the blind significance of culture and its loss of sense. The practical life and the theoretical praxis should be addressed to recover the belief in a meaningful world. This is the sole condition by which we can find a rational principle of self-determination helpful to build up our identity according to our experience.

"Already Husserl, nutring some hints at the inventive nature of the mind, rejected radically any identification of cognition so understood with the 'constitution' in which, according to him, the objectivities are formed and devised; the referential dependence of the cognitive processes (understood as constitutive) on any assumed 'referent' lying outside of the cognitive process is disclaimed. (. . .) Objectivity as such is shown by Husserl to be precisely the effect of constitution".[8] In *Erfahrung und Urteil* Husserl "attempted to show how the last instance of *dynamis* that he invokes, (. . .) is par excellence constitutive *and* opens upon the all-embracing and fundamental 'ontopoietic schema of the self-individualization of life' ".[9] According to this analysis, in Husserlian vocabulary *ontopoiesis* corresponds to *dynamis*. It is always a power that makes actual the outburst of our life. But we wonder if there is a blind side of *ontopoiesis*, also from an Husserlian perspective. In effect, if this power is a strong and creative one, is it possible to find its irrational part? Could it be enlivened by irrational sentiments like ancient *ate* or *menos*? In the following paragraphs we seek to go through the rationality of this power.

"ONTOPOIESIS" AND *LOGOS*

In the paragraph seven of *Experience and Judgment* Husserl claims that before every movement of cognition, the object of cognition is potentially present as a *dynamis*. The Aristotelian term is employed here to explain the sense of the creative and rational movement which is before every epistemological act. As we saw before, the logical rationality of this power is strictly connected to entelechy and teleology. These two terms denote a synthetic continuous movement from potency to act, from *dynamis* to entelecheia, from potency of act to the object of knowledge. As some studies remarked, the creative power of *dynamis* can be explained *as a kind of impulsive intentionality* (Husserl, Ms E III 5) addressed always to a *telos*.[10] Intentionality is a very rich and complex word by which we may intend the act of consciousness to refer to an object. Thus, for Husserl every spur of consciousness seems to be intended to a form of self-creation and determination, e.g. to an ethical entelechy and teleology. This could mean the following statement: the creative *dynamis* of subjective life is rational since it represents an ethical power referred to a *telos*, that is to a *summum bonum*.[11] Accordingly Husserl's *ontopoiesis* could point to a principle which enlivens human world and rules all subjective creative acts thanks to its teleological direction, establishing a sort of equipollence between action and aim. A *poietic* act should be rational whenever it is addressed to an ethical goal. In the following paragraph, we would verify this inference.

SPIRITUAL ACTS

It is not clear if Husserl retains *dynamis* as rational at all. In Husserlian work it is possible to detect many references regarding the general idea of this concept. Indeed, whenever he raises the issue of active and passive intentionality, he paves the way for a questioning in *ontopoiesis*. Nevertheless, here we would like to focus just on the sixth chapter of Husserliana XXXVII, where Husserl draws the question of intentionality by the topic of spiritual acts and their motivations.

This chapter is entitled "Die eigentümlichen Entwicklungsgesetzlichkeiten des geistigen Sein. Das Reich der Motivation" For Husserl, the link between legality, ontopoiesis and motivation passes through the crucial concept of spirit. In fact it is the spirit the main subject of Husserlian *dynamis* and legality represents the key of its rationality. The spiritual life makes up the main path by which we can explicate the acts of human soul (Husserl 2004, pp. 105–106). In effect Husserl divides human soul in a spiritual and natural life. While the latter encompasses passive and a-subjective life, the former is an active form of life. Natural life is a sort of not-controlled life, dominated by instincts, emotions and habits. On the contrary spiritual life is the result of all the decisions made by a subject and its intention. Husserl calls spiritual all the acts of an intentional subject. Thus spiritual acts are those acts which belong to the pure rational and intentional side of consciousness. He wrote: "Die eigentümlichen Wesen alles Geistigen führt zurück auf das Wesen der Subjekte aller Geistigkeit als Subjekte von intentionalen Erlebnissen" (Husserl 2004, p. 104). In this text Husserl intends for *Geist* the pure, active, personal and explicable part of consciousness. It is a meaningful word which stands for the subjective life spoiled from all the empirical acts accomplished by persons and it encompasses all the intentional lived of consciousness. "Bewusstseinstätig leben, d. i. in diesem Bewusstsein von irgendetwas Bewusstsein haben, von diesem Etwas bald affiziert sein und eventuell den Affektionen passiv nachgeben, bald aber sich aktiv dazu verhalten, dazu in eigentlichen Ichakten Stellung nehmen, theoretisch oder praktisch" (Husserl 2004, p. 104).

From this point of view spiritual life can be active or passive. In fact it can be an active or passive movement of a personal subject. For passive movement Husserl means that movement which has not a genesis. "Geistige Kausalität oder Motivation ist daher etwas durchaus Verständliches und steht in jedem Schritt unter Wesensgesetzen, denen gemäß alle geistige Genesis, prinzipiell gesprochen, durch und durch verständlich zu machen ist" (Husserl 2004, p. 108). Spiritual activity implies a chosen motivation (Husserl 2004, p. 108) which compels us to deed. Rationality of our deeds lays on our motivation to deed.

Verständlich im Geiste ist alles, was eine geistige Genesis hat, alles im Geiste, was motiviert auftritt, also auf ein Motivierendes verweist. Damit ist gesagt, dass es auch Unverständlichkeit geben kann. Ich erinnere (. . .) etwa ein Knall, ein sinnliches Datum überhaupt hereinbricht. Der Knall tritt auf und verläuft im Bewusstsein, aber im exakten Sinn hat er keine 'Genesis' (Husserl, 2004, 109).

Motivation is rational when it has an intentional or explicable genesis. Husserl asserts that we have two kinds of motivations: rational and irrational. The latter is the one which cannot be immediately explained because it is empathic and affective. It is felt and not chosen. On the contrary, the former has always a rational aim and root which makes the act legitimate and rationally founded (Husserl 2004, p. 105). Accordingly we can understand the power of our action when we choose to comply that *motus* following a specific direction. Every motivation is understandable since it has a genesis.

Moreover, every subject is a permanent flow of lived. "Das Ich-Sein ist beständiges Ich-Werden" (Husserl 2004, p. 104). In this sense being a subject means becoming a subject, that is a person with different ways of being or relationships within surrounding world (Husserl 2004, p. 102). "Die Subjektivität baut sich in ihrem passiven und aktiven Bewusstseinsleben ihre Umwelt auf, die ist, was sie ist, vermöge der immer neuen intentionalen Charaktere (. . .). In diesem Prozess entwickelt sich zugleich das Ich selbst als Personalität relativ bleibenden und doch immerfort ich wandelnden Habitus, seinen Charakter mit den verschiedensten Charaktereigenschaften, bleibenden Kenntnissen, Fertigkeiten usw" (Husserl 2004, p. 105). The origin of a personal spiritual life comes from its capability to deed according to an intention which has a reckonable genesis. Thus, for Husserl every spiritual act can be rational, whenever it is actively motivated according to a genesis. Indeed, the spiritual subject can choose what to do and its activity is the result of a choice accomplished by a practical, logical or a axiological reasoning.

The equipollence posed by Husserl between rationality and spiritual *dynamis* is accounted for the subjective reasoning. According to the philosopher, our spiritual life can be *logikos*, just when it is grounded on a legitimate movement. "Motivationen der Vernunft (. . .) stehen selbst unter Fragen der Vernünftigkeit und Unvernünftigkeit, der Rechtmaessigkeit oder Unrechtmaessigkeit, und das in dem verschiedenen, durch die Grundart der betreffenden Akte und Aktsetzungen vorgezeichneten Sinn, also nach dem Sinn der Schönheit als ästhetischen Rechtmäßigkeit, der theoretischen Wahrheit als logischen Rechtmäßigkeit und ebenso der ethischen" (Husserl 2004, p. 112). The reason of an act is consistent with the motivation which compels the subject to act. The motivation can be moved

222 SUSI FERRARELLO

by an active or passive association of ideas. Following Husserlian example I can choose that it will be a good day because I saw the high level of barometer. My theoretical decision is motivated by an active association of ideas. The personal actions of the spiritual part of subjective life can be always explained and understood by the reason of the act (Husserl 2004, pp. 103–104). "Jede geistige Tatsache ist motiviert. (...) Geistige Tatsachen erklären heißt, nach Aufklärung ihres eigenen geistigen Gehaltes, also nach Analyse und Aufweisung ihres 'Sinnes', die in der einzelnen oder sozialen Subjektivität liegenden Motive" (Husserl 2004, p. 106) Thus every causality of act is linked up to a specific issue of reason (*Vernunftfrage*). The rightness of an act is founded on a reasoning belonging to the same field of the act. The rationality of a motivation is given by this reason itself. As Husserl wrote, if the act is a theoretical one its fundament is given by the pure idea of truth. The comparison between the actual deed and its fundament is possible thanks to a pure reason and its contents. In the example of a barometer and the judgment "it is going to be a good day!", the evident fundament of this proposition is given by the theoretical reason which aims to the truth and by the right connection between facts and ideas.

"Zu jeder spezifischen Aktkausalität gehört die Vernunftfrage; d. h. jede solche Kausalität kann ich in die Form einsichtiger Begründung überführen, in der in den Begründungsgliedern etwas Neues auftritt (...) Kurz jede Aktmotivation unter Fragen der Vernunft" (Husserl 2004, pp. 112–113) Considering what has been highlighted up to this point, the question we pose is the following: How can an act find its rational justification? On what is it grounded?

AXIOLOGY, ONTOPOIESIS AND RATIONALITY

We might sum up what we reached in the previous paragraphs as it follows: for Husserl spiritual life represents the active and rational life of consciousness. It is ruled by a connection of motivations that holds an understandable and explicable sense. These connections exist when the motivations are legitimate, that is when they are founded on reason.

Now, we want to understand in what consists the rational justification of the act. According to Husserl, I can deed in a spiritual and rational way when I am motivated, that is when I follow a right idea, a rationally grounded idea. But when is an idea right or wrong, true or fault? Consequently when is an act rational or irrational moved? Let Husserl's words explain:

Vornherein muss man also darauf achten, dass Mittel und Endzweck (...) intentionalen Charaktere sind, Sinnescharaktere, die man befragen, die man aufwickeln kann, und diese Aufwicklung ist ein Hervorholen angezeigter, aber allererst zu klärender Akte und Aktmotivationen (...). Hingegen gehört zum Wesen eines Endzweckes, dass er vom Ich vermeint ist als in sich wert (...). Gewiss kann ein früheres Mittel zum Selbstzweck werden, aber dann nur in einer geistigen Motivation (...), ich erkenn, dass das Mittel in sich einen Wert hat, für den ich vorher keine Auge hatte (...). Also durchstreiche ich den früheren Zweck, ich gebe ihn als minderwertig auf und erstrebe das, was bisher Mittel war, um dessentwillen als eigenwertigen Zweck (Husserl, 2004, pp. 115–116).

The sense of a spiritual act and thus its rational motivation is enclosed in the relationship between medium and aim. "I am motivated to do something" means "I want to reach an aim by a specific medium". This relationship is ruled by the hierarchy of values: "I choose to swim instead of eating because the former end is on higher stage of values than the latter"; "I use a car than a bike to go to the swimming pool because now it is a better medium for me than the others". The rational fundament of the act is based on a reasoning articulated on the relationship between medium and aim. The reason's evidence about which Husserl talks, entails an axiological evidence. The sense of motivation and rationality is rightful when it is axiologically well founded and not when it is "ein blinder, irrationaler Instinkt" (Husserl 2004, p. 116). Every rational act is always motivated and active intention.

In this sense Husserl seems to reprove that we deed rationally when we deed following an ethical code. Rationality means axiology in a certain way. "Alle ethischen Fragen sind – he wrote – Rechtfragen, sind Vernunftfragen" (Husserl 2004, p. 116). An act seems to be legitimate when it is an axiological or, at large, an ethical one (Husserl 2004, pp. 109–110). We use the verb "to seem", because here it emerges an ambiguity. It is not clear, in fact, if the axiological reason encompasses generally all the reasons (included the logical reason too) in virtue of the structures of *Verflechtung* and *Parallelisierung* employed by Husserl's ethical lectures of 1914.[12] For example here, when he tackles the reasoning of medium and aim, he seems to consider the axiological reason interlaced with the logical one (Husserl 2004, p. 118). But, when he has to apply the results he reached, he bounds them only on the axiological domain of reason.[13] Until now Husserl seems to explain the rationality of motivation by the axiological relationship of aim and medium (which is at the basis of every kind of reason) and he seems to read the proposition we are going to cite as a sort of tautology, i.e. "Alle ethischen Fragen sind Rechtfragen, sind Vernunftfragen" (Husserl 2004, p. 116). An action is ethical when it is aimed to the accomplishment of a value and this value makes the act ethical. Moreover, we can infer from tautology that an ethical act (when axiologically well grounded) is rational and a rational act is ethical (when it is axiologically well grounded). Axiology, ethics and rationality are deeply linked up in a sort of identity.

In the same way, a spiritual act is motivated when the act accomplished is rightful, *e.g.* when it is founded on a value and it is moved by it (Husserl 2004, p. 113). Thus, it could be possible to answer the questions posed before by this path. The truth of a theoretical motivation, or the beauty of an aesthetical gesture or finally the correctness of practical deed lies in an axiological hierarchy, which is exploited by a subject in every connection between medium and aim. Every choice is rationally founded on a reasoning which entails this kind of connection.

Being nearly impossible to explain rationally the meaning of the evidence, Husserl sweeps the reign of axiology in order to establish the meaning of a rational choice. Every value can be a cultural product, that can be just lived by the human community. Nevertheless, it can be the rational fundament of our spiritual life, along with all the acts and motivations we can conceive, when it is posed as a medium or aim of an intentional act according to a specific (personal) hierarchy. In a certain sense the justification of our practical life comes from rational values which bear the wisdom of our society.

The natural life is based on values that are not always intentionally comprehended. "Natur ist das Reich der Unverständlichkeit. Das Reich des Geistes aber ist das Motivation" (Husserl 2004, p. 107). The layers of spontaneous and passive or a-subjective life represent the layers of irrational and instinctive creative *dynamis* of subject. These layers can be called antiquity, since they are all that occur before every kind of analysis or questioning. It is not possible often to recover this antiquity, because it is moved by an irrational kind of motivation, "Motivation der niederen, der passive dort affektiven Geistigkeit" (Husserl 2004, pp. 107–108). Indeed, in the layers of our past we can detect the actual reason of those values which account for the motivations of our acts. This antiquity is described by Husserl on one hand as a complex of intersubjective layers resulting from our choices and lives, on the other hand as a natural, passive and asubjective life which remains out of our comprehension (Husserl 2004, p. 105).

The main path that allows us to approach the creative life of a subject is that of a spiritual and active life. Therefore the *dynamis* of *ontopoiesis* is *logikos* when it is legitimated by motivation, *e.g.* when it is ruled by an axiological relationship between medium and aim in which the correctness (or not) of the choice accomplished lies.

CONCLUSIONS

Therefore *ontopoiesis*, that is the energetic flow whereby subject builds up itself, can be always clarified because a great part of this energy can be founded. All the acts can be brought under the normative domain of a scientific axiology.

Albeit the reign of passivity is not taken into consideration, Husserl retains that it is always possible to explain what is passively lived. Every causality is rationally founded, also the passive one. When there is an evident foundation, there is an appropriation which let the subject show the reason of what he lives. The foundation of the immediate correctness of a proposition compels to understand all the historical layers that make the life of everyone understandable. The truth consists just in this ability to explain the value posed in every human goal.

Ontopoiesis is *logikos* when it can be disclosed by motivations of spiritual acts. Namely, the creative power coming out from the energy of our spiritual life can be investigated by the axiological *motus* of our will. This movement on its turn is a product of antiquity which can be just partially explained. The mystery of human nature lies exactly in the antiquity of natural life that provides the fundament of our acts. Yet, it cannot be always grasped because of the scope of our consciousness. The reason of antiquity and the power of beingness gives origin to an endlessness work of comprehension and creation.

La Sapienza University, Rome, Italy
e-mail: ferrarello.susi@tiscali.it

NOTES

[1] See: A.T. Tyminiecka, *Logos and Life* (Netherlands: Springer, 2000); Louchacova, 2006, Springer, pp. 289–331. D. Verducci, "The Human Creative Condition between Autopoiesis and Ontopoiesis in the Thought of Anna-Teresa Tyminiecka", *Analecta Husserliana*, LXXIX, 2004, pp. 3–20; F.J. Varela, *Un know-how per l'etica* (Bari: Laterza, 1992).

[2] See: Dodd, in *The Greeks and the Irrational* (Firenze: Nuova Italia, 1997), p. 13.

[3] A.-T. Tyminiecka, op. cit., pp. 39. 256.

[4] Ibid., p. 39.

[5] Ibid., p. 259.

[6] See: E. Husserl, *Erfahrung und Urteil* (Hamburg: Felix Meiner Verlag, 1985), p. 74, A.-T. Tyminiecka, op. cit., p. 259.

[7] A.-T. Tyminiecka, op. cit., p. 101.

[8] Ibid., p. 190.

[9] Ibid., p. 216.

[10] See: N. Ghigi, "Creativity in Husserl's Impulsive Intentionality", *Analecta Husserliana*, LXXXIII, 2004, pp. 553–564; Nenon, *Phenomenological Approaches to Moral Philosophers* (Dordrecht: Kluwer, 2002), p. 255.

[11] Namely we refer to the telos of God and Love as it is remarked by U. Melle, "From Reason to Love; in *Phenomenological Approaches to Moral Philosophers* (Netherlands: Kluwer, 2002) pp. 229–248.

[12] See on this point: my "Ethical Project and Intentionality in Husserl", *Analecta Husserliana* (Dordrecht, Holland: Springer, 2009), pp. 161–177.

[13] See: E. Husserl *Logische Untersuchungen*. Erster Teil. *Prolegomena zur reinen Logik*. Text der 1. und der 2. Auflage, Halle: 1900, rev. ed. 1913, hrsg. von Elmar Holstein (The Hague: Martinus Nijhoff, 1975); E. Husserl *Vorlesungen über Ethik und Wertlehre, 1908–1914*, hrsg. von Ulrich Melle (The Hague: Kluwer, 1988); E. Husserl, *Einleitung in die Ethik 1920–1924*, hrsg. von Hennig Peucker (Dordrecht/Boston/London: Kluwer, 2004). Hua XIX, p. 41, Hua XXXVII, pp. 112–113.

REFERENCES

Dodd, J. 1995. *The Greek and the irrationality*. Firenze: Nuova Italia.

Ghigi, N. 2004. Creativity in Husserl's impulsive intentionality. *Analecta Husserliana* LXXXIII: 553–564.

Husserl, E. 1985. *Erfahrung und Urteil*. Hamburg: Felix Meiner Verlag.

Husserl, E. 1975. *Logische Untersuchungen*. Erster Teil. *Prolegomena zur reinen Logik*. Text der 1. und der 2. Auflage, Halle: 1900, rev. ed. 1913, hrsg. von Elmar Holstein. The Hague: Martinus Nijhoff.

Husserl, E. 1988. *Vorlesungen über Ethik und Wertlehre, 1908–1914*, hrsg. von Ulrich Melle. The Hague: Kluwer.

Husserl, E. 2004. *Einleitung in die Ethik 1920–1924*, hrsg. von Hennig Peucker. Dordrecht/Boston/London: Kluwer.

Melle, U. 2002. Logic to the Person. In *Phenomenological approaches to moral philosophers*, ed. J. Drummond, 229–248. Dordrecht/Boston/London: Kluwer, 2002.

Nenon, H. 2002. *Phenomenological approaches to moral philosophers*. Dordrecht/Boston/London: Kluwer.

Tyminiecka, A.T. 2000. *Logos and life*. Netherlands: Springer.

Verducci, D. 2004. The human creative condition between autopoiesis and ontopoiesis in the thought of Anna-Teresa Tyminiecka. *Analecta Husserliana* LXXIX: 3–20.

TÕNU VIIK

ORIGINATING THE WESTERN WORLD: A CULTURAL PHENOMENOLOGY OF HISTORICAL CONSCIOUSNESS

ABSTRACT

The paper investigates Husserl's account of Greek Antiquity as the origin of Western civilization in order to explicate his notions of the "spiritual surrounding world" (*geistige Umwelt*) and the "spiritual objects" (*geistige Objekten*) that are the elements of this world. The spiritual objects are proposed to be interpreted as cultural forms that play a crucial role in meaning-formation processes. Whereas Husserl sees the spiritual objects as intentional objects of a special type, the paper proposes to pay attention to their functioning, as what Husserl calls "grasping sense" (*Auffassungssinn*), by means of which an intentional object is constituted. This leads to re-examining the notion of *noema* and reading it as a "spiritual sense" that is shared by the members of a common "spiritual surrounding world", i.e., to reading *noema* as a socially shared cultural form that makes an object to be identified as an object of a certain type within a particular community. Thus *noema* is not the object as it is intended, as suggested by the East-coast interpreters, but a socially shared sense which belongs to the symbolic structures of a culture, and which makes the object to be intended *as* something meaningful in a given social context. In the end of the paper these findings will be applied to Husserl's own attempt to make sense of such a spiritual object as the unique character of European culture.

Antiquity is not a thing. It is specific way of thinking about Western (or European) civilization, of Western philosophy, literature, arts, economics, war craft, etc. Antiquity is also a cultural horizon for the European Renaissance and modernity. And what is more, antiquity can be seen as providing (and at the same time hiding) the specific nature of Western civilization if viewed as its historical origin. An origin is not just any starting point; it is the source of what comes from it, and as such, it determines its essence. At least this is how Antiquity is understood by Husserl in his "Vienna lecture", held in May 1935 under the title "The Crisis of European Humanity and Philosophy".

This paper investigates Husserl's account of Greek Antiquity as the origin of Western civilization in order to explicate his notions of the "spiritual surrounding world" (*geistige Umwelt*) and the "spiritual objects" (*geistige Objekten*) that are the elements of this world. I will interpret these objects as cultural forms that play a crucial role in meaning-formation processes. If Husserl sees the spiritual objects as intentional objects of a special type, then I propose to pay attention to their functioning as what Husserl calls the "grasping sense" (*Auffassungssinn*), by means of

227

A.-T. Tymieniecka (ed.), Analecta Husserliana CX, 227–240.
DOI 10.1007/978-94-007-1691-9_19, © Springer Science+Business Media B.V. 2011

which an intentional object is constituted. This leads to re-examining the notion of *noema* and reading it as a "spiritual sense" that is shared by members of a common "spiritual surrounding world". Thus *noema* is not the object intended, but a socially shared sense which belongs to the symbolic structures of a culture – a cultural form that makes the object to be intended *as* something meaningful in a given society. As a consequence, an object will be identified as an object of a certain type according to the typification that is commonly held in a given society. In the end of the essay I will apply these findings to Husserl's own attempt to make sense of such a spiritual object as the unique character of European culture.

Husserl sees the uniqueness of Europe as being based on new type of cultural ideals that were discovered by Ancient Greek philosophers and on a new type of attitude towards life that was formed on the basis of these ideals. But before we go into Husserl's account of European culture, let us clarify a few terms he is using here. The uniqueness of Western civilization is to be found, according to Husserl, in the European *geistige Gestalt*, translated as "a spiritual shape" of European culture.[1] Husserl explains the notion of *geistige Gestalt* by the concept of *Umwelt* – the "surrounding world". As he puts it in his "Vienna lecture", *Umwelt* is not the "objective world", nor the world of mathematical sciences and physics, but the world of valid realities (*geltende Wirklichkeiten*) for the subjects belonging to a particular historical cultural community. Thus for example "the historical surrounding world of the Greeks is not the objective world in our sense but rather their 'world-representation', i.e., their own subjective validity with all the actualities which are valid for them within it, including, for example, gods, demons, etc."[2] Further he describes *Umwelt* as something essentially spiritual (*geistig*):

"Surrounding world" is a concept that has its place exclusively in the spiritual sphere (*geistige Sphäre*). That we live in our particular surrounding world, which is the locus of all our cares and endeavors – this refers to a fact that occurs purely within the spiritual realm (*in der Geistigkeit*). Our surrounding world is a spiritual structure (*geistige Gebilde*) in us and in our historical life.[3]

Elsewhere the notion of *Umwelt* is not defined as being something purely "spiritual", but is seen as consisting of both material and spiritual entities. Here, however, Husserl talks about a spiritual *Umwelt* that can be seen in my view as a layer of a wider concept of *Umwelt*. What does the word *geistig* mean in these contexts? The English translation of this adjective has usually been "spiritual" in philosophical texts, and this is also David Carr's choice here, but clearly Husserl is not talking about something ethereal, or pertaining to religious otherworldly matters. Rather, Husserl is talking about a set of representations and typifications commonly held in a society. That explains best how our *Umwelt* is **present in us**, i.e., in each individual belonging to a society, as the *geistige Gebilde*, "the spiritual structure". All social representations exist in no other way than in the minds of individuals, yet they are not private fantasies or subjective particularities, but exist as objectively valid in a given community, and as pre-given for the individuals born into this community. As Husserl claims, if social representations include acting gods and demons then there **are** gods and demons in the *Umwelt* of a particular society. But the *Umwelt* is *geistig* not because it includes collective representations about religious matters, but it would be *geistig* even if its elements did not include any representations of religious deities. Thus the term *geistig* refers here to any type of collectively held idealities that are real and valid in a given society. Husserl makes it clear that *geistige* phenomena have a historical existence, which means that they are created by particular individuals in a particular point of time, after which they can become "communalized", institutionalized, and possibly spread to other cultural *Umwelten*. Thus the adjective *geistig* refers also to this cultural and historical character of the collectively held idealities.

However, translating *geistig* as "cultural" is complicated in this text, because Husserl also uses the term *Kultur*, and in some contexts (but not always) he differentiates between *geistige* and *kulturelle* phenomena: the terms *Kulturgestalt*, *Kulturgebilde*, and *Kulturform* designate the "real", materialized, and institutionalized social activities in which *geistige* phenomena are brought to the level of praxis, whereas *geistige Gebilde* and *geistige Gestalte* designate the collective representations themselves – commonly shared ideas, ideals, norms, and other elements forming the *Umwelt* that "has its place exclusively in the spiritual sphere".[4] Thus for example Husserl distinguishes between philosophy as a *geistige Gebilde*, and philosophy as a cultural formation (Husserl uses *Kulturgestalt, Kulturgebilde, Kulturform*). The first refers to the ideas discovered by philosophers, the second to the real deeds of historically particular individuals who practiced philosophy in their real historical life, and discovered and developed these ideas in their particular "vocational communities".[5] Thus the first term refers to idealities discovered by philosophers, and the second to the real historical forms of practicing philosophy, creating and communicating these idealities in real life. In the second volume of *Ideas* Husserl discusses the examples marriage, friendship, student union, and parish community (*Gemeinde*) as cultural institutions within which we can distinguish between the level of everyday social praxis and the level of spiritual essentialities.[6]

Perhaps the most well-known discussion of the nature of the "spiritual" elements of cultural *Umwelt* comes from the "Origin of Geometry" where they are named idealities (*Idealitäten*) – as in the "Vienna lecture", but also spiritual products (*geistige Erzeugnisse*), ideal products (*ideale Erzeugnisse*), ideal objectivities (*ideale Gegenständichkeiten*), and spiritual formations (*geistige Gestalten*).[7] The use of words here suggests that *geistig* is a synonym for ideal. What kind of ideality is it, and what kind of ideal objects is Husserl talking about? Put shortly, it is again the ideality specific to the products of culture. For as Husserl explains, they do not exist as private conscious representations of a singular individual,[8] but are available and objectively given for everyone within a particular spiritual *Umwelt*, yet their objectivity does not derive from their empirical existence (i.e., from the fact that they can be given to us in a form of empirically existent physical things). Rather, they possess a specific "'ideal' objectivity (*'ideale' Objektivität*)... proper to a whole class of spiritual products (*geistige Erzeugnisse*) of the cultural world (*Kulturwelt*), to which not only all scientific constructions (*Gebilde*) and the sciences themselves belong, but also, for example, the constructions (*Gebilde*) of fine literature".[9]

In the second volume of *Ideas* Husserl distinguishes between three types of objects; (1) the "real" objects, or the objects of nature, (2) purely ideal (*ideale*) or spiritual objects (*Geistesobjekten*), such as works of literature and music,[10] and (3) "spiritualized objects" (*begeistete Objekte*) that are both real and ideal,[11] such as a printed book that "contains" a literary work or, to use a modern example, a CD which "contains" music. Thus there are two types of cultural objects according to Husserl besides the natural or "real" objects: first, pure idealities, or purely spiritual objects that can form purely ideal/spiritual formations (*Gebilde*), such as scientific conceptions; and second, "spiritualized objects" and institutionalized forms of social praxis that instantiate pure idealities.

This dichotomy between pure symbolic idealities and materialized social phenomena coincides with the main structuralist insight of the social theories of the twentieth century about the existence of symbolic networks or cultural structures that give shape to social life and all cultural artifacts. Starting from Durkheimian "forms of classification"[12] social scientists have discussed the nature of cultural idealities that give form to the empirically particular social world. Thus social psychologists claim that our actions and thoughts, individual and collective self-identification, decision-making, and habitual life-styles – are all structured by nets of social representations, stereotypes, and interpretive schemes. Max Weber called the social idealities simply ideas. Anthropologist Clifford Geertz prefers to call them symbols. I will propose to name them cultural forms, because they are cultural constructions that have been created in the course of historical cultural praxis, but once created they structure the understanding of the world and social action in a given society.

Karl Popper and John Eccles, in their book *The Self and its Brain* present a view about the nature and ontological status of the "ideal objectivity" that is specific to the purely ideal objects of the cultural world. They draw a distinction between three different ontological domains; the world of physical entities (World 1), the world of mental states (World 2), and the world of the products of human mind (World 3).[13] The elements of World 3 bear a strong similarity to Husserl's notion of ideal objects, for they also include scientific theories, contents of the works of art, etc. The point of making the objects of World 3 a separate ontological domain is to insist that World 3 cannot be reduced to Worlds 1 and 2, even though the elements of that world obviously can take the form of material bodies, and can be become the contents of human mind. However, Popper and Eccles suggest that they have independent objectivity of their own, for "as World 3 objects, they may induce men to produce other World 3 objects and, thereby to act on World 1,"[14] and, "they may have, objectively, consequences of which nobody so far has thought, and which may be discovered".[15]

It seems that Husserl, at least in the "Origin of Geometry", is in agreement with these features about the objective idealities explicated by Popper. In addition to that, Husserl emphasizes a specific feature of cultural idealities that he calls their "singular uniqueness". Thus for example the Pythagorean Theorem does not come into existence each time anew when it is uttered, expressed, used, or thought of, but its existence is singular and precedes its particular expressions and applications (except perhaps when it was expressed for the first time). "It is", Husserl argues, "identically the same in the 'original language' of Euclid and in all 'translations';

and within each language it is again the same, no matter how many times it has been empirically uttered."[16] Husserl notices that in fact language is thoroughly built up from such "ideal objects", as for example the "the word 'lion' occurs only once in German language; it is identical throughout its innumerable utterances by any given persons."[17]

Similarly, when Husserl discusses the ideal nature of *geistige* phenomena in the manuscripts to his lecture series on passive synthesis, he also mentions that language is made up of these ideal formations that have the characteristic of singular uniqueness: "In a treatise, in a novel, every word, every sentence is singularly unique, and it cannot be duplicated by a repeated reading, be it aloud or to oneself."[18] For obviously we distinguish between the treatise itself and the manifold of its uttered reproductions and written documentations. And it is because of this distinction, Husserl argues, that we are able to say, for instance, that these particular editions and printed books are *of one and the same* work.[19] The same applies to non-lingual spiritual products of the cultural world, as for example to the Kreutzer sonata:

Even if the sonata itself consists in sounds, it is an ideal unity, and its sounds are no less an ideal unity; they are not for instance physicalistic sounds or even the sounds of external, acoustic perception; the sensuous, thing-like sounds, which are only really available precisely in an actual reproduction and intuition of them. Just as a sonata is reproduced over and over again in real reproductions, so too are the sounds reproduced over and over again with every single sound of the sonata in the corresponding sounds of the reproduction.[20]

Thus we may conclude that when Husserl talks about spiritual or ideal formations of a common surrounding world, he means intersubjectively accepted and objectively valid idealities that are produced by human beings in the course of history, and thus stem from a particular psychic existence in some individual, yet they are relatively independent from their consequent subjective and objective manifestations. They are *geistig* in a sense that they constitute ideal contents of the empirically sensible expressions and ideals of social praxis. In this sense, *geistig* means the same as ideal, but not as a standard of perfection, as in the expression "this is an ideal home", but ideal as opposed to something materialized or embodied, and therefore multiplied. They are intersubjectively valid and pre-given from the point of view of an individual, and yet they are historical products that have their empirical origin – their first occurrence in an individual mind of somebody, – the event of which we are most often unable to track.

The ideal elements constitute, as we saw above, the "spiritual sphere" of the surrounding world, or as put in the "Vienna lecture", the surrounding world itself. In the "Vienna lecture" Husserl claims that the surrounding world is a wholly spiritual phenomenon, but in other texts the surrounding world is seen as the world that includes both objects of World 1 and World 3. In the *Ideas II* and elsewhere Husserl claims that *Umwelt* also contains other subjects, as well as subjectivities of a higher order, – "social subjectivities" (*soziale Subjecktivitäten*) or what is the same, communities of subjects of different levels.[21] However, we are still entitled to talk about a specific "spiritual sphere" of *Umwelt* that is constituted by spiritual idealities. Numerous thinkers before and after Husserl have suggested a concept for the repository of such symbolic idealities that constitute cultural structures, such as

the "collective consciousness" of Émile Durkheim, "collective memory" of Maurice
Halbwachs, "cultural memory" of Jan Assmann, "collective unconscious" of Carl
Jung, and, of course, their forerunner, *Volksgeist* and "objective spirit" of Herder and
Hegel. Therefore it is no coincidence that Husserl talks about the spiritual *Umwelt*
in connection with the cultures of nations. In the "Vienna lecture" he mentions the
"spiritual space" (*das geistige Raum*) of a nation, which forms the spiritual *Umwelt*
of a national society as a whole.[22] A short discussion of the surrounding world of
a nation can also be found in a Husserl's manuscript from 1933 where he talks
about a "national surrounding world" (*völkische Umwelt*) and mentions even a "surrounding world of fatherland" (*vaterländische Umwelt*) that each nation possesses.
The national surrounding world is defined here as generatively accumulated common validities constituting the whole sense of the being (*Seinssinn*) that is valid for
everyone among national fellows (*Volksgenossen*).[23]

Now that we have gained some understanding of the nature of cultural idealities
and the surrounding world constituted by these, we can return to Husserl's claims
about the uniqueness of European civilization. As said above, this uniqueness is
to be found in the *geistige Gestalt*, the "spiritual shape", which is specific to the
Western world and which influences the whole cultural formation (*Kulturgestalt*)
of Europe. Needless to say, the "spiritual shape" of Europe cannot be defined
geographically.[24] Thus Husserl says that the United States belongs to Europe,
whereas some nations and cultures that are actually situated within the geographical
domains of Europe, do not; he names Eskimos, Indians and Gypsies.[25] European
culture is trans-national: Each European nation may well have its own national
Umwelt, but "the European nations nevertheless have a particular inner kinship of
spirit (*Verwandschaft im Geiste*) which runs through them all, transcending national
differences", and in this sense Europe provides the consciousness of the common
homeland of all Europeans.[26]

The uniqueness of European culture can be recognized by the representatives
of other cultures, as well as it can be felt by Europeans themselves, according to
Husserl, as a "spiritual *telos* of European humanity" (*das geistige Telos des europäischen Menschentums*).[27] It does not mean, of course, that this *telos* occupies all
Europeans all the time, or that it is the main goal of all of its cultural institutions.[28]
It is just the essential ideal of European culture as a whole. This *telos* was discovered and established by the Ancient Greeks in the seventh and sixth century B.C. in
the course of activities that they called philosophy. From that time on, it has created
of "a new sort of attitude of individuals toward their joint *Umwelt*".[29] Husserl describes instituting this new attitude as a cultural revolution – as a "transformation of
the whole praxis of human existence."[30]

What happened there in Ancient Greece that can be seen as a creation and institution of a unique spiritual shape of European civilization? What kind of spiritual
telos did the Greek philosophers discover? – It was, as Husserl tells us, the discovery
of cultural idealities of a new type, namely the infinite cultural forms:

The spiritual *telos* of European humanity ... lies in the infinite (*Unendliches*), in an infinite idea (*un-
endliche Idee*) toward which, in concealment, the whole spiritual becoming (*geistige Werden*) aims, so to
speak.[31]

No other cultural formation (*Kulturgestalt*) on the historical horizon prior to [Ancient Greek] philosophy is in the same sense a culture of ideas (*Ideenkultur*) knowing infinite tasks, knowing such universes of idealities (*Universa von Idealitäten*) which ... bear infinity within themselves. ...[32]

Let us recall that each culture has a spiritual *Umwelt* that consists of cultural forms of all sorts, such as collective representations of deities, social norms, etc. And even mythic cultures have, as Husserl says, certain "linguistically structured 'knowledge' of the mythical powers" that govern the world according to a particular spiritual *Umwelt*, – the knowledge that is cultivated among priesthood.[33] What was specific about the idealities produced by Greek philosophers that make the European spiritual *Umwelt* different among all others was their infinite and otherworldly nature. The idealities of all cultures prior to Greeks, and of all other civilizations besides European until today are finite in a sense that they are drawn from the particular life-world itself; the "ends, activity, trade and traffic, the personal, social, national and mythical devotion – all this moves within the sphere of its finitely surveyable surrounding world".[34] All within a surrounding world of a traditional culture is taken for granted "... with its traditions, its gods, its demons, its mythical powers, simply as the actual world".[35] But the Greek philosophers, starting from the idealization of magnitudes, measures, numbers, figures, etc. (that became first applied to cosmology, and thus the first non-mythical accounts of it were created)[36] discovered a whole sphere of infinite idealities that formed as if a parallel world that differs from the empirical world in the same way as Plato's world of ideas differs from the world of shadows. The latter one is finite, yet non-persistent and constantly changing, the first is unchanging, eternal, and universal. Based on these universal idealities "the new question of truth arises: not tradition-bound, everyday truth, but an identical truth which is valid for all who are not blinded by traditions, a truth-in-itself."[37] These otherworldly universal and in this sense infinite idealities soon became applied to the other areas of life, including ethics and politics. "Hence", Husserl argues, "there are, for us Europeans, many infinite ideas ... which lie outside the philosophical-scientific sphere (infinite tasks, goals, confirmations, truths, 'true values', 'genuine goods', 'absolutely valid norms'), but they owe their analogous character of infinity to the transformation of mankind through philosophy and its idealities."[38]

Thus according to Husserl the uniqueness of European culture consists in discovering a specific non-empirical universality and in attempting to yield all aspects of empirical life to it. It was the discovery of the theoretical gaze, a new "purely theoretical attitude" (*rein theoretische Einstellung*) that replaced the religious-mythic attitude of all previous cultures.[39] This was achieved due to the infinite nature of the new cultural forms. And it is precisely as infinite that they function as *logoi* of the whole life of European culture, so that their embodiment has become the unachievable (in the sense of not being able to reach completion) *telos* of all of the cultural life, including its ethical life and politics. And indeed, even European politics today attempts to be grounded on the infinite. Thus when we go to war we do it not just in order to accomplish some particular results – empower a regime and establish another. Rather, or at least this is what we say, we go to war for infinite ideals, such as freedom or justice. And, it is commonly accepted and expected that our wars should have such grounds. Those whose wars are not based on such grounds do not belong to the spiritual surrounding world that defines "us".

Thus the task of the meaningful relating of particular deeds and individual thoughts to infinity has become an automated task of meaning-formation of Europeans – the Europeans who are, of course, defined "spiritually", i.e., by the cultural structures of their surrounding world. By means of the infinite nature of our cultural forms we cross the line between infinity and mundane finiteness. Being infinite in themselves, these cultural forms are applicable to particularities that are finite, and they make them infinite on the level of how they are perceived. We could say that our cultural forms constitute a surplus of infinity that comes to define *as what* the particular and finite is perceived. This is how a statue of god, or a crucifix, can become more than just a material finite shape. And this is also how a war can be launched in the name of the eternal peace.

Today when we are used to be much more critical of such claims about the exclusivity of Western civilization, we need to take notice that Husserl cannot be accused of claiming that the European culture *is already* based on universal truths. Rather, he claims that it is a cultural ideal of Western civilization to attempt to do so in all spheres of life. In real cultural life, he says, it is an infinite task.[40] Thus in a way he is a cultural relativist – he sees Western civilization as having its culture-specific and historically contingent beginning that establishes cultural forms that distinguish the European spiritual world from all others. At the same time, it is true that he sees the West as the only civilization that attempts such universality (and therefore he claims that what is called Indian or Chinese philosophy is essentially different from the Greek one)[41] – a claim that can be easily criticized. However, Husserl does not attach any axiological superiority to the idea of the uniqueness of Western culture.

Let us now turn to Husserl's theory of meaning in order to prepare ourselves for the phenomenological analysis of Husserl's claim about the uniqueness of European culture. We know already from the *Logical Investigations* that the intentional object (*intentionale Gegenstand*) transcends the very act of experiencing it (*Erlebnis*), as well as the immanent contents (*immanente Inhalte*) of this act.[42] This is because of the following: what we intend, or the intentional object, is essentially different from the sensational content (*Empfindungsinhalt*) that is literally contained in the corresponding act of experience.[43] In other words, the immanent contents of consciousness are not what we are conscious of; or what is the same – the appearing of the thing (*Dingersheinung*) is not the thing which appears (*erscheinende Ding*). While things appear (*erscheinen*) to us, the appearing itself does not appear (*erscheinen*), but we live through (*erleben*) it, not being thematically conscious of it.[44]

Thus there is a basic phenomenological distinction between what appears and the processes within individual consciousness that provide for this appearance. These processes make the appearance possible, or in Husserl's vocabulary, they **constitute** the intentional object. Now, what is the nature of these processes? – The ability of consciousness to be a consciousness *of* something, i.e., the ability to constitute intentional objects, is based on various kinds of syntheses that operate on its immanent contents of consciousness and produce various unities.[45] One of the most important effects produced by these syntheses is the constitution of the identity of an object, within which various visual, auditory, tactile, olfactory, gustatory, and other sensations, remembrances, future projections and expectations, and any other particular contents of consciousness, are brought together *as* being sensations, remembrances, and projections of one and the same intentional object. Thus the function of synthesis is to produce the effect of different appearances of a thing to be the appearances of one and the same thing, and as its result an intentional object is constituted.

If we look more closely into the nature of these synthetic processes we see that in the act of appearing *of* a thing, its meaning (Husserl uses both *Sinn* and *Bedeutung*) plays a decisive role in these processes, i.e., meaning seems to be a decisive element in creating the synthetic unity and thereby constituting an intentional object. Thus Husserl writes in the *Cartesian Meditations*:

The "object" of consciousness, the object as having identity "with itself" during the flowing subjective process, does not come into the process from outside; on the contrary, it is included as a sense (*Sinn*) in it – and thus as an *"intentional effect" produced by* the synthesis of consciousness.[46]

Here and elsewhere Husserl seems to suggest that the object's identity, as well as its intentional objectivity, is produced by means of its meaning.[47] It does not imply that the intentional object is reduced to its meaning, for we do not experience the meaning of an object, but the object itself.[48] Rather, Husserl argues that meaning constitutes the identity and objective validity of the experienced object, because there is no other way of being conscious *of* something than being conscious of it *as* something. And the creation of this *"as"* is the function of meaning, as Husserl suggests.

Up to this point there seems to be a general agreement among commentators about Husserl's theory of meaning, but we need to go a little further into the details. In the 5th Logical Investigation Husserl offers us an account of how exactly the consciousness *of* something by means of its meaning is achieved:

We concede that such a [sense-complex (*Empfindungskomplexion*)] is experienced (*erlebt*) in the act of appearing, but say that it is in a certain manner "interpreted" ("*aufgefatβt*") or "apperceived", and hold that it is in the phenomenological character of such an animating interpretation (*beseelende Auffassung*) of sensation that what we call the appearing of the object consists.[49]

In the *Logical Investigation* Husserl calls the element of consciousness that performs the function of unification of appearances *as* the appearances *of* one and the same object the matter (*Materie*), or interpreting sense (*Auffassungssinn*). Later, most notably in the *Ideas*, a similar function is taken over by the concept of *noema*. Husserl gives us several explanations of this concept, which has caused a lot of controversy among interpreters. Two sides have been taken about the nature of *noema*; one party of interpreters, the so-called East Coast position hold by Gurwitsch, Drummond, and others, sees *noema* as the intentional object itself, simply considered from the phenomenological point of view, i.e., as it is intended. The other party of interpreters, the so-called West Coast interpretation hold by Føllesdal, Dreyfus, Smith, and McIntyre, sees *noema* as an intermediary entity which mediates the act's relationship to the intentional object. The latter interpretation enables us to see the connection between *noema* in the *Ideas* and the *Auffassungsinn* in the

Logical Investigations. In the *Ideas* Husserl indeed introduces the notion of *noema* in connection with meaning-bestowal (*Sinngebung*) that produces the object that is "meant" (*"gemeinte" Gegenstand*);[50] and uses often the word meaning (*Sinn*) to define it. In the case of perception, for example, *noema* is its perceptual meaning (*Wahrnehmungssinn*), and in other types of acts, such as remembering, judging, or liking the "noematic correlate" of the act is to be seen as sense (*Sinn*) in the extended meaning of the word.[51] Husserl also claims that the core of full *noema* is formed by its objectifying sense (*gegenständliche Sinn*).[52] Dagfinn Føllesdal summarizes the function of *noema* as following: (1) *noema* is a generalization of the notion of meaning. (2) It is that by virtue of which an act is directed towards an object; i.e., it is the objectifying device (the device constituting the objective validity) of an intentional object; and (3) *noema* is responsible for the self-identity of an object constituted in a complex act.[53]

Thus *noema* is not a part of the physical thing, nor a part of the intended object as intended, but that which "animates" the intended object by forming its identity, and by the same move constituting that *as what* the object is perceived. Now, I wish to claim that the Føllesdalian interpretation works best if we connect Husserlian account of meaning-bestowal with the concept of pure spiritual idealities, or cultural forms. This connection is most obvious in the case of an act of perception of cultural object, or "spiritualized" material objects, as Husserl called them. Husserl himself did not systematically work through this idea, but let us look at his own example of dice.

What are the phenomenologically observable processes behind the perception of such a thing as dice? – Obviously there have to take place all the timely and spatial, internal and external, as well as kinesthetic, syntheses of the sensuous contents that are given to me looking at the different surfaces on my side of the object. In the course of such synthetic activities, Husserl claims, I constitute a self-identical object including its horizonal potentialities that are not yet actualized in perception. But how do I know that the object before me is what we call "dice"? How do I know that such a word, and consequently such a concept is applicable to this thing here? For something like dice is a cultural object; and my knowledge of such word and concept must also have a constitutive effect in recognizing this object as dice, and not just as a cube with black dots on it. We must distinguish between dice as this object here – an object that is both spiritual and real – a spiritualized object, as Husserl says, and the dice in a purely spiritual sense that functions as the grasping sense of this object as dice. It must be precisely this spiritual dice that forms the "spiritual sense" that animates the sensuous appearances, fuses with them and unites them into this particular object – the dice.[54] Otherwise I would at best recognize the object before me as a cube; or perhaps even not, if the notion of a cube is required for recognizing something as cubical. Therefore, in order to complete the analysis we need to make a step Husserl did not make: we need to transcendententalize the notion of purely spiritual objects; and to view them as "grasping senses", or what is the same, the cultural forms. As cultural forms, these purely spiritual objects function as *noemata* – as symbolic surplus of meaning by virtue of which an object is identified *as* that object.

As a *noema* dice is a cultural form that functions not as an intentional content of experience, but as a transcendental figure that belongs to the "spiritual sphere" of ideal objects of a culture that form a "spiritual structure" present in all of us – those who can recognize this something as a dice. For something can be a dice only for the community of subjects for whom this word has an identifiable meaning – subjects who share a common spiritual *Umwelt*. And what is more, as the *noema* of the dice belongs to the traditions of the society, the identification that it enables is automatic in most cases. Thus, with transcendentalizing ideal objects we arrive at a cultural phenomenology with its new account of meaning-formation. Husserl himself was perhaps on his way towards revisions of his phenomenological project in this direction, as his manuscripts about generative phenomenology and intersubjectivity suggest, but there is no room here to discuss this trajectory of his thought.

In conclusion I propose to return to his account of the uniqueness of Western civilization in order to illustrate the transcendental function of cultural forms. What happens if we apply Husserl's own theory of meaning-formation, adjusted to the analysis of spiritual objects (i.e., if we read *noema* as a cultural form) to Husserl's own history of Western civilization. What is the *noema*, or the cultural form, of this story? It must be that element of the story that causes it to make sense, i.e., the element that constitutes the identity and meaning of the story as a whole.

As we saw, Husserl argued that the uniqueness of Europe is founded on a particular historical phenomenon – the discovery of a purely theoretical attitude by the Ancient Greek philosophers: "The theoretical attitude" as he puts it, "has its historical origin in the Greeks."[55] A particular historical event has become the origin of the culture that was then – in 1930s when Husserl presented his lecture, and perhaps continues to be now, in crisis. What does it mean for something to have an origin? How does having an origin differ, if it does, from a simple starting-point? Obviously having an origin particularizes and historicizes the phenomenon by giving it spatial and temporal coordinates. But that could be accomplished by any starting point. The question of an origin goes further than that; it establishes the ground for a phenomenon, and sees it as grounded on it. Being grounded, however, does not just belong to the past. The ground is there as long as the phenomenon that is grounded by it is; that is, the ground functions as a non-historical and timeless essence of the phenomenon that is itself historical and particular. Being able to see and comprehend the ground – and this is what Husserl accomplishes in his lecture – gives us the essence of the phenomenon that we are dealing with.

Thus Husserl himself established the "interpreting sense" by finding the origin of Western culture – the origin that defines the essence of Europeanness. It is, of course, difficult to know whether cultures and civilizations have origins and essences, or whether these essences can be discovered by philosophers, but we know for sure that they can be created and successfully presented in our (world-) historical narratives. And if these narratives become widely accepted, then these essences will become commonplaces in cultural surrounding worlds, even if only retroactively attributed to the real historical beginnings.

An origin thus construed starts to function as an automated interpretative machine in the historical consciousness of a narrator, as well as in the consciousness of the listeners. In other words, positing an origin forms an active center of meaning-formation, but once posited, it starts to function as a meaning-creating agency of its own right, and as such, it determines the meaning of the story, as well as the meaning of what is narrated about – *as* revealed in this story. Thus originating a phenomenon on something means turning the origin into a meaning-automaton of a historical narrative, the procedure of which is a typical "spiritual" feature of European historical consciousness. For cultural forms are not simply what we think about, but that by means of which we make sense of what we think about. As we know, the historical narrative with its origin defined as Antiquity has long ago acquired a normative status within the Western spiritual surrounding world. We will never reach any pure presentation of this cultural form, however, because something like an origin can only be presented in terms of what is already originated. The originating activity itself will remain hidden. Applied to our case, it means that we can only approach the essence of European uniqueness from the perspective of the narrated consequences of it, and in this sense it is these narratives that give the unique European "spiritual shape" its real birth. But what we can discover is the transcendental mechanism of this birth – which is not something the Greeks did, but something that Husserl accomplished in his account of it.[56]

Department of Philosophy, Tallinn University, Tallinn, Estonia
e-mail: tonu.viik@ehi.ee

NOTES

[1]   See Edmund Husserl, "The Vienna lecture" ("Die Krisis des europäischen Menschentums und die Philosophie" – Abhandlung nr. 3 in the *Husserliana* vol. 6), pp. 318–319/272–273. The number(s) before slash refer to the page numbers in Edmund Husserl, *Die Krisis der europäischen Wissenschaften und die transzendentale Phänomenologie*, ed. Walter Biemel, Husserliana 6 (The Hague: Martinus Nijhoff, 1954), pp. 314–348. The number(s) after slash refer to page numbers in Edmund Husserl, *The Crisis of European Sciences and Transcendental Phenomenology. An Introduction to Phenomenological Philosophy*, trans. David Carr, (Evanston, IL: Northwestern University Press, 1970), pp. 269–314.
[2]   Husserl, "The Vienna lecture", pp. 317/272.
[3]   Husserl, "The Vienna lecture", pp. 317/272.
[4]   Sometimes Husserl also mixes these terms when he mentions *Geisteskultur* or *Kulturgeist*.
[5]   See Husserl, "The Vienna lecture", pp. 321/276 and 333/286.
[6]   Edmund Husserl, *Die Konstitution der geistigen Welt. Text nach Husserliana, Band IV*, ed. Manfred Sommer, Philosophische Bibliothek (Hamburg: Felix Meiner, 1984), / *Ideas Pertaining to a Pure Phenomenology and to a Phenomenological Philosophy. Second Book. Studies in the Phenomenology of Constitution*, trans. Richard Rojcewicz and André Schuwer (Dordrecht, Boston, London: Kluwer, 1989), Zusatz § 51, pp. 31/210–211. From here on cited as *Ideas II*.
[7]   See Edmund Husserl, "The Origin of Geometry", pp. 368/356–357 and 373/363. ("Der Usprung der Geometrie als intentional-historisches Problem") – Beilage III in the 6th volume of Husserliana. *Die Krisis der europäischen Wissenschaft und die transzendentale Phänomenologie*, pp. 365–386./*The Crisis of European Sciences and Transcendental Phenomenology. An Introduction to Phenomenological Philosophy*, pp. 353–378.
[8]   Husserl, "Origin of Geometry", pp. 367/356.
[9]   Husserl, "Origin of Geometry", pp. 368/356–357.
[10] Husserl, *Ideas II*, pp. 74/255.

[11] Husserl, *Ideas II*, pp. 70/251.

[12] See Émile Durkheim and Marcel Mauss, *Primitive Classification* (Chicago: University of Chicago Press, 2007).

[13] Karl Raimund Popper and John C. Eccles, *The Self and Its Brain: An Argument for Interactionism* (London, New York: Routledge, 1993), 36–38.

[14] Popper and Eccles, op. cit., p. 39.

[15] Popper and Eccles, op. cit., p. 40.

[16] Husserl, "Origin of Geometry", pp. 368/357, changing the translation of *sinnlich* – sensibly for empirically.

[17] Husserl, "Origin of Geometry", pp. 368/357. See also the discussion of language in Husserl's manuscripts to the passive synthesis: "The word itself ... is ... an ideal unity that is not duplicated with its thousand-fold reproductions" 359/12. The numbers before and after slash refer to Edmund Husserl, *Analysen zur passiven Synthesis: Aus Vorlesungs-und Forschungsmanuskripten (1918–1926) (Husserliana, Band 11)*, ed. M. Fleischer (The Hague: Martinus Nijhoff, 1966). / Edmund Husserl, *Analyses Concerning Passive and Active Synthesis: Lectures on Transcendental Logic*, trans. Anthony J. Steinbock, 1st ed. (Dordrecht, Boston, London: Kluwer, 2001), from here on cited as "Analyses Concerning Passive Synthesis".

[18] Husserl, "Analyses Concerning Passive Synthesis", pp. 358/10.

[19] Husserl, "Analyses Concerning Passive Synthesis", pp. 358/10–11.

[20] Husserl, "Analyses Concerning Passive Synthesis", pp. 358–359/11. See also Husserl's discussion of real and ideal objects (*reale und ideale Gegenstande*) in the in the manuscript number 29 in the from 39th volume of *Husserlina*, where he says that the real objects have each their unique location in time and space (*raumzeitliche Lokalität*), but "the ideal objects also have the spatial and timely manifestations, but they can manifest themselves in several time-spatial places at the same time, and yet remain identically the same." – Edmund Husserl, *Die Lebenswelt: Auslegungen der vorgegebenen Welt und ihrer Konstitution. Texte aus dem Nachlass*, ed. Rochus Sowa, Husserliana 39 (Dordrecht: Springer, 2008), p. 298. From here on cited as Husserl, *Die Lebenswelt*.

[21] Husserl, *Ideas II*, pp. 26–30/205–208. See also Edmund Husserl, *Zur Phänomenologie der Intersubjektivität: Texte aus dem Nachlaß. Zweiter Teil. 1921–1928*, ed. Iso Kern, Husserliana 14 (The Hague: Springer, 1973), p. 209.

[22] Husserl, "The Vienna lecture", pp. 322/277.

[23] Husserl, *Die Lebenswelt*, Text nr. 35, pp. 345–349.

[24] Husserl, "The Vienna lecture", pp. 318/273.

[25] Husserl, "The Vienna lecture", pp. 318–319/273.

[26] Husserl, "The Vienna lecture", pp. 320/274.

[27] Husserl, "The Vienna lecture", pp. 320/275.

[28] Husserl, "The Vienna lecture", pp. 322/276.

[29] Husserl, "The Vienna lecture", pp. 321/276.

[30] Husserl, "The Vienna lecture", pp. 325/279 and 333/287.

[31] Husserl, "The Vienna lecture", pp. 320–321/275.

[32] Husserl, "The Vienna lecture", pp. 324/278–279.

[33] Husserl, "The Vienna lecture", pp. 330/284.

[34] Husserl, "The Vienna lecture", pp. 324/279.

[35] Husserl, "The Vienna lecture", pp. 332/286.

[36] Husserl, "The Vienna lecture", pp. 340/293.

[37] Husserl, "The Vienna lecture", pp. 332/286.

[38] Husserl, "The Vienna lecture", pp. 325/279. See also pp. 334/287: "If the general idea of truth-in-itself beomes the universal norm of all the relative truths that arise in human life, the actual and supposed situational truths, then this will also affect all traditional norms, those of right, of beauty, of usefulness, dominant personal values, values connected with personal characteristics, etc."

[39] Husserl, "The Vienna lecture", pp. 326–331/280–285.

[40] Husserl, "The Vienna lecture", pp. 336/289.

[41] Husserl, "The Vienna lecture", pp. 325/279–280.

[42] Edmund Husserl, *Logische Untersuchungen [1900–1901; nach Husserliana XVIII, 1975 und XIX/1–2, 1984]*, ed. Elisabeth Ströker, 1st ed. (Hamburg: Felix Meiner, 2009) / *Logical Investigations, Vol. 1–2*, trans. J.N. Findlay, New edition. (London, New York: Routledge, 2001), V, §11, p. 387/Vol. 2, p. 99.

[43] Husserl, *Logical Investigations*, V, §14, pp. 395–397/Vol. 2, pp. 103–104.

[44] Husserl, *Logical Investigations*, V, § 2 pp. V, 359–360/Vol. 2. p. 83.

[45] Edmund Husserl, *Cartesianische Meditationen: eine Einleitung in die Phänomenologie*, ed. Elisabeth Ströker (Hamburg: Felix Meiner, 1995) / *Cartesian Meditations: An Introduction to Phenomenology*, trans. Dorion Cairns, 10th ed. (Dordrecht, Netherlands, Boston, MA: Kluwer, 1995), § 17–18, pp. 41–48/39–46.

[46] Husserl, *Cartesian Meditations*, §18, pp. 44/43; translation modified.

[47] In the *Logical Investigations* Husserl also suggests that consciousness produces an intentional relationship, and thereby constitutes an intentional object, by means of its meaning. See I, §13, pp. 54–55/Vol. 1, p. 198; and § 15, p. 59/Vol. 1, p. 201.

[48] See Husserl, *Logical Investigations*, I, §34, pp. 108/Vol. 1, p. 232.

[49] Husserl, *Logical Investigations* V, §2, pp. 360–361/Vol. 2, p. 84, translation modified.

[50] Edmund Husserl, *Ideen zu einer reinen Phänomenologie und phänomenologishen Philosophie. Erstes Buch. Allgemeine Einführung in die reine Phänomenologie*, ed. Elisabeth Ströker (Hamburg: Felix Meiner, 2009) / *Ideas Pertaining to a Pure Phenomenology and to a Phenomenological Philosophy. First Book. General Introduction to a Pure Phenomenology*, trans. Fred Kersten (The Hague, Boston, Lancaster: Martinus Nijhoff, 1983), § 88, pp. 202/213–214. From here on cited as *Ideas I*.

[51] Husserl, *Ideas I*, §88, pp. 203/214.

[52] Husserl, *Ideas I*, § 91 pp. 210/221–222.

[53] Dagfinn Føllesdal, "Noema and Meaning in Husserl," in *Philosophy and Phenomenological Research* 50 (Autumn 1990): pp. 263–271.

[54] Husserl, *Ideas II*, § 56 h, pp. 69/250.

[55] Husserl, "The Vienna lecture", pp. 326/280.

[56] This research was supported by the European Union through the European Regional Development Fund (Center of Excellence CECT).

MAGDALENA PŁOTKA

THE RECOVERY OF THE SELF. PLOTINUS
ON SELF-COGNITION

ABSTRACT

According to numerous interpretations, Neoplatonism was a recovery of the spirit
of man and of the spirit of the world. The philosophy, whose founder was Plotinus,
influenced German classical philosophy as well as phenomenology considerably.
For Plotinus, the "spirit of the world", i.e., *Logos* is real, objective being, and also
forming principle, and principle of explanation. Additionally, it is causal principle
of unity and organization, and according to this aspect, the being of *Logos* is universal creative activity (*ontopoiesis*). Following Plotinus, it is the soul of the world, and
as such it underlies reality. All beings – insofar as they participate in *Logos* – are
able to contemplate. This applies specially to man who, exiled from the Absolute,
has to return to it. Human restoration leads only through contemplation. The latter is the process directed to unity and identity between being and cognition. Due
to the contemplation, the cognizing subject identifies itself with the cognized object. According to Plotinus, insofar as acts of cognition are intentional, namely they
are directed towards external objects, unity between knower and known object cannot occur in the case of the cognition of external world. Such an unity is possible
only in the case of self-cognition. When human mind knows itself, it attains the
unity between object and subject, and the identity between being and knowing is
established.

According to Hans Meinhardt, the German historian of philosophy, Georg Wilhelm
Friedrich Hegel was to say that Neoplatonism "has discovered spirit of the man
and spirit of the world" (Gatti 2006, p. 23). However, before Hegel Plotinus' philosophy, as well as philosophical theories of many other Neoplatonists had been
regarded as the theories which deformed the original thought of Plato for a long
time. Nevertheless, since the 18th century mostly in Germany Plotinus and his
philosophical system has been appreciated as the independent and autonomous
philosophy of its own unique value.

The influence of Plotinus' philosophical ideas upon the German thought seems to
be apparent. One can even hazard the guess that the German philosophy has its roots
in Neoplatonic thought. Indeed, while exploring modern and contemporary German
thought, one can find many various references to Plotinus; the Neoplatonic concept
of "being in the world" might be compared to the Martin Heidegger's claim that
we encounter ourselves as immediately and unreflectively immersed in the world
(Thomson 2010). Also, there is similarity between Plotinus' and Heidegger's concepts of time. Additionally, Plotinus' question about the possibility of freedom in

A.-T. Tymieniecka (ed.), Analecta Husserliana CX, 241–249.
DOI 10.1007/978-94-007-1691-9_20, © Springer Science+Business Media B.V. 2011

the determined material world resembles a question which underlies Fichte's philo-sophical system, whose primary task was to explain how freely agents can at the same time be considered as a part of the world of causally conditioned material ob-jects (Breazeale 2006). Moreover, Plotinus' observation that man is able to develop himself only as being temporal is parallel to Schelling's claim that eternal potentiali-ties have to become temporal in order to fulfill and realize (Schelling 2000). Finally, we could validly and convincingly maintain that Plotinus' concept of the spirit of the world, i.e., Logos, anticipates Hegelian concept of the Absolute Spirit.

Although the problem of Plotinus' influences concerns the German philosophy in general, this article asserts that such influences can be seen within the problem of self-cognition in particular. Inasmuch as Fichte, Schelling, Hegel and Heidegger tried to express human experience of self-cognition, also Plotinus referred to the problem significantly. Therefore, the aim of this article is to explore Plotinus' idea of self-cognition. The problem of self-consciousness or self-cognition is the specific problem of modern and contemporary philosophy. The idea of Réne Descartes that knowledge about self could be the basis of all knowledge has found its developments in later theories of self-cognition in Fichte, Schelling and Hegel (Halfwassen 1994, p. 5).

According to Jens Halfwassen, late medieval theories of intellect (theories of Dietrich of Freiburg, Nicolas of Cues and Master Eckhart) anticipate idealistic theories of subjectivity. Nevertheless, the medieval theories have their sources in antique philosophy, namely in Aristotelian and Neoplatonic metaphysics of spirit (Halfwassen 1994, p. 5). One can assume the idea of self-cognition takes central place within Plotinus' philosophical system, and hence, it helps to explain not only human ambiguous position in the world, but also the metaphysical structure of the universe.

The very first paragraphs of *The Enneads* present the bundle of questions concern-ing human nature. However, Plotinus does not assume what exactly human nature is. Rather his point of departure is the mere observation of particular human feel-ings, thoughts, desires and pains.[1] All of these mental acts are human, nonetheless, can man be the compound of these mental acts, or rather is he something more than his mental acts? While considering relations between mental representations of the objects and ourselves, Plotinus poses the question: Whether the intellect while cognizing its mental representations cognizes itself simultaneously.[2] The issue is important for Plotinus in his formulation of the crucial question concerning self-cognition. If the answer to the question was affirmative, it would mean that the concept of self can be defined as a collection of mental events. But, does the man identify himself with his own mental states?

In order to solve the puzzle, Plotinus describes the following thought experiment: "Suppose the hypothetical thinker to be considering any group of mental acts, any possible content for the consciousness (. . .). Now, since the thinker is not a separate substance apart from his own thoughts, the mental states of this thinker are in some sense a part of the thinker" (Rappe 2006, p. 263), but still, they are not identified with him. Plotinus emphasizes that one should distinguish between mental acts as contents, and "the sphere". The latter is for Plotinus the metaphor of consciousness, which contains mental events as its contents. Thus, Plotinus insists that behind mere mental states there has to be a subject or substance. Why is he so certain about the existence of the subject? The fact that hypothetical thinker is able to relate to his own mental states and cognize them as well, guarantees that there is such a subject behind the mental events. As Sara Rappe points out, "the person, qua knower, or subject of consciousness, will identify with the sphere, rather than with any of its contents" (Rappe 2006, p. 266). "I am not my own mental states" – Plotinus could have said.

However, such a view lays itself open to the charge of infinite regress. If we assume the existence of a certain observer who relates to his mental events, in consequence, we state the necessity of the next observer who relates to the observer perceiving his mental events, and so in *infinitum*. The argument has its sources in the sceptic tradition, namely, it has been formulated by Sextus Empirist. Nonetheless, Plotinus does not seem to solve this sceptic puzzle satisfactorily. He only says that in order to refute the sceptic argument, one has to assume self-cognition, namely one must assume that at least intellect cognizes himself. Thus, Plotinus' question is not whether man is able to cognize himself, but rather he asks how is self-cognition possible?

Plotinus' discernment between self and his mental states leads to the question about the self-identity.[3] Such a lack of self-identity arises from the distance between the subject and his own mental events. Let us notice, that the problem of the lack of human self-identity has its sources in the constitution of human nature. Plotinus says, that since the human being is a kind of compound of his *substance* and *distinctive feature*, he cannot identify himself. Therefore, according to Plotinus, human being is not self-identified with his own substance, what means that he is not merely the substance.[4] Plotinus (1991, p. 524) contrasts human nature with the One: Whereas the latter is what it is, and it does not differ from itself and does not differ as the substance, human nature, on the contrary, is not undistinguished, rather it differs as such from itself. But, one may ask, why is not human nature undistinguished? Plotinus (1991, p. 4) replies that if it were undistinguished, why would it need a cognition or desire? Any kind of the act of cognition or desire damages the internal, united and integrate structure of the self, and therefore, man cannot be undistinguished in himself, and as the compound he cannot identify with himself. The Plotinus' account of man, as the compound of substantial identity and distinctive feature leads to the explanation of what human nature is: Since the unity of man is permanently disturbed by external acts of cognition or desire, and since the disturbance is specific for man, namely it defines man, human intellect is essentially ecstatic (Plotinus 1991, p. 4). Hence, the ecstasy defined as the intentional mental act directed toward the external empirical objects, is crucial for being a man.

Plotinus considers the problem of ecstasy while explaining the Aristotelian theory of perception, for which the concept of passive intellect is its main notion. It is worth to notice that Aristotle treats perception as the case of interaction between two elements: objects capable of acting and capacities capable of being affected (Shields 2008). Let us remind that according to Aristotle, human intellect is such a "capacity capable of being acted", namely it is the mere passivity, which is actualized by its object (capable of acting). Hence, the Aristotelian intellect becomes active only if it confronts with its object. In other words, the intellect is an active power only in its acts of cognition. The process of cognition consists then in receiving forms of the object by intellect; Stagyrite uses in the context the metaphor of a seal impressed in wax to explain this concept. However, Plotinus rejects definitely such a conception (Plotinus 1991, p. 329). Instead, he presents four arguments against the Aristotelian theory of impression. First of all, Plotinus points out that to be able to receive such an imprint the soul would have to be in some way material, and of this there can be no question. Secondly, when we perceive an object by means of sight, we see where the object is, and we direct our power of vision to that point; it is clear, Plotinus says, this is how the perception takes place. Thirdly, Plotinus notices that the soul looks outside just because there is no impression in it, and it takes on no stamp. If it did it would have no need at all to look outwards, for it would already possess the form of the object. Finally, Plotinus claims that of the impression theory of sense-perception was correct, it would mean that we do not see the objects themselves but only some sort of images of them (Blumenthal 1971, pp. 70–71).

In Plotinus rejection of the Aristotelian foundations of psychology, we might find reemphasis on an active aspect of human intellect. Again, Plotinus stresses that human intellect is defined by the acts of ecstasy. If we accepted the Aristotelian theory of cognition, how could we explain the ecstatic acts of the human soul? Plotinus (1991, p. 329) says, that the soul observes what is outside, and not impressions inside it, because they are not there.

While exploring the concept of human ecstatic acts, Plotinus describes nature as undistinguished and self-identified. Such a nature lives in unity and eternity, and it does not move. As Andrew Smith (2006, p. 198) suggests, "eternity remains in unity", what also suggests "rest". Let us remind that the idea of eternity as a being in rest has been provided by Plato's *Timaeus* (Smith 2006, pp. 199–200). Indeed, Plotinus follows Plato when he says that time is an image of eternity.[5] Nevertheless, so far as Plotinus points out that nature has to become temporal in order to develop itself (and cognize itself as well), his vision of time and eternity differs from Plato's view. Thus, whereas the Platonic man raises up from temporal empirical being towards eternal ideas, the Neoplatonic man moves in the opposite direction: from eternal unity he descends towards empirical (and temporal) world. Descent from eternity is some kind of motion, therefore, so far as rest corresponds to eternity, motion corresponds to time (Smith 2006, p. 199). Thus, the moment of the nature's descent is also the moment in which time has come to existence. In other words, ecstatic acts are the source of time (Plotinus 1991, p. 227).

Plotinus rejects the Aristotelian definition of time as the measure of motion. According to Smith, "the doctrine of Aristotle is deemed inadequate precisely because it commences from and does not rise above an empirical analysis of time, an attempt to find an adequate account of how time operates rather than to ask what it is" (Smith 2006, p. 197), whereas Plotinus hopes for answering the question concerning the essence of time. Aristotle states that time is the measure of movement of heavens circuit. Such a movement would never cease, and it seems to be a good candidate for identification with time. Thus, time is measure of sunrises and sunsets.

Let us notice that the concept of time as a measure of heavens circuit movement has been maintained by Plato and his followers as well. Therefore, as a Platonist Plotinus refers to this idea if time. Nevertheless, he proposes his own view.

Plotinus' discussion with Aristotle's concept of time begins by rejecting the claim that time is movement of heavens circuit. First of all, he observes that movement can be regular as well as irregular, and he asks how is it possible to measure something which is not regular (Smith 2006, p. 207)? Moreover, he notices that the movement of heavens circuit can lapse, but time cannot. According to Plotinus, if the heavenly circuit should cease to move (and hence all physical movement cease) even its rest would be in time (Smith 2006, p. 211), and this rest would be measured by soul.

Plotinus' conclusion is the thesis that time is not a movement of the world, but rather it is a movement of the soul. Precisely, time is the life of the soul. According to Plotinus, time exists on two levels; on the one hand, it exists on the level of soul's life, on the other hand, it can be perceived in the physical world, when worldly things exist "in time". And since world exists in time, and since time is soul's life, as Plotinus concludes, the world exists in the soul (Smith 2006, p. 210). Thus, unlike Aristotle and Plato, Plotinus shows that time is internal to the soul, not external. He stresses that "we should not imagine the time as something being outside the soul, and similarly, we should not imagine the eternity as something <out there>" (Plotinus 1991, p. 227).

However, in Plotinus' view time is not only the life of the soul, but also it has its origin in the soul. Plotinus explains that as soon as nature desires "something more" than presence and stillness, it has made itself temporal. It is so, because, according to Plotinus, only being in time guarantees an authentic human experience. As Plotinus says, understanding what time is helps us to understand what we are (Smith 2006, p. 210). Hence, only in its ecstatic acts, the soul undergoes the changes, and within these changes it becomes temporal. In consequence, within becoming temporal, the soul creates the empirical temporal world as well.

While remaining in the unity and rest nature does not desire anything, and hence, it is self-sufficient. And the crucial question is: Why does nature want to disturb its unity and stillness by its ecstatic acts? And why does nature want to abandon its eternity and become temporal? According to Plotinus, the source of the soul's descent as well as beginning of time is nature's desire of mastering itself and belonging to itself. In order to do that, it has decided to achieve "more than presence" and has set itself in motion (Plotinus 1991, pp. 227–228). According to Blumenthal, "the soul must descend (. . .), but it does so by its own dynamism: it comes down by reason of its power to organize subsequent being, starting from an impulse of its own free will" (Blumenthal 1971, p. 5). Therefore, the source of the soul's descent is some "restless power", as Plotinus says, inside nature, and due to this power, the nature wants to spread itself in ecstatic acts.

Let us notice that this movement of nature can be regarded as a metaphysical explanation of human freedom. Georges Leroux, while considering the concept of freedom in Plotinus' thought poses the question: "Does the soul descend voluntarily, that is, does it freely move toward the lower states of its realization, and in particular toward the body?" (Leroux 2006, p. 295). But it seems that it would be better if we claimed that soul moves toward the lower states of its realization, because of its freedom. In other words, the process of emanation is entirely free process; the soul emanates and thus it moves towards lower and external states. Such movement is also the manifestation of freedom.

Plotinus' emphasis on the ecstatic character of the soul aims at the understanding of what the human being is. As he points out, this ecstatic property of man is not the property of man considered as a whole compound, but rather it is a property of mere intellect. Therefore, as Plotinus puts it, our intellect is our truest self (Blumenthal 2006, p. 96). This intellect is defined as διανοια, the real human intellectual capacity, the power of reasoning and judgment, with which Plotinus often says we are to be identified with Blumenthal (1971, p. 43). It may thus be regarded as the meeting place of the sensible and intelligible worlds (Blumenthal 1971, p. 111), and this is the psychic level when human concept of the self is being constituted.

Plotinus shows that in order to see and understand our intellect as our truest self, one should purify himself of all desires, thoughts, memories and material body. After such a purification, he would see himself as a pure and immortal intellect (Plotinus 1991, pp. 336–343). Hence, the first step of self-cognition is to recognize oneself as the intellect. In order to make this thesis clear, Plotinus creates the second part of his "hypothetical thinker" thought experiment: let us remind that hypothetical thinker was supposed to consider all his mental acts and contents of consciousness: "No matter how diverse the causes that initially produced these elements in the external world, as for the contents of the sphere considered solely as objects of thought, it is true to say that their productive cause is singular, namely, the hypothetical thinker himself" (Rappe 2006, p. 263). This is the very crucial moment in Plotinus' work, because he claims that we are able grasp the reality as it appears in our consciousness. And if we concentrated on our consciousness events, it would turn out that our consciousness is the "productive and efficient cause" of its contents. Nevertheless, it does not mean that the empirical material world is somehow dependent on our consciousness. Plotinus does not maintain anything like this. He only says that there are two ways of perceiving the world: as the macrocosm and the microcosm. "The macrocosm is a publicly available world, inhabited and experienced by countless sentient beings, each with a diverse perspective. The microcosm is that same world, seen from within the confines of an individual consciousness" (Rappe 2006, p. 262).

Since Plotinus claims that consciousness contents can be individuated in a complete independence of empirical objects, this thought experiment might be interpreted as a kind of internalism: mental states have their only cause and source in thinking intellect. However, how Plotinus can claim both that the human intellect in his very nature follows external objects in cognition, and the cognized world is just the totality of consciousness contents? Let us notice that Plotinus makes use of special notions of "externality" and "internality", which are crucial to his concept of self-cognition. He tries to show, as Rappe puts it, "how the soul constructs a (...) sense of self when it conceives the world as outside of the self; (...) the thought experiments reveal a way of conceiving the world as not external to the self" (Rappe 2006, p. 265). Thus, the world is not external to the intellect, it is rather internal: worldly objects are perceived as the contents of consciousness. Therefore, since the world is internal to the human intellect, the latter cognizes himself in his ecstatic acts.

Since borders of myself are simultaneously the borders of the world, self-cognition would be cognition of the world, which is identified with the self. If we look closer to the Plotinus' notions of internality and externality, we might ask whether there is any kind of external world in a strict sense, totally independent from the intellect. Plotinus states that the world of matter is such a world, because matter would never become internal to the intellect. Matter, as the last emanation from the One cannot be regarded as any being, because the latter, for Plotinus, is only that what is intellectual. On the contrary, matter is the end of the intellectual world, and therefore, it should be regarded as a nothingness. Plotinus compares matter to the mirror: the same as the mirror is indispensable for reflections, matter is indispensable for reflections of real beings. Matter as the mirror is not visible itself, it is only visible due to its reflected images of real intellectual beings (Dembińska-Siury 1995, p. 54).

The concept of the self which is identified with the mere intellect is exactly a result of Plotinus' doctrine of matter. It is so, because the statement applies to human body as well: since the human body and its organs are material, they cannot be regarded as the parts of the self. While describing the process of perceiving, Plotinus notices that we perceive only the external objects. But he asks about perception of the internal processes of an organism. Do we perceive our bodily experiences as internal to ourselves or rather external? Plotinus distinguishes power responsible for the perception of external objects from the power of perceiving what goes on within us. Plotinus talks of the power of internal perception. However, all sensation is of externals because the affections of the body which such a faculty cognizes are also external to the soul (Blumenthal 1971, p. 42). Thus, according to Plotinus, every time we experience any kind of "bodily disorder", we used to experience it as if it came from outside (Plotinus 1991, p. 309). Therefore, the body is not a part of myself, but the part of the external – material world (Plotinus 1991, p. 367). "I am not my body, I am only my intellect" – Plotinus might say.

The specific notion of externality in Plotinus' thought is a result of habitually identifying with the body (Rappe 2006, p. 265). Let us stress, following Rappe, that "gradually the boundary that separates self and world is erased, when the demarcations of selfhood are no longer around the body, but around the totality of any given phenomenal presentation" (Rappe 2006, p. 265). In consequence, "every cognizable fact about the knower's identity as subject is converted to the status of an external condition: body, personality, life history, passions, and so forth" (Rappe 2006, p. 266). Within Plotinus' works, these qualities have received the status of mere modifications of the self. Behind the modifications, there is an authentic self. Cognition of the authentic self is for Plotinus the proper self-cognition.

However, having established our self as the intellect, Plotinus goes one step further and asks about the principle of the unity of the self. Our intellect has been defined as discursive potency, namely as διανοια. Moreover, since its movement has been defined as circular which means that the intellect moves from intelligible rules (Νουσ) to the sensible world and back, it is not united and thus not one. In consequence, the intellect has been also defined as Dyad: it is duality of a cognizing subject and a cognized intellect, it is also indefinite and unlimited. Intellect's position between intelligible and sensible world, as well as its other attributes are precisely the reasons of deficiency of its unity. Therefore, Plotinus poses the question about the grounds of self-unity: On what grounds do we cognize ourselves as one?

Let us emphasize that relation to ourselves is being constituted in reciprocity of thinker and thought. The unity of self-thinking is not absolute unity, because, as Plotinus says, the unity in multiplicity is primary the multiplicity. Thus, Plotinus' aim is to introduce some kind of the third element which would unite thinker and thought in an act of self-cognition. It has to be the principle of both unity and multiplicity, and as such, it would be the ground of unity of the self in self-cognition (Halfwassen 1994, p. 9).

Plotinus answers that we perceive the unity of ourselves in the light of Νουσ (Halfwassen 1994, p. 22). How do we discover presence of Νουσ within us? Plotinus shows two ways of our participation in Νουσ: firstly, Νουσ is the power which unites multiplicity of our thinking, namely it unites variety of λογοι. And secondly, we become Νουσ through intellectual insight. According to Halfwassen, there are two concepts of self-cognition which are joined to these two ways of participation in Νουσ. Therefore, self-cognition can be regarded either as the cognition of the essence of discursive thinking, or as an intellectual self-insight which relies on intellectual turn to Νουσ with complete omitting discursive potencies of intellect (Halfwassen 1994, p. 22). Plotinus definitely chooses the second option. Thus, the man does not cognize himself as a discursive thinking which is aware of its reception of external truths. Preferably, not only he cognizes himself as a principle of his own unity, but also while participating in Νουσ he ceases to be indefinite and unlimited.

To sum up, let us stress that Plotinus claims that the very nature of human being consists in ecstatic acts. Because of these intentional acts, directed towards external objects, the man cannot be self-identified. Thus, transgression describes human condition in the world, and it derives from freedom. While transgressing his unity and self-identity, man becomes temporal. Plotinus would agree that only being in time helps the man to develop and cognize himself. Therefore, in order to cognize himself, the man has to be in time. Since ecstasy is the intellectual property, Plotinus claims that intellect is the human truest self. Plotinus' "hypothetical thinker" thought experiment has led him to the conclusion that the world is internal to the man. This applies to the body as well, which is just a part of external and empirical world. And as far as the man is able to recognize himself in his pure intellect, and as far as he knows that the world, time, his body, memories, personality and mental events are only modifications of himself, and he is something behind all these qualifications, then he would cognize himself. This pure intellect has been defined by Plotinus as διανοια, nevertheless the principle of its unity is not himself, but Νουσ understood as an intellectual intuitive insight.

Department of the History of Ancient and Medieval Philosophy, University of Cardinal Stefan Wyszyński, Warsaw, Poland
e-mail: magdalenaplotka@gmail.com

NOTES

[1] "Pleasure and distress, fear and courage, desire and aversion, where have these affections and experiences their seat?" (Plotinus, 1991, p. 3).

[2] "Are we to think that a being knowing itself must contain diversity, that self-knowledge can be affirmed only when some one chase of the self perceives other phases and that therefore an absolutely simplex entity would be equally incapable of introversion and of self-awareness?" (Plotinus, 1991, p. 364).

[3] According to Blumenthal, there is another explanation why Plotinus had problems with answering the question "who we are": "Our soul does not descend completely, but a part stays up in the intelligible world" (Blumenthal, 1971, p. 6).

[4] "This is a compound state, a mingling of Reality and Difference, not therefore reality in the strictest sense, not reality pure. Thus far we are not masters of our being; in some sense the reality in us is one thing and we are another. We are not masters of our being" (Plotinus, 1991, p. 524).

[5] "For Plotinus himself one important and central element of this is the linking of eternity with the unchanging and transcendent intelligible world and time with the physical world of becoming. Clearly Plato lies partly behind this" (Smith, 2006, 196).

REFERENCES

Blumenthal, Henry J. 1971. *Plotinus' psychology. His doctrines of the embodied soul.* The Hague: Nijhoff.

Blumenthal, Henry J. 2006. On soul and intellect. In *The Cambridge companion to Plotinus*, ed. L. Gerson, 82–104. Cambridge: Cambridge University Press.

Breazeale, Dan. 2006. *Johann Gottlieb Fichte.* [Online] Stanford, California: E. N. Zalta, ed. *The Stanford encyclopedia of philosophy (fall 2009 edition).* http://plato.stanford.edu/archives/fall2009/entries/johann-fichte/. Accessed 12 Mar 2010.

Dembińska-Siury, Dobrochna. 1995. *Plotyn.* Warsaw: Wiedza Powszechna.

Gatti, Maria Luisa. 2006. Plotinus: The Platonic tradition and the foundation of Neoplatonism. In *The Cambridge companion to Plotinus*, ed. L. Gerson, 10–37. Cambridge: Cambridge University Press.

Halfwassen, Jens. 1994. *Geist und Selbstbewustssein. Studien zu Plotin und Numenios*, Mainz, Stuttgart: Akadmie der Wissenschaften und der Litteratur, Franz Steiner Verlag.

Leroux, Georges. 2006. Human freedom in the thought of Plotinus. In *The Cambridge companion to Plotinus*, ed. L. Gerson, 292–314. Cambridge: Cambridge University Press.

Plotinus. 1991. *The Enneads* (trans: from Greek by McKenna, S.). London: Penguin Books.

Rappe, Sara. 2006. Self-knowledge and the subjectivity in the "Enneads". In *The Cambridge companion to Plotinus*, ed. L. Gerson, 250–274. Cambridge: Cambridge University Press.

Schelling, Wilhelm Johann. 2000. *The ages of the world.* Albany, NY: SUNY Press.

Shields, Christopher. 2003. *Aristotle's Psychology.* [Online] Stanford, California: E. N. Zalta, ed. *The Stanford encyclopedia of philosophy (winter 2008 edition).* http://plato.stanford.edu/archives/win2008/entries/aristotle-psychology/. Accessed 12 Mar 2010.

Smith, Andrew. 2006. Eternity and time. In *The Cambridge companion to Plotinus*, ed. L. Gerson, 196–216. Cambridge: Cambridge University Press.

Thomson, Iain. 2010. *Heidegger's aesthetics.* [Online] Stanford, California: E. N. Zalta, ed. *The Stanford encyclopedia of philosophy (spring 2010 edition).* http://plato.stanford.edu/archives/spr2010/entries/heidegger-aesthetics/. Accessed 12 Mar 2010.

CEZARY JÓZEF OLBROMSKI

SOCIAL CONNOTATIONS OF THE CATEGORY OF THE «NOW» IN THE LATE WRITINGS OF EDMUND HUSSERL VS. J. DERRIDA AND B. WALDENFELS

"The concept of time, in all its aspects, belongs to metaphysics, and it names the domination of presence. Therefore we can only conclude that the entire system of the metaphysical concepts, throughout its history, develops so-called 'vulgarity' of the concept of time [...], but also that an other concept of time cannot be opposed to it, since time in general belongs to metaphysics' conceptuality."[1]

ABSTRACT

The author analyses the late Husserl's phenomenology of time giving a new interpretation of the «now» which is based on statement that the «now» should be expressed by non-temporal terms. According to the author, this process of temporal devoid is present in the very late Husserlian considerations on *lebendige Gegenwart* and this process is threefold. The third level of freeing the «now» from temporality is "being of the form of the pure non-temporal «now»". Derrida's temporality of origin discloses the simultaneousness of objective ontology and objective consciousness. Dialectics of conversion of subjectivity into temporality, which is present in Derridean philosophy, requires a direct and an original insight in the difference. The Husserlian solution of the problem is reduced by the author to an explanation of the «now» as a noun. According to the author, this interpretation overcomes Derridean apories. Also, the paper shows the basic significance of the category of the «now»—that is devoid temporality on the most basic level—in the constitution of the consciousness ot time. Double character of the «now»—temporal and non-temporal—is a source of a cognitive tension but also it is a level of sociality.

The philosophy of pure consciousness—the Husserlian phenomenology of time—is strictly related to the notion of time as the core of the consciousness. This paper shows the basic significance of the category of the «now» in the constitution of the consciousness of time. The most essential issues of this topic are presented in the analysis of the consciousness. *Lebendige Gegenwart* is described as a cognitive tension released by the depiction of the constitution of the flow of time. This flow is temporalized within a-temporal surroundings.

A.-T. Tymieniecka (ed.), Analecta Husserliana CX, 251–265.
DOI 10.1007/978-94-007-1691-9_21, © Springer Science+Business Media B.V. 2011

A   P O L E M I C   A G A I N S T   D E R R I D E A N
A N T I - P R E S E N T I A L I S M

A pre-social and a primordial sphere is not a result of a reflection although it means that it is not a domain of intersubjectivity. In other words, the origin or the genesis of the transcendental "I" cannot create itself. Jacques Derrida explains this problem in the introduction to his book *The Problem of Genesis in Husserl's Philosophy.*

> Without recourse to an already constituted logic, how will the temporality and subjectivity of transcendental lived experience engender and found objective and universal eidetic structures?[2]

The eidetic reduction and the transcendental reduction lead us to the suspension of our knowledge about facts. This suspension conducts us to define the internal consciousness of time on an eidetic level. Thus, the temporality is a point of the phenomenological arrival. There are the phenomenological rudiments.

Derrida argues that the phenomenology of time ought to stay on a non-temporal level in its attempts of taking up temporality—it seems to be directed against Husserl's intentions to portray the phenomenology as a dialectics of temporal "moments" between phenomenology and ontology.

According to Derrida, the Husserlian phenomenology is reduced to a dialectical depiction of temporal "points" in its relation to the phenomenological and ontological background. Husserl's source temporality *a priori* synthesises the existence of time with the constituted sense of time. Husserl does not intend to discuss the problem of temporality any further because he considers it to be an eidetic structuralization, and additionally points out at the non-temporality of this problem. In other words, one may recall an emblematical opinion of Derrida: Husserl is still a prisoner of the classic tradition. This tradition reduces an individual to the isolated cases of the universal history of the universal conception of man. In this configuration, it seems that temporality manifests itself in an actual eternity existing within *periechon*—a container like this would include an internal consciousness of time.

Husserl is the first philosopher to change the grammatical qualification of the category of the «now». He defines the «now» as a noun. The «now» is a noun not only as a specific term of philosophy of time, but also as a part of speech, in which we ask a question "what?" not "when?". The «now» is not a noun because it is a category which is added to our philosophical vocabulary. The Husserlian «now» is a noun because, substantially, it answers the question "what" or "who"? For this reason, we cannot find any contradictions in the evolution of the «now» in the works of Husserl. What we can see is only how he shifts in the categorizing of the «now». In my opinion, one of the most important breaking points in Husserl's work is giving up his diagrammatic depiction of the theory of time. The category of the «now» which is constituting time is a background of an intentional act which is characterised retentionally and protentionally. Giving up a retentional↔protentional time is not actual but methodological.

According to Husserl, the flow of time is represented by a sequence of the consecutive and successive points of time. In his theory of time, the future is later than the past, the past is earlier than the «now». The past and the future, on the one hand,

and the «now», on the other, do not possess the same nature: the «now» is not a border between the past and the future, but the only present time of the creator of time. The main difficulty lies in the fact that this sequence cannot be characterised in a temporal terminology for two reasons.

(1) The «now» (also in the retentional↔protentional setting) is the smallest "unit" that the consciousness constitutes.[3]
(2) Consciousness cannot measure constituted time by means of the «now» defined as category of time.

The «now» does not answer to question "when?". Well, the «now» must answer other questions than "when?". According to Husserl, (in his definition of the «now» in the retentional↔protentional setting as well as in *lebendige Gegenwart*) the «now» answers to question "what?". Let me use a birth of individual consciousness as an example (supposing that an individual is not the eternal monad). I am not taking into consideration the time as the factor which is constituting my universal sense of the world—the sense which relates to my retention–protention, to my consciousness of the flowing time, as well as to the socialised and the inter-subjective time. I am only interested in a feeling of time in its specific «now».

Let's analyse the problem of the actual phenomenon (a subjective aspect) and *a priori* nature of consciousness (an objective aspect). One of the main objectives of Husserl is to try to define as well as to precede an experienceable—but not yet predicative—way of *Zeitigung*: it can be called temporality independent from consciousness. Derrida, who was inspired by Husserl's phenomenology of time in *The Problem of Genesis in Husserl's Philosophy,* is searching atemporal *a priori* which is an underlying foundation of phenomenology. However, it does not imply returning to a substantial originality of the subject in relation to consciousness. In other words, Derrida claims that what an origin is—is not substantial.

Jacques Derrida's pre-predicative absence—analysed in a temporal context—becomes complicated in the retentional↔protentional context of category of the «now» and becomes complicated in the atemporal infinity in Husserlian understanding. Pre-predicative origin of Derrida's philosophy differs from a-temporal ones and becomes limited to *lebendige Gegenwart* origin of Husserl phenomenology of time. Derrida's temporality of origin discloses the simultaneousness of objective ontology and objective consciousness. Dialectics of conversion of subjectivity into temporality, which is present in Derridean philosophy, requires a direct and an original insight in the difference. The dialectics of being and sense goes hand in hand with the dialectics of being and time. The essence of this issue lies in the fact that primary temporality of passive pre-constituted being is more important than immanent temporality of consciousness. That is, primary temporality of passive pre-constituted being is mixed with being and thus it precedes every phenomenological temporality, which is a background of this pre-constituted being.

We can quickly notice the bipolarity of such structure:

(1) (a) The existence in the *Nullpunkt* is the pure (pre-temporal) and unconditional reception of reality, and (b) we deal with a reference to the *Nullpunkt* as a basis of the interpretation. On the one hand, the consciousness is blind because it

does not known retentional↔protentional perspective, on the other hand, the consciousness outside retentional↔protentional time is the intentional correlate for the consciousness of time.

(2) The objectivisation of the first level takes place outside the time; the objectivisation of the second one takes place above the time. The second kind of the objectivisation exceeds the monolinear pattern of a sheer succession in «now» of the acts since each reference to primordial temporality supposes a continuity of an action. The action is deprived of a limited perspective of retention-«now»-protention and is potentially referred to a "future" by «now»; moreover, an action does not take place in the «now» noticed in the prism of *before*.

(3) (a) The consciousness (as a *being of the form of the pure non-temporal* *«now»*) is anonymous and is not non-individual (as only individual consciousness can enter the reality). The creation of the consciousness of internal time is a derivative process that leads to the consciousness, which is inherently atemporal—which means that the first «now» is recognised only into perspective of *before*. An experience of the first «now» is a temporal *unconscious*.[4] We can say so because the consciousness has not experienced the internal time in the retentional↔protentional perspective, the consciousness was not motioned in the objective time. Also, an experience of the first «now» is temporally *conscious* because the consciousness participates in reality in a pure way and this process takes place without the participation in the temporal character any «now». The consciousness as the pure *Einfühlung* of reality wins the memory of reality; and it wins the internal and temporal perspective of social communication. Simultaneously, the consciousness loses a part of its nature (namely—its atemporal character) as a result of the transcendental reduction and the pure consciousness appears as absolute). (b) There is an existential tension (in *being* *the form of the non-temporal «now» between before and after*), which appears at the moment when the consciousness recognises «now» in the context of the future. There is the existential tension between non-temporality in pre-cognition and cognition into perspective of retention-«now»-protention, between *before* and *after*. The «now»—as a basis for the temporal «now»—exists and the *before* and the *after* fix its borders.

We may therefore say that the temporal «now» is the product of the intentiveness to the non-temporal «now», that it is essentially and necessarily an identifying synthesis. Time is a result of individual *Zeitigung*. The temporal «now» is a result of constitution of the pre- and beyond-temporal «now». But this can only be possible because the retentional↔protentional structure constituting time in the proper sense, and mental living as inherently temporal, is objectivated as the identical time at each intermediary level of constitution. According to Kersten, the process of "self-temporalization", the process of "self-constituting" of transcendental mental living as past, present, and future in the manner described does not, however, reconstitute itself or multiply itself.[5] That is to say, that at the level of the oriented constitution peculiar to time, transcendental mental life is transcendentally temporalized, with the identical structure of a transcendental intensity to time. Given schema of a transcendental mental life-process with respect to process, as a whole is objectivated as an unflowing frame consisting of future, present, and past. The current extent flows through this frame so that the relation of any portion of the extent to each part of the frame changes continuously. The tense of the posited characteristic of each portion changes continuously from "will be later", to "will be soon", to "is", to "was recently", to "was earlier", to "was still earlier" etc. The change in tense of the positioned characteristics of the extents is a consequence of the flow out of the future, through the present into the past. If it is not the case, the mental life-processes would be nothing but a continuous recurrence, hence would provide no basis for building up the real and the objective world within which mental life-processes find them. It is the condition for my transcendental life. However, the change/flux in tense is only a necessary but not a sufficient condition for existing in the world. It is true, but the mental construction of time, or in other words, the transcendental mental living which constitutes «now», disappoints when we try to define the pure «now». This Husserlian construction does not take into consideration a pure concept of flowing time. The unity of an enduring extent of any mental life-process is possible only in so far as it presents itself in the correlation with something identical presented as well as through a multiplicity of different temporal extents continually changing in the orientation and tense. The consciousness of the internal time relates to the present (the consciousness of time and its reference to the wider, retentional↔protentional context is built by the sense of «now») but in the contrary—the social time is built by the reference to the past and the past experience. The centre of gravity of immanent temporality moves into the past. But the past, although being temporal, does not impose its own temporalization on the «now». The «now» constitutes the temporality into the perspective of the past, and the «now», as a moment, cannot be separate from time, because the pre- and temporal «now» does not answer to the question "when?".

Let us consider the following question: is an ideal sphere—which is purposefully given by a genetic interpretation of what we recognise as a sphere of objective validity—temporal or a-temporal? If it is indeed a temporal and original sphere the subjectivity cannot be simultaneously constituted in the present. If it is temporal it is historical and psychological. In that case the constitution is reduced to the formal norms.

This kind of temporality in an original sphere in Husserl's *Logische Untersuchungen* and in *Philosophie als strenge Wissenschaft* is more clearly showed than in *Vorlesungen zur Phänomenologie des inneren Zeitbewusstseins*, but we can notice a discontinuousness in reply to this question. There is a difference between the objective and subjective temporality. The objective temporality depends on a temporal constitution taking place in the individual (constitution) of time. This objective temporality can be only accomplished when consciousness constitutes its beginning in a temporal sense. An attempt at finding the beginning in the opposite direction—in terms of the becoming in an ontological sense—does not bring the required results apart from the necessity of being. In other words, Husserlian *lebendige Gegenwart*—in contrast to Derrida's dialectics of difference—is given as a source and it is constitutive.

From the beginning of Husserl's analysis of time and his study of phenomenology in *Philosophie der Arithmetik*, his theory has become a call for searching for the secondary basis of the transcendental philosophy. Derrida changes this "Platonic" method of a philosophical investigation and finds it in the dialectics of genesis. The "ineradicable" *aporiae* of the transcendental *Schein*, which Derrida radicalises, shows that the evidence is always given in person as something.[6] The opposition between the transcendental and the mundane, non-presence[7] and presence is the non-*arché* and non-*thelos* origin. According to Lawlor,

the metaphysic of presence is a discourse that presupposes a sense of being, the sense as presence.[8]

This unfortunate anticipation in the sphere of dialectics gives as the beginning without the beginning; in other words, it presents itself as a dogmatism of presence. A change of Husserlian ontic presence for the presence as origin, is in fact, only a verbal transubstantiation. Derrida did not only fall into a temporal presence, but also lost his sight of self-evidence in time. Husserl transcendental method led to some language difficulty to express self-evidence. Derrida—being convinced of the impossibility of self-evidence—has accepted the method of a dialectical and recurrent approach to self-evidence. In other words Derrida, has combined the metaphysic of presence with the self-evidence by means of an infinite chain. One end of this chain spreads out in the subjective evidence, the other one vanishes in a quasi-sensitive and infeasible self-evidence. Certainly, the difference between the radical discontinuity (and subjective retention) and the objective time, which exists without any intervention of a subject, is of no importance.

Husserl distinguishes between the psychological, objective (*sic!*) and phenomenological understanding of time. This rudimentary statement put in the context of consideration about non-conditioned foundation of phenomenology is quite surprising. However, if you have in mind the socially conditioned Waldenfels' concept of *Zwischenreich* this statement suddenly becomes entirely clear.[9] Is seems that the most important argument against Derridean metaphysic of presence can be explained by the fact that Derrida assumes that the empiricalness is dialectically mixed by the source juxtaposition of the ontic continuity with the temporal discontinuity. If Derrida treats the temporality as—*activeness*—derivative of intentionality and, at the same time, as—*passiveness*—subject of sensual perception, it falls in *aporiae*. Every experience of the external world processes in internal stream of time consciousness, which has not beginning and the end. Derrida gets bogged down in details of time, it means that he loses the beginning and the end of retentional↔protentional time. But to get bogged down in details and to know that there is no beginning and the end, these are two different matters. The same starting point—Husserl and Derrida consider in what way individual act of consciousness, so to say specific and limiting temporality, can be a grounds of depiction of infinity of time—leads to so much different results. Husserl in point of view of individual consciousness (the late Husserl) extends this schema to temporal horizon of the participation of latent monads, while Derrida writes:

What does this flux of lived experience mean, taken in its infinite totality and nevertheless distinct from every piece of lived experience in particular? It cannot be lived as infinite. On the other hand, its infinity cannot be constituted from finite lived experience as such.[10]

Let us examine the Husserlian senses of infinity. Husserlian phenomenology of internal time uses the term "infinity" as at least a threefold meaning.

(1) *Infinity is an extension of the protentionality of the «now»*. In this sense infinity is a synonym of the lack of knowledge about future events. A homogeneous tone of ticking of the grandfather clock if finished, that is—it can be separated in the retentional↔protentional «now», and simultaneously—this is why it can be separated—the tone emphasises the infinity that does not exit because on the basis of the tone, one cannot know what will be continued. This part of the analyse can be characterised by the term of the atemporal and unmeasurable infinity.

(2) According to Husserl, *infinity is a fulfilment of the retentionality of the «now»*. The constituted time is not an interval time. Retentionality is a total reflection of what had occurred in the finished past. The perspective of the past is not described in terms of remembering [*Erinnerung*]—and remembering specific things in the past does not possess the characteristic of infinity. According to Husserl, a latent monad becomes an active monad. It can be interpreted in the way that it fixes the temporal caesura, or a "moment" which is adequate to (and in) time, in which this "transition" was accrued but—for the sake of the actual state of the monads (and their reference to the acts in the «now»)—it is necessary to the past. A similar situation happens with the infinity of the "past", which has a border called the «now». The consciousness in the «now» is a non-thematic consciousness of the infinity of the monads.

(3) According to Husserl, *infinity is (in) the «now»*. The «now» is not a moment but a beyond-temporal lack of time. In that sense the «now» is infinitive as well as atemporal in a manner of the phenomenological time that does not have a temporal value, or which can be used in the physical calculations where infinity is not a temporal infinity. What is in time is subjected to time, what is equipped with the change and an aftermath is a basis of the constitution of the immanent time. The misunderstanding is caused by Derrida's argument that Husserl tries to define the phenomenology of time by means of temporal categories. Derrida leads his own argument in the same way as he treats infinity as temporal.

"The «now»"—"no longer than the «now»"—"not yet the «now»", are the three fundamental *modi* of the phenomenological time. The «now» is the punctually inexpressible *modi* of time; the «now» is additionally specified in the retentional↔protentional context. According to Husserl, retention and protention do not have any temporal extension recognition of the cardinal importance of the «now» that seems obvious. Thereby self-identification of the consciousness originates in the experience of the flow, in which a retentional fall into the past takes place. The category of the «now» is not only an original impression but also an entity that includes an individual and actual interest of a subject—a limited horizon of experience by *lebendige Gegenwart*.[11] In this context *lebendige Gegenwart* becomes a temporal present of the «now» and it is expressed as invariability. The «now» given as non-reflective is anonymous. The anonymity of the «now» is identical with the impossibility associated with the specific temporal "place" of the «now» in time. The «now» is universal, the «now» is always.

Contrary to Husserl, Derrida claims that the constitution of time cannot be limited only to the passive synthesis which derives its own temporality from retentional guiding of the «now». He asks:

> What radical discontinuity is there between this already constituted past and objective time that imposes itself on me, constituted without any active intervention on my past? Husserl will not pose this fundamental question in the *Vorlesungen*.[12]

In what way the multiplicity of the experience of time can be reconciled with its immanent coherence?[13] If it is only the unity resulting from the multiplicity experience of time of individual consciousness it would be difficult to explicate in what manner the internal consciousness of time fulfils the condition of the source included in the infinite flow of time. The internal consciousness of time is finished and limited.[14] According to Derrida, the infinity of time is neither universal nor noematic in the internal experience. The question is if the pure time of a pre-predicative experience is a form of a completely non-determined «now» and the future. The "I"—as transcendence in lived immanence—cannot appear in a pure monadic *ego*. The "I" is between retention and protention, it is in the infinity reference to the past and the future and as a noetic and noematic ontic ground. According to Derrida, Husserl remains in the noematic temporality, the importance of which is constituted. The time of the lived immanence is the time that is reaching much deeper, because it is a time of individual consciousness.[15] This time is a time *for me*. This time is not contaminated by the empirical character of retentionality. In Husserlian phenomenology of time the "I" has got only access to an updated and non-original experience of history the in retentionality of act of constitution. In Derrida's criticism of phenomenology, the freedom as the basis of temporalization is not an abstractive and formal freedom, but it is a freedom that is essentially temporal by a direct reference to retentionality of time.[16] Husserl claims that the flow of time has a feature of an absolute subjectivity, what does not necessarily mean that he connects absolute subjectivity with absolute temporality. Derrida on the other hand, is not able to confine his consideration to this statement. He claims:

> Freedom and absolute subjectivity are thus neither *in* time nor *out* of time. The dialectical clash of opposites is absolutely 'fundamental' and is situated at the origin of all meaning; thus, it must be reproduced at every level of transcendental activity and of the empirical activity founded thereon.[17]

Derridean criticism of Husserlian phenomenology of time includes a false interpretation of Husserlian dislocation of epistemological sense of the immanence.

In the Husserl's early phenomenology of internal consciousness of time, time is described as retentional (in the past of the actuality of the present «now»). The retentional «now», in a temporal life of the "I", makes it possible for the reflective incorporation of intentional acts to happen. Derridean anti-presentialism is based on a recognition that the origin of time is non-present but temporal. Husserl claims that the core of time lies in the non-temporality which is identical with the «now». Derridean proposal, contaminated by untranslatability of terms, as well as Husserlian *lebendige* (also *stehende*) *Gegenwart* remains within the limits of a classic philosophical tradition. There are no solutions for the fundamental problem of time in the «now».

Husserl started to analyse the term of *lebendige Gegenwart* at the beginning of 1930s. This term—seemingly ignoring a retentional↔protentional context of the "punctual" «now» (as Derrida clams)—is a final solution of the problem of the constitution of time. The procedure of uncovering of the life of the transcendental "I" lies in the *lebendige Gegenwart*.

HUSSERLIAN PERSPECTIVE OF PLATO'S
METAX Ý—I C H - S P A L T U N G

The depiction of the constitution of time which was shown above does not explain adequately the constitution of time and temporalization [*Zeitigung*]. It seems important to differentiate between the passive and the active temporalization of consciousness of time. Husserl tries to put a bigger stress on this difference by using the notion of the separation of the "I" [*Ich-Spaltung*]. This notion refers to the term of common presentness [*einfühlende Vergegenwärtigung*] being a circumstance of temporalization of the stream of the consciousness of the Other. It consists of the separation of individual consciousness on the "I"-subject and the "I"-object. The first mentions of this statement can be found in Husserl's notes descended from 1930. Later on Husserl writes about the "I" as a subject in the context of directness [*Zentrierung*] to the whole relived life of the conscious "I".[18] The consciousness of time is on the border between these two kinds of the "I" which connects what was given with what is retentional in the context of the actuality of the present «now». As I tried to show above, there is a cardinal difference between the first and non-retentional «now» and the «now» in the retentional↔protentional context. Finding these parts or aspects of identity is dynamic. It could be said that the "I"-object is always taken under consideration and reflected after the «now». The presentness of this "I" is a secondary presentness but it does not mean that it has secondary significance. My interpretation of Husserlian *Ich-Spaltung* is very similar to the Waldenfels' interpretation of ancient *pathos*. In his statements given during the conference entitled "Actuality of Husserl Thought" (held on 22nd of November 2003) Waldenfels claims:

we understand *pathos* of astonishment which appears on the border between what we know and what gives us new optic of depiction and which is not non of these former ones.[19]

This is not Husserlian *nuns stans* but it is *nunc distans*. According to Husserl,

"I" is beyond-temporal. Obviously, there is no sense that 'I' is treated as temporal. "I" is beyond-temporal—it is a pole of reference to the temporal, it is a feature of a subject. (author's translation)[20]

and

"I" in its original primordiality is nothing temporal—it is constant as living modally original presentness in present. (author's translation)[21]

Waldenfels puts the carriages before the horses and depicts this temporal diastase in the context of the "I". The "I" is analysed in the context of the first reflection on the "I"-object. The temporal separation of time in the "I" permanently starts from the beginning—from a temporal diastase.[22] *Ich-Spaltung* shows the second aspect of the constitution of time which overlaps with the constitution of time known from *Vorlesungen zur Phänomenologie des inneren Zeitbewusstseins*.

The constitution of time based on a primal impression, retention and protention is not enough to describe a variety of the world experience. All what can be described as such experience can be depicted in universal horizon of the world in which outside and inside horizons are contained. Also, a question comes to mind—how the complex horizon of the world can be created in the transcendental subject.

According to Husserl (1933), the transcendental ego is beyond time. Its is an atemporal being which is a carrier all of different kinds of time (primordial, intersubjective, immanent, objective and so on). The ego is an original source of all temporal modalities. There are not objects that are put in time but only appearing of objects which are strictly connected with their temporalization. In this context Husserl refers to the notion of a passive synthesis. He claims that temporalization in transcendental subjects (as primal impressions, retentions and protentions) is original passive occurring without active participation of the transcendental ego. In comparison, in his early writings Husserl claims that retention embraces only a very close horizon of the «now» directed into the past, belonging to the *lebendige Gegenwart*. According to Husserl who depicts the notion of the passive synthesis as a part of a constitution without any participation of the transcendental *Ego*, a pre-predicative unity is created and it refers to the immanent world and the self-reference of the "I". This is an anonymous process which is a phenomenon based on the consciousness of creation of the transcendental subjectivity. In this interpretation, Husserl treats the original synthesis of the original consciousness of time of the transcendental subject as something that is beyond the subject. The core of this depiction is the notion of the style of the world [*Weltstil*] and sedimentation [*Sedimentierung*]. Sedimentation means that the subjective sense is deposited in a phenomenon due to flow of time. Sedimentation has got an influence on the retentional modification of an original impression and protentional intention of expectation. In his *Die Krisis der europäischen Wissenschaften und die transzendentale Phänomenologie*,[23] Husserl describes the world as being a temporal modality that had transformed from a static and structural analysis to the genetic and dynamic depiction of immanence. He shows in his genetic phenomenology that in the background of the experiences structure of the subjective sense lies the original structures of temporal relations which—in his universality—depict the existence of the immanent world. The only sense of genetic phenomenology is drawing out intentional implications of horizons and giving the sense of conscious experiences.

The his late writings Husserl puts a strong emphasis on the issue of time. In the centre of his analyses of time are temporal horizons. Every horizon describes *a priori* presentness in its genetic effect by the sedimentation of the sense. For Husserl, a temporal horizon and its sedimentations of the sense are the connotations of the past and historicalness of the transcendental subject. He names these connotations "monads". An individual sense of temporality, namely, the internal relation with transcendental temporalily gives us the sense of a monad. There is no succession in the flow of time or any unity of any coexistence of temporal places or moments.[24] Although we can find in Husserlian *Vorlesungen zur Phänomenologie des inneren Zeitbewusstseins.*

In the object there is duration: in the phenomenon, alteration. Thus we can also sense, subjectively, a temporal succession where, objectively, we must confirm a coexistence.[25]

and

The break in qualitative identity, the leap from one quality to another within the same genus of quality at a temporal position, yields a new experience, the experience of variation; and here it is evident that a discontinuity is not possible in every time-point belonging to an extent of time.[26]

The original time is not time, it is a previous stage of tie as a form of coexistence.[27]

In staying flow takes place the first self-constitution of ego as temporal flowing constant unity. (author's translation)[28]

According to Husserl, *lebendige Gegenwart* is a multiplicity of phases of non-successive retentions and protentions. It is a continuous and a flowing change. Simultaneously, this flow is a non-temporal and a non-spatial constancy. Also, a reduction to the *lebendige Gegenwart* is a strictly transcendental. This reduction gives us a possibility to reach the transcendental *Ego* as an anonymous being. The Identity of the "I" is not the identity of something that remains in time but it is some kind of constancy of finite functioning in the temporalized time.

Identity of "I" is not simple identity of duration—it is a pole of "I"—and when «in everyness of staying of pole of "I"» also «will be as» the constituted, it also remains only the unity—it is called identity of executor [of "I"].

The identity of the transcendental "I" is covered for a philosophical reflection. The reflection stops before the original "I" and it reaches only the "I"-object.[30]

According to Held, the question about a manner of being of the transcendental "I" tied with time issue is validated. However, Held indicates an aporiae of the Husserlian depiction of time. On the one hand, *lebendige Gegenwart* is finally functioning "I", namely, it is an atemporal and constant "I" in the flow of time, on the other hand, the transcendental "I" is anonymous and possible to be depicted only on the pre-predicative level of cognition. There are two opposite aspects of *lebendige Gegenwart* which give the notion of the transcendental "I" if connected. The anonymity of the transcendental "I" means that it is not directly connected with any "place" and any "moment" of time. The Husserlian *nunc stans* of the transcendental "I" is an expression of the universal dimension of temporalization of change and succession. It is everywhere and nowhere at the same time.[31] Everywhere is an atemporal constancy and nowhere is nothingness which is understood as the anonymity of atemporality of time places.[32] The unity of the flow of temporal experiences can be defined as a Kant's idea in which the transcendental "I" constitutes its time as something to which "I" is getting closer—I am using a baroque expression at the moment—the actuality of the presentness of the present «now» (Held).

Also, can one say that the transcendental "I" constitutes the flow of time? To the contrary, focusing on temporal reality of *lebendige Gegenwart* as *nunc stans* could be a stage toward a recognition of the original and passive character of time which has no reference to the constitution of time as an activity of the transcendental "I". Temporalization of the original flow of time—as the first transcendental stage of an activity—is primarily an act of the transcendental "I". This is a new outlook of phenomenology of time given by Husserl in the middle of 1930s. This phenomenology of time is based on the primordial *Ego*. *Ego*—in atemporality of constancy of the flow of the constituted time, moves its centre of gravity from an individual subject to a monad and co-presentness. According to Held, the most important notion is *nunc stans* used by Husserl has three different meanings. It means

(1) *lebendige Gegenwart* or
(2) staying "I" or
(3) the habitual "I".

It is very difficult to verbalise the idea of staying flow [*strömend–stehenden*] of "I". According to Husserl, a connection of these opposite terms indicates the main position of the «now». The «now» as non-retentional and non-protentional notion is given by Husserl as a reference to the stream of consciousness. In other words, the question is: does the «now» include a simultaneous *continuum* of original content of the consciousness and *continuum* of depiction ? According to Husserl,

I am as flowing present but my being-for-me is constituted itself in this flowing present.[33]

Husserl's twofold depiction of time consists of the realisation of the constitution of time as (1^O) the stream of consciousness constituted in the manner of a temporal unity fixed by retentionality and (2^O) as a reference to the appearing objects in the context of time and beyond directly given continuum temporal duration, change, and succession. There are no two independent streams of consciousness in the Husserlian phenomenology of time but two aspects of the epistemological relation of the complementation of the consciousness of time. The apperception of the object proceeds in a dynamics and in the flow of stream.

Philosopher and the Chair of Theory of Politics, Lublin
e-mail: ktpkul@yahoo.com

N O T E S

[1]   Jacques Derrida, "Ousia and Grammē: Note on a Note from Being and Time" [in:] Jacques Derrida, *Margins of Philosophy*, The University of Chicago Press, Chicago 1982, p. 63.
[2]   Jacques Derrida, *The Problem of Genesis in Husserl's Philosophy*, The University of Chicago Press, Chicago — London 2003, p. 2.
[3]   The smallest notion is used in its epistemological sense, not in the sense of temporal duration.
[4]   Also, I refer to the interpretation of Bernet's understanding of the term "unconscious". According to him, "[...] the Unconscious as the presence of the non-present is first of all a matter of the particular

type of act-intentionality, called 'presentification' (*Vergegenwärtigung*), which characterizes the acts of fantasy, memory and empathy. It can be shown that the possibility of these acts is ultimately grounded in the temporal structure of inner consciousness and that a correct phenomenological understanding of the Unconscious is first opened up through the analysis of this inner consciousness. Then, of course, the essence of the Unconscious can no longer be understood on the basis of the mere absence of inner perceptual consciousness. Instead, its appearance, and thereby its phenomenologically determined essence, results from the possibility of another form of inner (time-) consciousness, namely, the reproductive form. Husserl himself did not develop this new phenomenological understanding of the Unconscious any further, although he prepared for us all the means for doing so. It was never difficult for Husserl to think about the possibility of the presentification of something non-present because he always understood consciousness as the subjective achievement of intentional apperception and appresentation and never as the mere presence of sense data. In his early work in the *Philosophy of Arithmetic* and the *Logical Investigations*, presentifications were still understood as inauthentic, that is, non-intuitive forms of *thought*. According to this theory, which withdraws from the intuitive or authentic thought but is presentified by means of a sign that functions as a surrogate or by an image that represents by similarity. It is not surprising therefore that initially, and up to and including the 1904/05 lecture on the *Main Issues in the Phenomenology and Theory of Knowledge*, 5. Husserl conceived of the acts of sensuous presentification, like memory and fantasy, as types of pictorial consciousness (*Bildbewusstsein*). He could base this view on an already extensive exploration of perceptive pictorial consciousness, one with which he continued to occupy himself with later, especially in the connection with the analysis of aesthetic pictorial consciousness. Thus, fantasy and memory were forms of pictorial consciousness in which an inner pictorial image (later called by Sartre the '*image mentale*') takes place in a physical perceptual picture. A past occurrence or an unreal fantasy-world would therefore come to appearance in a present depiction without thereby forfeiting its absence from this present." Rudolf Bernet, "Unconscious consciousness in Husserl and Freud", *Phenomenology and the Cognitive Sciences 1*, p. 331.

5   Cf. Fred Kersten, *Phenomenological Method: Theory and Practice*, Kluwer Academic Publishers, Dordrecht — Boston — London 1989, pp. 269, 273.

6   Cf. Leonard Lawlor, *Derrida and Husserl, The Basic Problem of Phenomenology*, Indiana University Press, Bloomington, IN 2002, p. 21.

7   Author use notion "presence" as peculiar kind of present. Presence is taken out from the presenceness as a peculiar aspect of the present ontologically "later" than the presentness.

8   *Ibidem.*

9   Cf. Bernhard Waldenfels, *Das Zwischenreich des Dialogs, Sozialphilosophische Untersuchungen in Anschluss an Edmund Husserl*, Martinus Nijhoff, The Haag 1971, *passim.*

10   Jacques Derrida, *The Problem of Genesis in Husserl's Philosophy*, The University of Chicago Press, Chicago — London 2003, p. 95.

11   Cf. Edmund Husserl, *Zur Phänomenologie der Intersubjektivität, Texte aus dem Nachlass, Dritter Teil: 1929–1935*, Martinus Nijhoff, The Haag 1973, p. 174; Bernhard Waldenfels, *Das Zwischenreich des Dialogs, Sozialphilosophische Untersuchungen in Anschluss an Edmund Husserl*, Martinus Nijhoff, The Haag 1971, pp. 149–151, 204–205.

12   Jacques Derrida, *The Problem of Genesis in Husserl's Philosophy*, The University of Chicago Press, Chicago — London 2003, p. 56.

13   Cf. Jacques Derrida, *The Problem of Genesis in Husserl's Philosophy*, The University of Chicago Press, Chicago — London 2003, p. 95.

14   *Ibidem.*

15   Cf. Jacques Derrida, *The Problem of Genesis in Husserl's Philosophy*, The University of Chicago Press, Chicago — London 2003, p. 96.

16   Cf. Jacques Derrida, *The Problem of Genesis in Husserl's Philosophy*, The University of Chicago Press, Chicago — London 2003, p. 65.

17   Jacques Derrida, *The Problem of Genesis in Husserl's Philosophy*, The University of Chicago Press, Chicago — London 2003, p. 65.

264   CEZARY JÓZEF OLBROMSKI

18 Cf. Edmund Husserl, *Manuscripts*, C 3 III, p. 1 (1930) [quoted after:] Klaus Held, *Lebendige Gegenwart. Die Frage nach der Seinsweise des transzendentalen Ich bei Edmund Husserl, entwickelt am Leitfaden der Zeitproblematik*, p. 4.

19 *Actuality of Husserl's Thought*, conference organized by Polish Phenomenological Association, The Institute of Philosophy and Sociology of Polish Academy of Sciences, and Goethe-Institut Warschau.

20 "Das Ich ist unzeitlisch. Natürlisch hat es keinen Sinn, das Ich als zeitlich zu betrachten. Das Ich ist über-zeitlich, et ist der Pol von Ich-Verhaltungsweisen zu Zeitlichem, er ist das Subject, das sich zu Zeichem verhält." Edmund Husserl, *Manuscripts*, E III 2, p. 50 (1920 or 1921), [quoted after:] Klaus Held, *Lebendige Gegenwart. Die Frage nach der Seinsweise des transzendentalen Ich bei Edmund Husserl, entwickelt am Leitfaden der Zeitproblematik*, p. 117.

21 "Das Ich in seiner ursprünglichsten Ursprünglischkeit ist nicht in der Zeit—hier der beständing als lebendige urmodale Gegenwart sich zeitigenden gezeitigten Gegenwart." Edmund Husserl, *Manuscripts*, C 10, p. 21 (1931), [quoted after:] Klaus Held, *Lebendige Gegenwart. Die Frage nach der Seinsweise des transzendentalen Ich bei Edmund Husserl, entwickelt am Leitfaden der Zeitproblematik*, p. 117.

22 *Actuality of Husserl's Thought*, conference organized by Polish Phenomenological Association.

23 Cf. Edmund Husserl, *Die Krisis der europäischen Wissenschaften und die transzendentale Phänomenologie, Ergänzungsband, Texte aus dem Nachlass 1934–1937*, hrsg. von R. N. Smid, *Husserliana*, Vol. 29, Dordrecht 1993.

24 Although, the reference to the spatial exemplifications is not ontologically correct reference.

25 Edmund Husserl, *On the Phenomenology of the Consciousness of Internal Time*, p. 8.

26 Edmund Husserl, *On the Phenomenology of the Consciousness of Internal Time*, p. 91.

27 Cf. Klaus Held, *Lebendige Gegenwart. Die Frage nach der Seinsweise des transzendentalen Ich bei Edmund Husserl, entwickelt am Leitfaden der Zeitproblematik*, Den Haag 1966, pp. 115, 135.

28 "In ständingen Strömen vollzieht sich «überhaupt erst» die Selbstkonstitution des Ego als «zeitlish–» strömend verharrender Einheit." Cf. Klaus Held, *Lebendige Gegenwart. Die Frage nach der Seinsweise des transzendentalen Ich bei Edmund Husserl, entwickelt am Leitfaden der Zeitproblematik*, Den Haag 1966, p. 135.

29 "Die Identität des Ich ist nicht die bloße Identität eines Dauernden, sondern die Identität des Vollziehers—das ist der Ichpol—und wenn «der in der Jeweiligkeit verharrende Ichpol» schon auch «als» eine Dauereinheit konstituiert «wird», so bleibt es «doch» *ein einzigartig Eigenes, was da Identität des Vollziehers heißt.*" Edmund Husserl, *Manuscripts*, C 10, p 28 (1931), [quoted after:] Klaus Held, *Lebendige Gegenwart. Die Frage nach der Seinsweise des transzendentalen Ich bei Edmund Husserl, entwickelt am Leitfaden der Zeitproblematik*, Den Haag 1966, p. 118.

30 Cf. Klaus Held, *Lebendige Gegenwart. Die Frage nach der Seinsweise des transzendentalen Ich bei Edmund Husserl, entwickelt am Leitfaden der Zeitproblematik*, Den Haag 1966, p. 139.

31 Cf. Klaus Held, *Lebendige Gegenwart. Die Frage nach der Seinsweise des transzendentalen Ich bei Edmund Husserl, entwickelt am Leitfaden der Zeitproblematik*, Den Haag 1966, p. 143.

32 *Ibidem.*

33 "Ich bin als strömende Gegenwart, aber mein Für-Mich-Sein ist selbst in dieser strömenden Gegenwart konstituiert." Edmund Husserl, *Manuscripts*, C 3 III, p 33 (1931), [quoted after:] Klaus Held, *Lebendige Gegenwart. Die Frage nach der Seinsweise des transzendentalen Ich bei Edmund Husserl, entwickelt am Leitfaden der Zeitproblematik*, Martinus Nijhoff, The Haag 1966, p. 115.

REFERENCES

SOURCES

Husserl, Edmund. 1928. Vorlesungen zur Phänomenologie des inneren Zeitbewusstseins. hrsg. von Martin Heidegger, *Jahrbuch für Philosophie und phänomenologische Forschung*, Band IX, pp. 367–490.

Husserl, Edmund. *Die Krisis der europäischen Wissenschaften und die transzendentale Phänomenologie, Ergänzungsband, Texte aus dem Nachlass 1934–1937*, Reinhold Nikolaus Smid (Ed.), Husserliana XXIX, Dordrecht/The Hague: Kluwer, 1954 [II 1976, III 1993].

Husserl, Edmund. 1973. *Zur Phänomenologie der Intersubjektivität. Texte aus dem Nachlass. Dritter Teil, 1929–1935*, hrsg. von Iso Kern, The Hague: Martinus Nijhoff.
Husserl, Edmund. 1991. *On the phenomenology of the consciousness of internal time* (trans: from German by Brough, John Barnett). Dordrecht/Boston/London: Kluwer.

RESOURCES

Bernet, Rudolf. 2002. Unconscious consciousness in Husserl and Freud. *Phenomenology and the Cognitive Sciences* 1:327–351.
Derrida, Jacques. 1982. Ousia and Grammē: Note on a Note From Being and Time. In *Margins of Philosophy*, ed. Jacques Derrida (trans: Bass, Allan), 29–67. Chicago: The University of Chicago Press.
Derrida, Jacques. 2003. *The Problem of Genesis in Husserl's Philosophy* (trans: Hobson, Marian). Chicago/London: The University of Chicago Press.
Held, Klaus. 1966. *Lebendige Gegenwart. Die Frage nach der Seinsweise des transzendentalen Ich bei Edmund Husserl, entwickelt am Leitfaden der Zeitproblematik*. Den Haag: Martinus Nijhoff.
Kersten, Fred. 1989. *Phenomenological Method: Theory and Practice*. Dordrecht/Boston/London: Kluwer.
Lawlor, Leonard. 2002. *Derrida and Husserl, The Basic Problem of Phenomenology*. Bloomington, IN: Indiana University Press.
Waldenfels, Bernhard. 1971. *Das Zwischenreich des Dialogs, Sozialphilosophische Untersuchungen in Anschluss an Edmund Husserl*. The Haag: Martinus Nijhoff.

SECTION V
COGNITION, CREATIVITY, EMBODIMENT

POUND, PROPERTIUS AND *LOGOPOEIA*

"My job was to bring a dead man to life, to present a living figure"[1]

ABSTRACT

Ezra Pound's "Homage to Propertius" is an unusually free translation of selected poems by the Roman poet Propertius which has generated a fruitful debate about the translator's task. Among the qualities Pound meant to find in Propertius, and consequently strove to recreate, was *logopoeia*, "the dance of the intellect among words". Tantalizing though it sounds, this definition remains somewhat vague, as does Pound's other references to the concept. The present paper seeks to clarify the meaning of *logopoeia*, which is done by first revisiting Pound's own statements and then juxtaposing the opinions of previous scholars. The scholars chosen include classicists as well as scholars on both Pound and Laforgue, the French 19th century poet who was Pound's initial inspiration for the concept. The conclusion reached is that *logopoeia* is not to be understood as locally limited wordplay, as some classicists have assumed, but rather as a more general detached attitude towards the language used which often includes an element of irony and humour.

When modernist poet and literary critic Ezra Pound finished his "Homage to Sextus Propertius" in 1917,[2] it represented something quite new in the modern use of the classics. Twelve poems were offered as translations from selected poems by Propertius, a Roman poet of notorious difficulty who had until then been little appreciated outside the ranks of classicists, but whose dense imagery and tortuous syntax seemed to have much in common with the developing modernist aesthetics.[3] Besides the unorthodox choice of author, the main novelty of the collection lay in the approach taken to the task of translation. Rather than trying to mirror the idiom of the ancient language as closely as possible, the aim of traditional translation, Pound sought to give the text a modern flair in a process that has been labeled "creative translation".[4] The precise nature of the approach, as well as the level of success achieved, has been the subject of much controversy. The present study, however, deals with one famous particular quality which Pound meant to have discovered in Propertius and consequently strove to recreate. That quality is *logopoeia*, which was never satisfactorily defined by Pound himself and consequently has sparked off a debate of its own. In the following a clarification of the term's meaning is sought by first revisiting Pound's own statements and then juxtaposing a number of later views. A main aim of the latter part is to integrate insights developed within fields normally kept apart: responses from classical scholars with an expert knowledge

269

A.-T. Tymieniecka (ed.), Analecta Husserliana CX, 269–278.
DOI 10.1007/978-94-007-1691-9_22, © Springer Science+Business Media B.V. 2011

of Propertius, Pound-scholars and scholars working with the French poet Laforgue, whose relevance will soon become clear.

Beginning now with Pound himself, his most extensive definition of *logopoeia* is to be found in the essay "How to Read", originally published in the *New York Herald Tribune* in 1929:[5]

Logopoeia, "the dance of the intellect among words", that is to say, it employs words not only for their direct meaning, but it takes count in a special way of habits of usage, of the context we expect to find with the word, its usual concomitants, of its known acceptances and of ironical play. It holds the aesthetic content which is peculiarly the domain of verbal manifestation, and cannot possibly be contained in plastic or music. It is the latest come, and perhaps most tricky and undependable mode.

In *ABC of Reading* in 1934 he elaborates:[6]

You take the greater risk of using the word in some special relation to "usage", that is, to the kind of context in which the reader expects, or is accustomed, to find it. This is the last means to develop, it can only be used by the sophisticated. (If you want really to understand what I am talking about, you will have to read, ultimately, Propertius and Jules Laforgue).

Tantalizing though the catchy "dance of the intellect among words" sounds, the two passages do not make it entirely clear what Pound has in mind with the concept, and it is this which has generated the scholarly debate. In the following I shall first review the response of two classical scholars, whose opinions I shall find to be inadequate. Then I shall proceed to a third classicist, whom I shall find to have a more convincing view. I shall find support for his view in central scholars within the field of Pound studies, and finally in work done on the French poet Jules Laforgue (1860–87), who is mentioned together with Propertius in the quotation just above.

The first classical scholar I take a look at is Mark Edwards.[7] He finds the term *logopoeia* to be "quite unacceptable",[8] although sadly not explaining why this is so. Further, he argues that the concept is in any case not as unique as Pound makes it out to be since there has already been done quite a lot of work on what he calls "intensification of meaning", both in Propertius and in other classical poets. Developing a list of various subcategories, he finds as the third kind of "lexical ambiguity" "cases where the straightforward effect of a word is enhanced by consciousness of another meaning or a common association". This, according to Edwards:[9]

is true "logopoeia" – use "of habits of usage, of the context we *expect* to find the word, its usual concomitants, of its known acceptances" – and I think some fairly certain instances can be found in Propertius, though I am not sure that they justify Pound's lavish praise of him.

The second classical scholar I take a look at is Niall Rudd.[10] Having quoted Pound's definition, he first remarks:[11]

This is all very general, and (understandably enough) those who have tried to elaborate the concept theoretically have not always made it clearer.

Rudd then performs a learned analysis of specific instances where the Propertian original shows novelty in the use of the Latin language and whether or not Pound in his translation seems to perceive and respond to this. His conclusion[12] is that Pound often does and that

It was surely this novelty, in its various manifestations, that Pound had in mind when he spoke of *logopoeia* – a term which might be translated as "creativity in language".

Leaving Rudd, I arrive at John Patrick Sullivan, the classical scholar who has worked most extensively on the relationship between Pound and Propertius, resulting in a fundamental 1964 monograpy on the subject.[13] *Logopoeia* is, as could be expected, given extensive treatment,[14] and I find that Sullivan's discussion improves upon those of Edwards and Rudd in two ways. Firstly, he draws into the discussion the French poet Jules Laforgue, who is mentioned together with Propertius in the second quotation from Pound above, but is conspicuously absent in Edwards and Rudd. Secondly, Sullivan takes a broader view of the concept, finding in it not merely a narrow play with words, but rather a general attitude on the author's part:[15]

Logopoeia is not, as one might immediately think, simply "wit" of the Augustan or even metaphysical kind (even though Rochester is in the direct line of the metaphysical tradition). Nor is it the sort of verbal ambiguity analyzed by William Empson or the very rhetorical "wit" we normally associate with Tacitus. It is something more subtle than these. It is much more a self-conscious poetic and satiric attitude which is expressed through a certain way of writing.

As support for this claim, he quotes Pound's great contemporary T.S. Eliot, who in the preface to the *Selected poems* of Pound, says of the *Homage*:

It is also a criticism of Propertius, a criticism which in a most interesting way insists upon an element of humour, of irony and mockery, in Propertius, which Mackail and other interpreters have missed. I think that Pound is critically right, and that Propertius was more civilized than most of his interpreters have admitted.

On this basis, Sullivan's own definition of *logopoeia* becomes:[16]

I suggest then that *logopoeia* is a refined mode of irony which shows itself in certain delicate linguistic ways, in a sensitivity to how language is used in other contexts, and in a deployment of these other uses for its own humorous or satiric or poetic aims, to produce an effect directly contrary to their effect in the usual contexts. Thus magniloquence can be deployed *against* magniloquence, vulgarity *against* vulgarity, and poeticisms *against* poeticizing. *Logopoeia* is not simply parody, for it may even be directed against the poet himself, but a very self-conscious use of words and tone which would be requisite for parody. Despite its sporadic appearance in other periods it must strike us as an extremely "modern" style of writing – which may explain why Pound thought that it was the latest come and the most tricky to handle.

If one compares Sullivan's definition to those of Edwards and Rudd, a major difference is as mentioned the level at which *logopoeia* is thought to operate. For Edwards and Rudd it is a play with words on a level very close to the text, whereas Sullivan finds it to be a more general attitude towards the kind of language chosen.

When I find myself in support of Sullivan, it is partly because of a further comment on *logopoeia* made by Pound immediately after the definition in *How to Read*
quoted above:

> *Logopoeia* does not translate; though the attitude of mind it expresses may pass through a paraphrase.
> Or one might say, you can *not* translate it "locally", but having determined the original author's state of
> mind, you may or may not be able to find a derivative or an equivalent.

Key expressions are of course "attitude of mind" followed a little later by "state
of mind". Equally important in the present context, though, is in my opinion the
comment that *logopoeia* cannot be translated "locally". Sullivan does quote the addition, and, as will become clear below, he makes use of it later on in a critique of the
Pound-scholar Kenner's explanation of *logopoeia*. Here, however, I suggest that it
can be used as an argument against the views of Edwards and Rudd, whose closeness
to the text seems to lead to a focus on precisely "local" translation. Perhaps telling
is the fact that they both leave out the addition in their quotations from Pound's
passage.

SCHOLARSHIP ON POUND AND ON LAFORGUE

Leaving the classicists I now take a look at two other separate scholarly fields
that have concerned themselves with Pound's *logopoeia*. The first is scholarship
on Pound himself and the second studies of the French poet Laforgue, whose inclusion in the debate was mentioned as the first improvement of Sullivan above. As
will become clear, the results from both fields give support to the view of the term
developed by Sullivan. Moreover, some studies of Pound stress the point that the
phenomenon defies "local translation", which lends support to my own critique of
Edwards and Rudd.

The first Pound-scholar I take a look at is Hugh Kenner, who mentions *logopoeia*
twice. The first time is in connection with puns on the Latin.[17] *Logopoeia* is here
defined as "elaborate contextual wit" based on discovered parallels in the Latin.
Kenner quotes the passage about "local translation" and concludes that: "hence it
is useless to try to expose the dimensions of the Latin in which he is interested by
direct rendering". The second mention of *logopoeia* is in connection with a certain
quality in Pound's later *Cantos*.[18] Beginning with the *Homage*, Kenner first finds
that

> It is impossible to represent by quotation the enormous freedom and range of tone, the ironic weight, the
> multiple levels of tongue-in-cheek self-deprecation everywhere present in the *Propertius*.

Then he singles out as one of these devices "the ironic use of Latinate diction",
which he finds to exemplify *logopoeia* in the *Cantos*:

> If the reader, by frequenting the Propertius sequence, will acquire a sensitivity to the weight of Latin
> abstract definition in unexpected contexts, he will find it easier to see how large stretches of the *Cantos*,
> in which for reasons of decorum rhythmic definition is diminished to contrapuntal status, are organized
> as it were from the centre out, by stiffening and relaxing the texture of the vocabulary.

Reviewing Kenner's two references to *logopoeia*, the first is in fact criticized by Sullivan, who writes thirteen years later.[19] A major point in his critique is the passage that *logopoeia* does not translate locally, but as seen Kenner *does* make room for this passage in his explanation, making Sullivan's critique appear unjust at least in this respect. The second passage, moreover, seems to express a view on *logopoeia* that clearly comes close to that of Sullivan, presenting the quality as an attitude that has an element of irony and even self-mockery. Finally, one can note another place where Kenner sees a link to Laforgue in a formulation which again stresses attitude and humour:[20]

it is impossible, after Laforgue, to be unaware of a calculated excess of atmospherics, to miss risible implications.

Moving on from Kenner, I arrive at Donald Monk,[21] who turns out to give important support to the idea that *logopoeia* is about an attitude rather than local instances of verbal play. In discussing the concept,[22] he first points out that "*logopoeia* is necessarily much more a matter of tone than paraphrasable content". Then he quotes the passage on "local translation", on which he comments:

He is looking at an "attitude" or "state" of mind as his material, and cutting totally loose from any idea of "local" translation. *Propertius*, then, is already firmly a matter of atmosphere, not fact.

Earlier on he has stated that "it is unhelpful to quarrel with Pound on the level of local mistranslation".[23] This, indeed, seems to be precisely the level on which Edwards and Rudd have been found to operate, so that the evidence from Monk strengthens my present case against these two classical scholars.

The last Pound-scholar I turn to is Donald Davie. He has some rather extreme opinions, claiming for instance that "Pound's poem is *in no sense* a translation"[24] so that Sullivan's book is as a whole "vitiated by this assumption that Pound's dealings with Propertius are a model of what the translator's should be with his original".[25] Furthermore, he dismisses any significant relationship to Laforgue:[26]

It is true that Pound was later to claim that Propertius and Laforgue were two of a kind, and to define the kind as "logopoeic". But this is unconvincing, and irrelevant to the *Homage*.

As can be seen, the dismissal seems to include a rejection of the term *logopoeia*, but it is a pity that Davie does not offer any argument for his assertion. Instead, he makes an observation that may have relevance for the view of *logopoeia* argued here when he discusses an interesting passage in a letter to Thomas Hardy dated March 31, 1921:[27]

I ought – precisely – to have written "Propertius Soliloquizes" – turning the reader's attention to the reality of Propertius – but no – what I do is to borrow a term – aesthetic – a term of aesthetic *attitude* from a French musician, Debussy – who uses "Homage à Rameau" for a title to a piece of music recalling Rameau's manner. My "Homage" is not an English word at all. (...). I ought to have concentrated on the subject – (I did so long as I forgot my existence for the sake of the lines) – and I tack on a title relating to the treatment – in a fit of nerves, fearing the reader won't sufficiently see the super-position, the doubling of me and Propertius, England to-day and Rome under Augustus.

Pound, it is clear, expresses doubt about the title he has chosen for the *Homage*, and Davie shows that this may be understood as part of a more general uncertainty generated by the harsh critique the work had been met with. Personally, however, Davie has

come to suspect that the whole business about "the doubling of me and Propertius" is a rationalization after the fact, a fiction uneasily promoted by Pound to meet a parrot-cry for "contemporary relevance".

In other words one should expect that the actual title gives a better impression of Propertius' original perception of his project. If one re-reads the passage with this in mind, it becomes clear that the *Homage* is primarily about the recreation of an aesthetic attitude, which comes very close to *logopoeia* as understood by Sullivan and Monk in particular.

Leaving now scholarship on Pound for scholarship on Laforgue, one should initially note the mention of this French poet in the quotation from Pound's *ABC of Reading* above. The only denial of any true relationship is as just demonstrated to be found in Davie, who does not offer any argument. Better, then, to accept the majority view, which is that it was the encounter with Laforgue that made Pound first discover the quality he would then find in Propertius[28] and later label *logopoeia*:[29]

sometime after his first "book" S.P. ceased to be the dupe of magniloquence and began to touch words somewhat as Laforgue did.

At one point he was not quite certain that *logopoeia* was to be found in Propertius, but claimed that it was in any case undoubtedly present in Laforgue:[30]

Unless I am right in discovering *logopoeia* in Propertius (which means unless the academic teaching of Latin displays crass insensitivity as it probably does), we must almost say that Laforgue invented *logopoeia* observing that there had been a very limited range of *logopoeia* in all satire, and that Heine occasionally employs something like it, together with a dash of bitters, such as can (though he may not have known it) be found in a few verses of Dorset and Rochester. At any rate Laforgue found or refound *logopoeia*.

However, he seems always to have seen a close connection between Propertius and Laforgue, as is made clear negatively just below in the same passage:

Laforgue is not like any preceding poet. He is not *ubiquitously* (my emphasis) like Propertius.

The close connection between Propertius and Laforgue in Pound's thought means that it should be possible to gain further insight into his view of Propertius through a separate study of his Laforgue. In particular it should be possible to learn more about Propertius' *logopoeia* through studying that which Pound found in Laforgue. The full potential of this approach seems so far not to have been realized, for even in Sullivan little is said beyond the mention of Laforgue's name, and no study of *logopoeia* in either Propertius or Pound that I have come across makes use of scholarship on Laforgue. In the following I shall take a small step towards rectifying this situation by taking a look at two different Laforgue-scholars, and it will become clear that these have reached views on *logopoeia* that are surprisingly similar to those of Sullivan, Kenner and Monk.

The first scholar I take a look at is Warren Ramsey,[31] who treats the relationship between Pound and Laforgue without any mention of Propertius. Moreover, he neither himself mentions scholarship on Pound and Propertius, nor is he mentioned by

Sullivan or any other of the scholars above, so that he can give important independent support. As indicated already by the title of his book, his focus is on Laforgue as an ironist, and *logopoeia* is consequently presented as an ironic quality.[32] Laforgue's irony is based on "an attitude of detachment", and "logopoetic ironies" arise from incongruous oppositeness in the use of language.[33] On Laforgue's poetry he finds in general:[34]

The Latinisms that Laforgue relished – "alacre" from "alacer", "albe" from "alba", "errabundes" from "errabundus" – are regularly pressed into ironic service, and clash with colloquial vocables in the same or proximate lines.

In connection with certain poems of Pound he claims that:[35]

They represent a kind of intellectual discussion that can be pertinently described as "logopoeia". (. . .). The cliché, "march of events" is pressed into ironic service, according to characteristic Laforguian procedure.

If one takes a closer look at these statements, it is striking how similar they are to the views on *logopoeia* in Sullivan, Kenner and Monk. The stress on irony Ramsey has in common with Sullivan and Kenner, and the conception of *logopeia* as an "an attitude of detachment" is central to the whole discussion above. The focus on Latinisms he has in common with Kenner, and finally comes the incongruous oppositeness, which compares with a statement by Monk so far not quoted: "Juxtaposition is at the heart of *logopoeia*".[36]

Leaving Ramsey, a more recent treatment of the relationship between Pound, Laforgue and *logopoeia* can be found in Jane Hoogestraat. As to the nature of the concept, she has the following to say:[37]

With remarkable consistency in his definitions of *logopoeia* and his criticism of Laforgue, Pound distinguishes between ordinary irony and the irony he discovers in Laforgue, and he takes care to emphasize the particular qualities of the Laforguean ethos: a specific attitude of an identifiable speaking subject toward the language that subject employs.

A little later she writes:[38]

All the examples of *logopoeia* he alludes to or cites directly share a particularized ethos on the part of the poetic speaker: an extremely self-conscious, overintellectualized voice directed toward relentless social satire. The diction in this poetry ranges from the clichés of popular culture to abstract Latinate terminology from numerous nonliterary disciplines. This aspect of logopoeia, the sharp ethos which holds no subject immune from poetic ridicule and no language out of bounds for use in a poem, was a necessary and direct reaction to sentimentalized or bourgeois aesthetics.

Among the wealth of references to Laforgue is a comment on his poem "Complainte sur certains Ennuis":[39]

The speaker further questions whether his own ennui would be of sustained interest, achieving both a distance and a self-mockery that would be impossible in, say, a Baudelaire poem, or in the larger tradition Laforgue satirizes.

Finally, Hoogestraat has a single short comment on Pound's Propertius: "*logopoeia* and Laforgue operate in a fairly straightforward way behind *Homage to Sextus Propertius*".[40] In the present context one can of course only lament the absence of a further elaboration of this point.

Summing up, Hoogestraat's view gives ample support to the view of *logopoeia* that has by now been established. There is the fact that the concept is about a general attitude rather than local wordplay and the element of irony and humour. There is the play with Latinate language central to Kenner and the distance which allows for self-mockery emphasized by Sullivan. However, one should realize that Hoogestraat is not quite as independent a source as is Ramsey. The brief mention of Propertius can perhaps be overlooked, as can the small number of references to Davie and Kenner in the notes. Not to be overlooked, however, is the note which explicitly mentions Sullivan's book as an "excellent discussion of *logopoeia* and Pound's *Propertius*".[41] Although hardly independent, then, the important fact remains that Hoogestraat arrives at the same conclusion as Sullivan, and so there exists a quite recent study of Laforgue that gives support to the view of *logopoeia* in Pound and Propertius argued here.

SUMMARY AND CONCLUSION

The term *logopoeia* was introduced by Ezra Pound in order to describe a quality he meant to find in the Roman poet Propertius and consequently sought to recreate when translating him. Tantalizing though it sounds, Pound's own definition of the term as "the dance of the intellect among words" remains unprecise, and the present paper has aimed at a clarification by first revisiting Pound's other statements about the concept and then juxtaposing the views of a number of earlier scholars. The context being a Roman poet, it has been natural to begin with the views of three classicists, whereafter have come three Pound-scholars and two scholars on the French poet Laforgue, the contact with whom was Pound's original inspiration for the concept. The main line of argument has been that the two first classical scholars, Mark Edwards and Niall Rudd, are wrong in explaining the concept as isolated local instances of verbal play, a posititition against which Pound himself seems explicitly to warn. Rather, one should understand *logopoeia* as a general attitude towards the kind of language used, an attitude which moreover often involves an element of humour in the form of irony, satire or even self-mockery.

To explain *logopoeia* as an attitude is of course not to say that one does not need to approach the phenomenon at a local level as does Edwards and Rudd. To analyze in detail the use of single words in relation to the words around them must remain the necessary, indeed the only sensible, way of approaching a poem. My point here, however, is that such a word-by-word local analysis is just the first step towards a full study of *logopoeia*, which must take into consideration also how each individual case as well as all the cases taken together both relate to and contribute to the general attitude lying behind the poem. Particularly demonstrative of the exclusively local approach seems Rudd, who has been seen to find that Pound seems to perceive and respond to novelties in Propertius' use of the Latin language. In itself this analysis is splendidly done, and it throws much light on a particular aspect of Pound's skill as a translator. However, I am not so certain that Rudd is right in identifying this quality as *logopoeia*, nor that this kind of analysis has any potential for increasing our understanding of the concept.

Department of Linguistic, Literary and Aesthetic Studies, University of Bergen, Bergen, Norway
e-mail: Lars.Gram@student.uib.no

NOTES

[1]   Pound to Alfred Orage, in *The selected Letters of Ezra Pound: 1907–1941*, ed. D.D. Paige (New York: Harcourt Brace and co., 1950), p. 149.

[2]   Publication did not take place until 1919, perhaps because these witty love poems seemed unfit for printing before after the war, see D. Davie: *Studies in Ezra Pound* (London, 1991), p. 77.

[3]   "There is something in the Umbrian poet that appeals to the modern mind", writes classicist Georg Luck, "whether it is the rich texture of his imagery or rather the desire to avoid the banal and conventional at all costs", *The Latin Love Elegy* (London: Methuen, 2nd ed. 1969) p. 121. It should be noted, though, that the affinity with modernism can be overstated, as seems to be done by D.T. Benediktson, in *Propertius: Modernist Poet of Antiquity* (Illinois: Southern Illinois University Press, 1989), see the reviews of J.L. Butrica, *The Classical Review*, New Series, 40:2 (1990), pp. 266–8, and F.H. Mutschler, *Gnomon* Vol. 63/5 (1991), pp. 461–3.

[4]   J.P. Sullivan, in *Ezra Pound and Sextus Propertius: A Study in Creative Translation* (London: Faber and Faber, 1964), which remains a fundamental study of the subject. Invaluable for further research is also Sullivan's inclusion of the complete text of the *Homage* with the Latin original running in parallel. Those not acquainted with Latin can use the Loeb parallel text (G.P. Goold, *Propertius: Elegies*, Harvard, 1990) for a more literal translation, which can then be compared with the translation of Pound. A schematic overview of the corresponding passages is to be found in K. Kenner, *The Poetry of Ezra Pound* (Norfolk, CT: New Directions, 1951), pp. 150–1.

[5]   Now available in *Literary Essays of Ezra Pound*, ed. T.S. Eliot (London: Faber and Faber, 1954), pp. 15–40, present quotation from p. 25. N. Rudd correctly points out that the original year of publication, 1929, is over a decade later than the *Homage*, *The Classical Tradition in Operation* (Toronto: University of Toronto Press, 1994), p. 140. However, appreciation of the quality may come before the eventual theoretical elaboration. J. Hoogestrat finds that the influence on Pound from the French poet Laforgue, which will be seen below to be the origin of his thinking about the concept, began around 1914–15, which is just before the *Homage*, "Akin to Nothing but Language: Pound, Laforgue and Logopoeia", *ELH* 55/1 (Spring 1988), pp. 259–85, present point p. 265. W. Ramsey also goes back to 1914, although pointing to F.S. Flint rather than T.S. Eliot as the catalyst, *Jules Laforgue and the Ironic Inheritance* (Oxford, 1953), p. 204.

[6]   E. Pound, *ABC of Reading* (New Haven, 1934), pp. 37–8.

[7]   M.W. Edwards, "Intensification of Meaning in Propertius and Others", *Transactions and Proceedings of the American Philological Association* 92 (1961), pp. 128–44.

[8]   M.W. Edwards, op. cit., p. 128.

[9]   M.W. Edwards, op. cit., p. 137.

[10]   N. Rudd, op. cit., pp. 140 ff..

[11]   N. Rudd, op. cit., p. 140.

[12]   N. Rudd, op. cit. p. 146.

[13]   J.P. Sullivan, op. cit. The topic is taken up again in Sullivan's *Propertius: A Critical Introduction* (Cambridge, 1976), pp. 147 ff., which does not, however, contribute much new, consisting mainly of text taken directly from the earlier book.

[14]   J.P. Sullivan, op. cit., pp. 64 ff..

[15]   J.P. Sullivan, op. cit., p. 66.

[16] J.P. Sullivan, op. cit., p. 67.

[17] H. Kenner, op. cit., p. 149 n.

[18] H. Kenner, op. cit. pp. 158–60.

[19] J.P. Sullivan, op. cit., p. 103.

[20] H. Kenner, op. cit. p. 147.

[21] D. Monk, "How to Misread: Pound's Use of Translation", in *Ezra Pound: The London Years*, ed. P. Grover (New York: AMS Press, 1978).

[22] D. Monk, op. cit. pp. 80–1.

[23] D. Monk, op. cit. p. 73. The comment is admittedly made about another and more general quality in poetry Pound calls *melopoeia*, but since the point is that *melopoeia* does not translate either it should apply to *logopoeia* as well.

[24] D. Davie, *Studies in Ezra Pound* (London: Carcanet, 1991), p. 73, in a chapter originally printed in *Ezra Pound: Poet as Sculptor* (London: Routledge, 1965), pp. 77–101.

[25] D. Davie, *Ezra Pound* (New York: Penguin, 1975), p. 58 n. 20.

[26] D. Davie, 1991, p. 79.

[27] D. Davie, 1975, pp. 46–8.

[28] As mentioned in note 5, the discovery took place around 1914–5.

[29] E. Pound, *Selected Letters*, op. cit., p. 178.

[30] E. Pound, *How to Read*, op. cit., p. 33.

[31] W. Ramsey, op. cit., see particularly pp. 204 ff. for the relationship between Laforgue and Pound.

[32] W. Ramsey, op. cit., p. 135.

[33] W. Ramsey, op. cit., pp. 135–6.

[34] W. Ramsey, op. cit,. p. 138.

[35] W. Ramsey, op. cit., p. 207.

[36] D. Monk, op. cit., p. 85.

[37] J. Hoogestraat, op. cit., pp. 259–60.

[38] J. Hoogestraat, op. cit., p. 263.

[39] J. Hoogestraat, op. cit., p. 272.

[40] J. Hoogestraat, op. cit., p. 276.

[41] J. Hoogestraat, op. cit., p. 284 n. 34.

KIYMET SELVI

PHENOMENOLOGY: CREATION AND CONSTRUCTION OF KNOWLEDGE

ABSTRACT

This paper proposes a discussion about the creation and construction of knowledge through the phenomenological way of searching for meaning. Individuals continuously deal with creating meanings of their own lives. Each individual follows a unique way in order to create and construct meaning in any situation. The ability of learning that can be defined as a natural and inner intention of becoming self in the world can improve individual's learning. The learner, as a meaning maker, creates new knowledge of the whole life process. Constructing the meaning of a phenomenon is the individual's self-inquiry. Descriptions of concepts continuously change and new meanings of concepts are acquired. Self-inquiry about life can be described as the individual's self-learning. Creation and construction of new knowledge corresponds with the individual's ability of learning. Learning improves the capability of the individual as a self-creator and develops phenomenological understanding of life. Creation and construction of new knowledge is also concerned with individual's learning ability, creative capability, freedom, subjectivity, way of thinking and perception of a phenomenon.

Phenomenological investigation is a key method of searching for meaning of life. This search develops personality so that the individual is interested in not only materialistic aspect but also spiritual aspect of his/her personality. This search can also help the individual to form his/her own personality depending on the creation and construction of the meaning of the world. Phenomenological learning should motivate the individual to form his/her personality for searching and constructing the meaning of life. Self-inquiry about life can promote creation and construction of new knowledge. Meanings develop within the endless conscious and unconscious processes in which new knowledge and products are created. The process of creation and the results of phenomenological inquiry cannot include verifiable knowledge. This process and results occur uniquely and authentically because of the individual's self-interpretations of the world.

INTRODUCTION

Knowledge is the abstract of the individuals' experiences corresponding to searching for and capturing the meaning of the phenomenon. Individuals construct their own meanings in order to create their own knowledge. Knowledge can be created in different ways depending on the individualistic bases such as capability of intuition,

A.-T. Tymieniecka (ed.), Analecta Husserliana CX, 279–294.
DOI 10.1007/978-94-007-1691-9_23, © Springer Science+Business Media B.V. 2011

perception, imagination and creativity. Thus individualistic bases are established upon ready-made knowledge such as scientific studies, cultural heritages of humans, historical process, social rules and customs of societies. Individuals also observe phenomena in daily life in order to create and construct knowledge. Moreover, they criticize and analyze past events, current situations and future possibilities while creating and constructing knowledge. Following this process, individuals create their own philosophies and values; make their own choices and preferences in their own lives. According to Tuomi "Knowledge consists of truths and beliefs, perspectives and concepts, judgments and expectations, methodologies and know-how knowledge has to be extracted from its raw materials, and in the process, meaning has to be added to them" (1999, 110). As the individuals create and construct their own knowledge by using different tools from their lives, they become self-interpreters.

Individuals mostly create and construct knowledge by referring to their own inner and outer worlds. The inner and outer worlds of individuals change from one individual to another. It is very difficult to explanation the individuals' inner worlds due to the complexity of inner worlds. Inner world is comprised with the metaphysical world which involves mystical and secret issues for human comprehension. Moreover, there is no clear explanation of how the individuals create and construct knowledge by means of their own inner and outer worlds. In this context, knowing is essential for humans to become self-beings in their own lives. Human beings are always busy with creating and constructing the knowledge of phenomenon to catch the meaning of life. Furthermore, knowing enables formation of personality and self-being. It supports individual development and triggers creative capability of humans and this provides them with the opportunity of self-actualization. Dewey sated that ". . . knowledge, even the most rudimentary, such as is attributable to low-grade organisms, is an expression of skill in selection and arrangement of materials so as to contribute to maintenance of the processes and operations contributing life" (1958, 290). It means that all organisms have their own processes that they need to realize activities and exist as self-beings. Similarly, individuals need very high levels of human activity and creativity for creation and construction of knowledge.

Knowledge is an essential tool for organizing the individualistic and societal life. But, it is not easy to decide about what type of knowledge is needed. The type of knowledge needed may change depending on the lives of individuals. Bonnett (1999, 316) asks the question of "what kind of knowledge will best illuminate and equip us to deal with issues of sustainability?". This question is very important for creating and constructing knowledge in lives of individuals. Different types of knowledge can introduce different receipts for managing the life process. Individuals may need a certain type of knowledge in certain stages of their own lives.

The way of constructed and created of knowledge may change based on the shifts in dominant paradigms. It is known that positivist and qualitative research paradigms were dominant in the past whereas the qualitative research paradigm has been dominant for the last thirty years. Changes in the current research paradigms affect current research methods and this is called as paradigm shifts. A research paradigm introduces different ways of searching for meaning. A shift in a research paradigm may result from shifts in the types of knowledge and the ways of search-ing for meaning. "The research paradigm shift has to do with major shifts in the way of knowledge is constructed and created" (Campbell 2010). The research paradigm introduces different ways of creating and constructing knowledge. The aim of this paper is not explaining the past and the current paradigms and paradigm shifts. The individualistic ways of searching for meaning, the pathway for and stages of creating and constructing knowledge are discussed in this paper. This paper aimed at discussing only the creation and construction of knowledge based on the phenomenological way of searching for meaning.

THE WAYS OF SEARCHING FOR MEANING

Phenomenological understanding follows the hermeneutic methodology while cre-ating and constructing of new knowledge. The hermeneutic methodology is the way for catching new meaning of phenomenon based on the individualistic perceptions and experiences. Individuals become self-creators while creating and constructing knowledge by means of hermeneutic methodology and phenomenological under-standing so that every individual becomes a researcher and reflects his/her own meaning. Phenomenology has been adopted as the appropriate way of exploring "the essence of lived experience" in order to find a way of constructing knowl-edge (Campbell 2010). A phenomenologist who studies in different disciplines may create new research methods based on the experience that did not exist before. The method of searching for meaning can be differentiated based on the phenomeno-logical research paradigm. Phenomenological inquiry methods different from the ones applied in the present and past will be applied in the future owing to the fact that phenomenological search for meaning will be important for creation and construction of knowledge.

Phenomenology associates prior knowledge with everyday experience. Individuals interacting with phenomenon catch and construct new meanings based on their past and current experiences. The constructivist thought, that is interested in creation and construction of knowledge, prevailed in 1990s. Piaget, Dewey, Husserl, Kuhn and Vygotsky are the well-known constructivists who has tried to explain how individuals create and construct their own knowledge. According to Dewey, a learner actively constructs in his/her knowledge by means of his/her own learning experiences in his/her environment (cited in Morphew 2000). An individual utilizes his/her own subjective life and environment in order to catch and construct new meanings. In the process of creation and construction of knowledge, the individual may use his/her own subjectivities.

Subjectivity is the main source of knowledge for individuals. "Subjectivity is defined as naturalistic, anarchic and authentic human perceptions which are ab-stractions of the knowledge of life experiences" (Selvi 2009, 8). The subjective knowledge can be defined as the individual's first perspective in which no scientific test based on the positivist understanding is applied to individualistic perceptions. Thus, many of the artistic, scientific and creative studies root in the subjectivity of individuals. Some scientists and artists such as Leonardo da Vinci, Voltaire, Rousseau, Lessing, Hegel and Chernyshevsky viewed creativity as the subjectivization of the idea (Kurenkova et al. 2000). Subjectivity provides unique perspectives of individuals and this is a way of getting their authentic bases. These authentic bases enable creation of subjective knowledge that corresponds to the first phase of creation of scientific knowledge. Dewey pointed out a method of knowing and he called this method as the "introspection." That is totally different from the concept of observation. This method is an inquiry about the meaning of phenomena and it is totally different from the epistemological inquiry. Subjectivity is the main source of creating and constructing authentic and new knowledge as can be seen in Figure 1.

In Figure 1, it can be seen that different individuals can create different knowledge even if they can perceive the same phenomenon. The main question is why different individuals create different meanings and construct different knowledge when they perceive the same phenomenon. The answer to this question is that individuals become authentic and subjective self-beings while creating and constructing knowledge. Another question is what affects individuals' construction process of knowledge and why the knowledge constructed process differs from one individual to another. The answers to these questions can be related to the subjectivity of individuals. But, there is not an adequate answer to these questions.

Subjectivity of individuals may lead to differentiation in the process of creating and constructing knowledge.

In order to explain the subjectivity and authentic bases of creating and constructing knowledge, an example from daily life can be given. Many radio listeners may listen the same radio program and the same song x at the same time as seen in Figure 1. However, each listener's feelings, sensations about the same song, tastes of the same song and meaning he/she gives to the same song can be different from one another's. While listening or after listening the song x, one individual's feelings and imaginations and experiences related to the song x must be unique. For example I am a listener of the song x, I know just my inner situation, my own

Figure 1. Subjectivity and authentic knowledge

experiences, feelings and have knowledge and these can be different from other listeners' immediate experiences of the song x. Anyone's senses of the song x can not
be same with my senses of the same song. I have knowledge about the meaning of
my self-experiences. Only I have totally awareness of knowledge about my experiences related with listening to the song x. Each individual's experience of the song x
and creation and construction of his/her knowledge about it are private for him/her.
What happens when each individual listens to the song x and reaches new meaning
of the same song? Learning theories may give some answers to this question but,
these answers may not provide adequate description of phenomenological understanding. It is very clear that only individuals themselves are aware of the knowledge
of the song x and why their knowledge differs from others' knowledge. According
to Dewey (1958, 301–308), knowing occurrences of the only existence able to know
its "own" states and process that immediate and intuitive self-knowledge of the individuals. Each individual's knowing in its particularity can be explained based on
his/her own subjectivity.

<h2>THE PATHWAY AND LAYERS OF SEARCHING
FOR MEANING</h2>

Meaning can be actively created by means of individual's conscious perception of
it. The individual can create his/her knowledge based on his/her own social, biological and metaphysical being. These features of the individual can affect his/her
own meaning of the phenomenon. Thus, creativity and construction of knowledge
become very complex tasks for the individual.

Searching for meaning, composed of seven layers, has a complex structure that
is too ambiguous for individuals to comprehend and thus it is explained by means
of Figure 2. The Figure 2 is prepared to provide visual description and presentation
of the layers and pathway about individualistic ways of searching for meaning and
creating and constructing knowledge. The pathway of individualistic searching for
meaning may be explained in seven layers such as spirituality, will to know, intuition, perception, imagination, creativity and knowledge. These layers are discussed
briefly in this paper. These layers are ranked in a linear and curvilinear pathway as
can be seen in Figure 2. This pathway begins with spirituality and ends with knowledge but each layer can feed all the others layers. For example, if individual reaches
new knowledge of phenomenon, this knowledge can be feedback for the other six
layers. That is, the relationship is not only liner but also curvilinear.

Searching for meaning, creation and construction of knowledge can be explained as "learning" or "experience" of individuals. It is known that many learning
theories give some explanation of the forms, process and models of the learning.
Nevertheless, learning theories do not provide sufficient and adequate explanation
about the phenomenon of individualistic ways of learning. Therefore, new explanations about individuals' ways of learning are needed. The act of learning may
be explained in terms of layers as shown in Figure 2. Current forms and models of

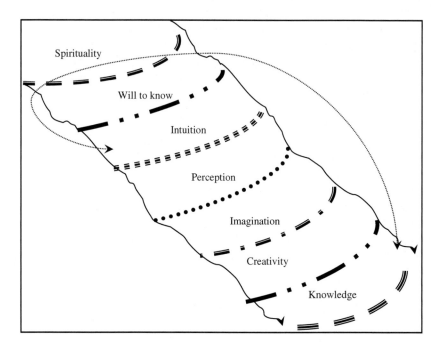

Figure 2.   The layers of searching for meaning

learning should be reanalyzed in order to improve pathways of learning or searching for meaning.

Spirituality of the individual is the first layer of the searching for meaning. Spiritual nature of the individual has a broader aspect that is not restricted with the individual's belief, values and understanding of mysticism. Spirituality may be composed of two courses. The first course is the inner process or inner side of individuals, that is mystical and spiritual aspect of individualistic life.. The second course is social experiences about individual's outer world including social rules, cultural heritages, and daily life experiences. The individual's inner world is the outcome of human soul and mind that is the individual intentionality becomes a self-being in his/her life. Inner process of the individual works based on his/her sense of knowing his/her existence. Spiritual nature of the individual can affect his/her self-value, ethical and moral aspects. Spirituality of the individual can explain metaphysical world for human. According to phenomenologist, phenomenological search develops personality so that the individual is interested in not only his/her materialistic side but also spiritual side (Cozma 2007). It means that individual should develop his/her materialistic and metaphysical sides to realize self-actualization.

Spirituality refers to super-natural power of humanities and it covers the knowledge about mind, body, intellect, mentality and soul of the individual. It is not easy for the individual to get this kind of knowledge from his/her spiritual world in his/her life. It is believed that the real knowledge about the universe is held in the spiritual world and the individual with a strong spiritual nature may capture a certain amount of real knowledge from his/her spiritual world. Although the individual becomes a spiritual being in his/her life, he/she cannot fully understand becoming a spiritual being. The first layer is so mysterious that it is concerned with ontological knowledge for the individuals. The first layer can be defined as the metaphysical knowledge about the nature of reality. Metaphysics deals with trying to understand the meaning and nature of the life and the reality about the universe. The individual is full of senses and eager to gain knowledge about his/her spiritual perspective. He/she is always ready to jump into spiritual world but capability of him/her may create barriers for him/her to touch or enter into his/her own spiritual world.

Spirituality is the power for individualistic and societal development. According to Saeed, "in the broader sense, spirituality is an inner uplift for the individual as well as society" (2008, 267). Spirituality is a necessity for the individual in the sense that it feeds up the mystic side of the human. The human is always concerned with mystic world and desires to know about it. This desire encourages the individual to create and construct knowledge for organizing his/her own life. Cozma sated that "…the man being interested not only about his material, but also about his spiritual welfare" (2007, 31).

The will to know is the second layer of the searching for meaning. The will to know is the individual's intrinsic power which can stimulate him/her to act for knowing. The individual's act for knowing activates inner process of the individual and this can be called as the will to know. The will to know is related to becoming a self-being in life and it promotes self-actualization of the self-being. And it is also that it can create energy in the life process of the individual. The will to know encourages the individual to acquire knowledge from his/her inner and outer worlds and to manage and accomplish his/her own life. The individual becomes a self-being by means of his/her own will to know and accomplishes his/her existence.

Davis (1995) stated that the individual's experiences and knowledge about world comes from the individual actively being in the world. It means that as the individual is as a biological creature, he/she has a tendency to act to know. The will to know has been a main topic of all philosophical and scientific studies beginning with the Aristotle. According to Tymieniecka (2004, 7) "philosophy and the other sciences have followed distinct but parallel paths, partly nourishing each other, partly promoting each other's progress." It means that the philosophy and sciences are deal with understanding individual's will to know in order to support self-actualization. The will to know is a tool for the fulfillment of both the individual and the others. Fulfillment of the self-actualization is the main goal for the individual and the will to know is the main force for it. Will to know can be defined as the energy that supports the individual's self-actualization. It is said that will to know comprises very important issues for philosophy and positive sciences.

Intuition is the third layer of the searching for meaning. It can be defined as the ability to acquire knowledge without inference, the use of reason or results. Intuition corresponds to the inner powers of individual and the individual may not need outer supports to know about phenomenon. The intuition has a mystical aspect that "looks inside" while focusing on the senses of the will to know. The intuition activates the individual to gain knowledge that may not be needed to justify. The intuition, a special observation through the mystical and the metaphysical world, activates innovative acts and creativity of the individual. It refers to the internal energy for perceiving the phenomenon to catch new and authentic knowledge. The ability of sensing of the phenomenon can be fostered by means of the intuition of the individual. It is connected with the spiritual nature of the individual and the spiritual nature of the individual is an inscrutable process, there is not sufficient knowledge to explain the process of intuition. The results of the intuition process can be seen as innovative and creative acts of individual but the process of the intuition can not be visible for individual. However, the intuition promotes the search for meaning and the self-learning in life.

The layers of spiritual nature, will to know and intuition are not clear issues for human understanding and they compose of the hidden capacity of the individual. These three layers are related to the nature and self-being of the individual. It can be very hard to explain how these layers affect the process of creation and construction of knowledge. These three layers composing a hidden space for materialistic world can be defined as the metaphysical world for the human being. It seems like there is a horizon between the first three layers and the last four layers. This horizon may occur in different places, as seen in Figure 2, depending on the power and vision of the individual. The layers of spiritual nature will to know and intuition may work unconsciously and spontaneously.

Intuitions and perceptions compose the source of data in phenomenological descriptions. That is, intuitions and perceptions are used to form phenomenological knowledge. The intuition leads individual to the object that will be described. Following this, the individual is consciously inclined towards the object and the process of perception begins. The intuition makes events and objects ready to be perceived. According to phenomenology, intuitions and perceptions provide the basis of knowing and the knowledge based on intuitions, perceptions and obser-vations should be reflected in appropriate forms. Phenomenological knowledge constructed by the individual becomes available by means of the phenomenolog-ical reflections. Phenomenological perceptions, phenomenological experience and phenomenological reflections comprise the parts of a whole.

The fourth layer is the perception of life to search for meaning. Perception can occur in two ways and the two ways have different patterns in the process of search-ing for meaning. The first way is related to the individual's internal sensations that inform him/her about developments in his/her body such as being trusty, walking, feeling hungry. The second way is related to the individual's external *sensations* that inform him/her about the world outside his/her body. These two ways provides senses based on which the individual can create and construct knowledge. The pro-cess of searching for meaning is connected with the first and the second ways. Both ways of sensing support the process of searching for meaning and interpretation of life and this means creation and construction of knowledge.

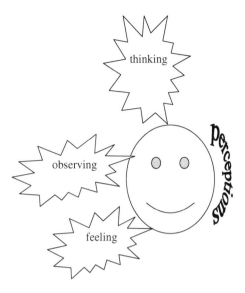

Figure 3.   The perception of searching for meaning

The fourth layer is the perception of life during the search for meaning as seen in Figure 3. According to Barbaras (2003, 160) "meaning of life is linked to possibility of perception from life; thus the meaning of this life will take shape in contact with perception." The ability of perceiving is the main force behind creating individualistic meaning of phenomenon. Creation of meaning results from individual's interpretation of his/her own perception. Perception may occur within the individual's inner world. If the individual reflects his/her perception of the phenomenon to other individuals, the others can understand what and how he/she perceives. Perception becomes the main gate between inner and outer worlds of the individual because it can provide the knowledge from external and internal worlds. The power of perception gives a chance for creating and constructing knowledge.

The fifth layer comprises the imaginative nature of the individual and imagination is a primary means of knowing about phenomenon as shown in Figure 4. It has been mentioned that there are seven layers in the pathway of searching for meaning. But there is a ambiguity about whether the creativity, the sixth layer, comes before the imagination, the fifth layer, or not. In this paper, there isn't any explicit answer to this question. However, it is only assumed that the imagination comes before the creativity.

Imagination is a kind of mental experience including reality and unreality and images. The individual doesn't need any equipment, any place, any action, anybody while imagining. Imagination is the colorful, enjoyable, creative and silent experience of the mind. It is referred as the freedom of human mind and the mental experiences of the individual and corresponds to the untouched and hidden gardens of human life. Imagination corresponds to the uniqueness of mental activity and

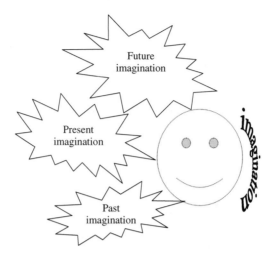

Figure 4.   The imagination of searching for meaning

depends on the individual capacity of spirituality, will to know, intuition and perceptions of life (Selvi 2006). Imagination is the ability of analyzing past, present
and catching future possibilities.

Depending on the time of imagination, imagination can be grouped as past imagination, present imagination and future imagination. Past imagination comprises the
imaginative experience the individual had in his/her past. It is also said that the
individual catches imaginative ideas and thoughts by means of his/her imaginative
past experience. Imaginative experience the individual had in the past forms the
basis of the present and future imagination. Past imagination can be defined as the
individual's history of searching for meaning in his/her life. Present imagination
is composed of the individual's immediate experience about phenomenon. When
the individual perceives a situation, he/she can imagine his/her perception. Present
imagination might enable solution of problems, changes in direction of the current
patterns and nourishment of the future imagination. Imagination of the future refers
to creativity of the individual and is mostly called as the creative imagination. Future
imagination is more important than other imagination types because it provides the
individual with the possibilities of searching for meaning in the future.

Capability of creativity is the sixth layer in the pathway of searching for meaning.
Creativity is not only a philosophical and aesthetical problem but also the main
problem of scientific study and knowledge. Creativity can be defined as the ability
to remember past, live in present and foresee future and create unique forms of
things and/or processes of becoming as shown in Figure 5. If an individual forms
frames of his/her past experience, he/she will find many solutions to problems and
show very creative acts (Selvi 2006). Nevertheless, mature creative experiences of
others become barriers to the individual's new creations and he/she does not act
creatively in his/her life. The main delusion about creativity is that only certain

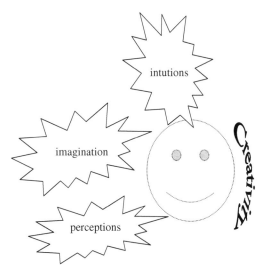

Figure 5.   The creativity of searching for meaning

individuals have creative capabilities. Every individual has the power of creativity but only certain individuals know how to reflect this power. Cozma stated that "... creativity not only in artistic achievement, but in self achievement self-fulfillment" (2007, 37). Self-actualization of the individual is composed of totally creative acts in his/her life and the capability of creativity is the main source of becoming a self being in life.

The individual has the power of creativity because he/she becomes a self-being in the creative process. Since the individual has creative experiences related to becoming a self being in life, creativity is the vital force for him/her. "Creativity was seen there as the human capacity hidden in an irrational depth of the human being-in intuition, will vital force, vital spontaneity the unconscious, etc." (Kurenkova et al. 2000). It means that creativity reflects the individualistic capability of catching new meaning of phenomenon. As the individual has the unique capability of creativity, he/she creates new and authentic ways, processes, products, methods, and ways of thinking, questions and answers to questions. Moreover, creativity comprises new and authentic reflections.

Creativity can support creation of new and authentic knowledge. Individual is creative in his/her whole life because creativity works very well in each stage of his/her life. Creativity provides the individual with the autonomy and freedom of creating and constructing of knowledge. In the course of creativity, creative imagination becomes more intense compared to other acts of the individual. This situation is very common in artistic studies in which knowledge is created and reflected. The search for meaning, corresponding to self-interpretation of the phenomenon must comprise creativity. Creativity reflects the uniqueness of the individualistic base, that is, the individual's own experiences in life.

The seventh layer of searching for meaning is the knowledge. It can be referred as the individualistic interpretations of life. Individual's self-interpretation of the phenomenon or life is unique to him/her. And, this can be the individual's aim of existing in the world, as a creative being. Self- interpretation is very important for the individual to actualize the aim of becoming a self-being. But it is known that the individual becomes aware of the phenomenon by means of others' interpretations of the world before his/her own interpretation. Since education system promotes the adults' interpretations of the world, the individual may not be to the self-interpreter of the world. Coucerio-Bueno sated that "educational theorists fail to highlight the aspects of human sensitivity and intelligence and from an early age, children become aware that the world is interpreted by adults" (2007, 373). This reality can establish barriers to the individualistic ways of creating and constructing knowledge through self-interpretation.

In the last century, the empirical model has been heavily criticized by, several prominent philosophers of knowledge, such as Bergson, James, Husserl, Heidegger, Mead and Merleau-Ponty. Criticisms of the empirical model have focused on the problem of objectivistic and empiricist knowledge from somewhat different directions (Tuomi 1999). These criticisms reflect that the individual can not be free to cerate and construct his/her own knowledge. Phenomenology clarifies the ways that individual constitutes his/her reality and life. The ways followed by an individual to create knowledge may not be applied by others. Thus the ways followed by the individual is unique to him/her.

The individual has the chance to self-actualization depending on what he/she experiences in his/her own life. Experiences of the individual refer to his/her total effort to search for meaning in the seven layers. The concept of "experience" connotes a very broad and complex process of human endeavor to create and construct knowledge. The meanings of "experience" and "learning" are the same. The individual becomes a self-creator within the pathway composed of layers of the search for meaning and the process of learning. The individual as a self-creator becomes an interpreter in life. At the end of the interpretation process, new knowledge can be created and constructed by the self. This process includes learning and creation and construction of knowledge.

The pathway and the layers of searching for meaning constitute a very discussible and hard topic for philosophers and scientists who give some explanation about searching for meaning as seen in Figure 2. In this pathway, the first, second and third layers might be mistier than the other layers for creation and construction of knowledge. It seems like there is a wall occurring as a frosted glass and it is called as a horizon after the first three layers of the pathway in this paper. This glass wall hides the layers of spiritual nature, will to know and intuition and the individual may see some reflections of the last four layers. However, it is hard to see what happens in the first three layers. The individual at the last four layers may catch some reflections if he/she has power or some senses such as intuitions but these reflections are too ambiguous for him/her. The reflections are the evidence of the fact that some layers come out of the individual's visions. The individual's awareness of the knowledge of some layers is hidden and a secret for him/her and there is some horizons or borders between the first three layers and the last four layers that activate the individual's perception to create and construct new knowledge.

The wall between two groups of layers might force the individual to search for meaning in order to realize moral development and self-actualization. If we look at the horizon between two groups of layers, we can not see what happens or what becomes before the horizon. But we know that even if our vision can not go beyond the horizon, many different things may occur there. The layers of spiritual nature, will to know and intuition can be hidden behind the individual's vision. The horizon is the changeable depending on the point of the individual's vision and perception. Individuals may have the capability of forming images behind this horizon. These images may turn to personal knowledge and experiences about metaphysical world. It can be said that the power of the individual's perception introduces new horizons for new searches of meaning of life.

DISCUSSION

Philosophy and science follow different pathways of creating and constructing knowledge and this has created many problems of fully understanding the phenomenon. Different ways of searching for meaning may lead to crises for the individual who follows an unnatural trend for creating and constructing of knowledge. Philosophers and scientists make their own explanations but try to avoid touching on each others' explanations. They also follow different pathways while creating and constructing knowledge. This attitude results in total differentiation in the ways of searching for meaning. According to Bolton (1979, 255), empirical research results and subject matter understanding become dominant in the educational area that includes too abstract knowledge existing in various forms. Whereas philosophical discourses and explanations are carried out by some philosophers, these philosophers are not forced to follow scientific research methods. This tendency has led to crises in the scientific studies as well as the field of philosophy. To cope with the crises of the scientific and philosophical studies, paradigm shifts have begun to be discussed.

Dominant paradigm of searching for meaning is based on the positivist understanding which is criticized in terms of the crisis of the science. The method of searching for meaning of phenomenon has shifted from the positive research paradigm to the qualitative research paradigm. Qualitative research paradigm is based on the descriptive analysis that relates to spontaneous and intentional experiences of the individuals. Qualitative research reflects the subjectivity of the individual in the process of creation and construction of authentic knowledge and it is related to phenomenology. Phenomenology has the potential to use different forms of creating new and authentic knowledge. According to Tymieniecka,

... the new philosophical paradigm, with actual transformations going on in scientific research, method, is course, there is possible, and has begun, a most illuminating dialogue between philosophy and science. I mean here the dialogue that phenomenology/ philosophy of life has begun with sciences of life (2004, 11)

Tymieniecka points out that scientific approach is in the way of great changes and the new paradigm mostly regards that the individual's subjectivity and authenticity comprise the way of creating and constructing knowledge. The new paradigm supports the human subjectivity that develops new and authentic meaning of phenomenon. This subjectivity also includes a high level of creativity and imagination. The pathway and the layers seen in Figure 2 reflect totally new and authentic creation and construction of meaning based on the subjectivity of the individual. The creation and construction of knowledge presented and discussed depending on the pathway in Figure 2 is also very similar to the phenomenological way of thinking and searching for meaning. It is said that phenomenology provides the opportunity to investigate the meaning of life. The individual constructs meanings in his/her life and guides his/her own actions and experiences depending on meanings he/she constructs. Phenomenology is concerned with the nature of the meaning of phenomena.

Creation and construction of new knowledge corresponds to the concept of phenomenological learning. Thus, learning can be defined as the creation and construction of the meaning of phenomenon. Ability of learning can help individual's learning that can be defined as the natural and inner intention of becoming a self-being in the world. Learning is a dynamic process in the individual's life and it is also the energy needed for becoming a self-being and the source of this energy is comprised of the individual's body and soul. This energy supports the internal and external conditions of the self. If learning doesn't find good supporters in the external world of the individual, his/her ability to learn can be damaged. A part of learning occurs in school and learning is planned and applied based on certain principles in the learning-teaching environment. As an educator I ask the question that how the ability to learn can be supported by means of the external learning environment, especially in school. It is known that creation and construction of new knowledge is related to learning ability of individual, freedom of individual, subjectivity of individual, individual's way of thinking and perception of phenomenon. Learning can also support the individual to construct his/her self-being on his/her own.

I struggled to find the meaning of the concept of "learning" while I was preparing the paper and this was really hard for me because I had some sense of or intuition about learning. But, it was not easy to catch, explain or reflect my own meaning of learning. I had just some sense of learning and I wasn't able to explain it. This was an ambiguous situation and it also disturbed me. I asked myself why I decided to write about this topic. I decided to leave the topic because I felt that my explanation was not clear for me and I wanted to search for meaning of the concept of "learning" while I was preparing my paper. But after I left the topic for a while, again I came back to the same topic, because this issue was unconsciously on my mind. This situation really disturbed me and I decided to write and find some descriptions of and solutions to my own problem about the concept of "learning." This process has taken almost five years of my life.

I had a sense that it was very important to reflect my perception of the concept of learning. However, sense of anyone else who has a perception of this topic should be important as much as my own senses. This sensation is based on the phenomenological perception of the phenomena. I have been brave to reflect this kind of phenomenological self perceptions. It is said that firstly phenomenological learning occurs and then phenomenological reflection comes out. Individual's learning needs to be reflected and discussed in order to catch the new meaning of the phenomena that is creation and construction of new knowledge.

Learning improves the behaviors of the individual as a self-creator and develops phenomenological understanding of life. Since searching for meaning is a new method of learning and gaining new knowledge, the individual becomes a creator of the new knowledge. Phenomenological investigation might be a key method of searching for meaning of life. This search develops personality so that the individual is interested in not only his/her materialistic side but also spiritual side (Cozma 2007). This search can help the individual to form his/her own personality. Phenomenological learning should activate the individual's self-creation to search and construct the meaning of life.

Construction of the meaning is related to learning that continually improves the meaning. Learner, as a meaning maker and a self-creator creates new knowledge of the whole life process. Constructing the meaning of phenomenon can be defined as a kind of self-inquiry and is related to descriptions of meanings that always change and reach new meanings. The meaning develops within the endless conscious and unconscious process in which new knowledge and process are created. The creation process and the results of phenomenological inquiry can not include verifiable knowledge. This process and these results occur uniquely and authentically because of the individual's self interpretations of the world.

The individual expresses his/her own ideas, intuitions, concerns, feelings, emotions, reasons, interests, desires, needs, aims, ideas, senses, thoughts, actions, intentions through self-interpretations of the meanings he/she learns. Self-interpretations of the meanings are completely creative self-knowledge about life. Self-interpretations help the individual to reach the unique self-knowledge depending on his/her self-basis that includes spiritual nature, will to know, intuition, perception, imagination and creativity of him/her. The meaning of phenomenological life is connected with the individual's ability to learn by using the phenomenological method and this can make the individual a self-creator.

The individual searches for catching the deeper meaning of his/her own experiences while applying the phenomenological method. Phenomenological method helps the individual to construct his/her own new knowledge. Learning is the individual's creative function that improves the creative potentiality of his/her life. Learning takes place in the individual's life composed of social, physical and mental situations. Learning has cognitive, affective and social dimensions. These kinds of multiple constituents can affect the individual's learning preferences which refer to individual's unique learning styles. The pathway and layers of searching for meaning are also related with the individual's learning styles. The individual's self-creation of the meaning can comprise all of the seven layers of searching for meaning. But these layers may not be sufficient to explain the search for meaning for creation and construction of knowledge. These layers must be improved by means of different discourses on the topic. These layers and the pathway of creating and constructing knowledge also need to be re-analyzed in a deeper and broader sense based on the pathway of individuals' searching for meaning.

Anadolu University, 26470 Eskisehir, Turkey
e-mail: kselvi@anadolu.edu.tr

REFERENCES

Barbaras, R. 2003. Life and perceptual intentionality. *Research in Phenomenology* 33:157–166.
Bolton, N. 1979. Phenomenology and education. *British Journal of Educational Studies* XXVII(3): 245–258.
Bonnett, M. 1999. Educational for sustainable development: A coherent philosophy for environmental education? *Cambridge Journal of Education* 29(3):313–324.
Campbell, B. 2010. Phenomenology as a research method. http://www.staff.vu.edu.au/syed/alrnnv/papers/bev.html. Accessed 30 Mar 2010.
Coucerio-Bueno, J.C. 2007. Education without paideia: A phenomenological view of education today. In *Analecta Husserliana: The yearbook of phenomenological research*, ed. A.T. Tymienecka, Vol. XCIII, 371–383. Dordrecht: Springer.
Cozma, C. 2007. *On ethical in the phenomenology of life*. Roma: Edizioni Eucos.
Davis, K. 1995. The phenomenology of research: The construction of meaning in composition research. *Jac: Rhetoric, Writing, Culture, and Politics* 15(1):1–7.
Dewey, J. 1958. *Philosophy of education*. USA: Littlefield, Adam & Co.
Kurenkova, R.A. et al. 2000. The methodologies of life, self-individualization and creativity in educational process. In *Analecta Husserliana LXVII*, ed. Anna-Teresa Tymieniecka, 195–205. Dordrecht: Kluwer.
Morphew, V.N. 2000. Web-based learning and instruction: A constructivist approach. In *Distance learning technologies: Issues, trends and opportunities*, ed. Linda Lau, 1–15. London: Idea Group Publishing.
Saeed, K.M. Spiritual experience and the foundation of education: A Tagorean approach. In *Analecta Husserliana: The yearbook of phenomenological research*, ed. Anna-Teresa Tymieniecka, Vol. XCV, 267–277. Dordrecht: Springer.
Selvi, K. 2006. Learning and creativity. In *Analecta Husserliana: The yearbook of phenomenological research*, ed. A.T. Tymienecka, Vol. XCIII, 351–369. Dordrecht: Springer.
Selvi, K. 2009. Phenomenological learning in our living reality. Paper presented in 59th International Congress of Phenomenology, 7 July 2009, University of Antwerp, Belgium.
Tuomi, I. 1999. Data is more than knowledge: Implications of the reversed knowledge hierarchy for knowledge management and organizational memory. *Journal of Management Information Systems* 16(3):107–121.
Tymieniecka, A-T. 2004. The pragmatic test of the ontopoeisis of life. In *Phenomenological inquiry*, ed. Anna-Teresa Tymieniecka, Vol. 28, 5–35. Watertown, MA: Evans & Faulkner Inc.

AYDAN TURANLI

PERSPICUOUS REPRESENTATION: A WITTGENSTEINIAN INTERPRETATION OF MARTIN HEIDEGGER'S VIEW OF TRUTH

ABSTRACT

Martin Heidegger criticizes the representational view of language and truth from the perspective of phenomenological ontology. Primarily he criticizes the presupposition that the content of an idea is an object it stands for and that judging is related to having a representation of an object in our minds, in our consciousness. Heidegger's critique of the representational theory of truth and language goes hand in hand with his critique of modernity. He thinks that in the West, thought about thinking resulted in a discipline of logic gathering special knowledge concerning a special kind of thinking, which is called logistics. Logistics is considered to be the only possible form of strict philosophy because it is integrated with the technological universe, which exercises power over other disciplines in our era. As a result of this, thinking is transformed into one-track thinking generating an absolute univocity. In this paper, I concentrate on Heidegger's critique of the representational view of language and truth by correlating it with that of Wittgenstein, and by focusing on what sense Heidegger's critique of the representational theory of truth is related to his critique of modernity. The first part of the article discusses the critique of the representational theory of truth, the second part of the article presents Heidegger's alternative and the third part deals with his critique of modernity.

Heidegger presents a different approach for an analysis of knowledge and truth one which does not require traditional distinctions. In both *What is Called Thinking?* and *Being and Time* Heidegger concentrates on the question of whether we can identify knowledge and truth in terms of assertion and judgment. He criticizes the view of truth that necessitates an agreement between a judgment and a fact, which presumes that judging is a "*Real* psychical process" and that which is judged is an ideal content (Heidegger 1962, 258–259 [216]). The basic question of the view, assuming that there is an agreement between the ideal content of judgment and the real psychical process, is "What is the ontological relation between an ideal entity as a fact and real psychical process?" (Heidegger 1962, 258–259 [216]).

Heidegger thus questions an established opinion in philosophy that assumes that an idea is called correct in case it conforms to its object. This correctness in the forming of an idea is equated with truth. For example, the statement "It is raining," is correct in case it directs the idea to the weather conditions. In terms of the contemporary cognitivist John Searle, the direction of fit is from words to the world or from the mind to the world in this case (Searle 1979, 3).

A.-T. Tymieniecka (ed.), Analecta Husserliana CX, 295–305.
DOI 10.1007/978-94-007-1691-9_24, © Springer Science+Business Media B.V. 2011

Heidegger states that judging is forming ideas correctly or incorrectly according to this established view. Forming an idea or a representation, on the other hand, is supposed to be having a representational idea of related objects in our minds or in our consciousness or in our soul (Heidegger 1968, 38–39). Hence, when we say "That tree is blossoming" our idea has, again in John Searle's terms, "the direction of fit" from words to the world namely toward the object of the blossoming tree. The crucial question here is whether the ideas inside us answer to any reality at all outside ourselves.

Traditionally, this event of forming ideas is correlated with the process taking place in the sphere of consciousness or in the soul, as some philosophers such as Descartes presume. Heidegger questions this cognitivist attitude by saying, "But does the tree stand 'in our consciousness,' or does it stand on the meadow? Does the meadow lie in the soul, as experience, or is it spread out there on earth? Is the earth in our head? Or do we stand on the earth?" (Heidegger 1968, 43).

In what ways does Heidegger criticize this concept of truth and what is his alternative? The next section discusses Heidegger's critique of the representational theory of truth.

CRITIQUE OF THE REPRESENTATIONAL THEORY OF TRUTH

Heidegger criticizes the representational theory of truth that assumes a true judgment regarding a fact corresponds to a real psychical process. His critique focuses on several implications of the representational theory of truth. One of the implications of the representational theory of truth is that it requires the definition of thinking and logic by means of propositions. Heidegger says, "When we ask our question 'What is called thinking?' ... it turns out that thinking is defined in terms of the λόγος. The basic character of thinking is constituted by propositions" (Heidegger 1968, 163). According to this view, which finds a clear expression in Plato and Aristotle, only the part of language dealing with propositions is important.

Heidegger's critique of the traditional understanding of truth goes hand in hand with his critique of a traditional view of language that implies that every sentence is proposition. In a Wittgensteinian manner, Heidegger underlines that every proposition is a sentence, but not every sentence is a proposition. In the *Philosophical Investigations*, Wittgenstein, like Austin, states that not all of our sentences are statements: there are questions, commands, exclamations, which do not fit into the form of statements (Wittgenstein 1967 §§ 23-25-27). Stating is just one form of telling there are other types. Heidegger says that the sentence "The moon has risen" when used as a part of a poem is neither a proposition, nor a sentence. He criticizes logo-centric view of language, which claims that stating and thinking through propositions are the most obvious things in the world (Heidegger 1968, 196).

Heidegger challenges not only the idea that thought or logic is possible by means of propositions, but also the idea that sentences have determined meanings. He implies that meaning changes depending on contexts. For example, "The moon has risen" may be used in its literal sense as an assertion and means exactly what it says, and can be correct if it is the case that the moon has already risen, or it may be used as part of a poem without carrying any literal sense of the sentence. However, its metaphorical, or indirect or sarcastic use does not show that it is not a part of thinking. Actually, the problem here transcends the distinction between literal and nonliteral uses of sentences. The problem here is, as Searle discusses in detail in his article "Literal Meaning," that even literal meaning is contextually dependent (Searle 1979, 117). Heidegger criticizes traditional understanding of logic that presupposes a causal connection between the units constituting a proposition. Like Wittgenstein and Austin, he emphasizes the pragmatic aspect of language by implying that stating is telling and it tells only in a context. For example, the copula "is" makes sense only in a referential whole and in a context. Hence, he presents a nonessentialist view of language by drawing an analogy between "moving within language" with moving on shifting ground, or moving on the "billowing waters of an ocean" (Heidegger 1968, 192).

Heidegger implies that when we predicate something of a thing, our attitude is affirmation. This is in contrast with a view that presupposes that an asserted proposition represents causal relations imposed upon us by nature. He says, "To predicate does not mean here primarily to express in speech, but to present something as something, affirm something as something" (Heidegger 1968, 162). This remark reminds us of the later Wittgenstein's discussion of seeing something as something. Similarity becomes obvious when we consider that instead of using the term "idea" Heidegger sometimes uses the concept "aspect" (Heidegger 1977, 20). Seeing something as something is related to aspect-seeing. Two points are significant here: one is in propositions predication or conjunction does not serve to represent the necessary relations between subject and a predicate. Therefore, Heidegger says, "Such presentation and affirmation is ruled by a conjunction of what is stated with that about which the statement is made. The conjunction is expressed in the 'as' and the 'about' " (Heidegger 1968, 162). The second point Heidegger stresses is that predicating something as something is affirmation. "Affirmation" requires human confirmation. Hence, a proposition is said to be correct not because it shows causal and necessary properties of the world having a reference to a decontextualized world, but because we affirm it. Like Wittgenstein, Heidegger emphasizes the context-dependency of meanings.

The second problematic aspect of the representational theory of truth is that it intellectualizes our existential position in the world. When we say "That tree is blossoming" we stand outside of science, and stand in front of the tree, it faces us. We stand face to face with the tree. Heidegger does not incline to define this as " 'ideas' buzzing about in our heads" (Heidegger 1968, 41). On the contrary, he interprets the Greek word "idea" in a different way. He states that the word "idea" in Greek means to see, face and meet -in other words to be face to face. When we are in that position, in front of the tree, we are out of the realm of science and philosophy. We are not in the realm of science because our connection to the world in this case cannot be described by appealing to psychology or cognitive science. We are not in the realm of philosophy because a traditional philosophical approach questions whether we are in a position to assert and justify the existence of the blooming tree standing in front of us.

Heidegger implies that our position in this case is immediate enough not to allow sophisticated explanation and justification. Hence, intellectualization or scientization of our immediate existence, or experience is useless.

Heidegger's concern regarding "the blooming tree" overlaps with that of the later Wittgenstein, who in *On Certainty* questions the attitude of skeptical philosophers that doubt the existence of a tree standing in front of them (Wittgenstein 1969 §§§ 349-350-352). Heidegger underlines the immediacy of our experience by denying the intellectualization of our existential condition.

In *On Certainty* Wittgenstein emphasizes that it is nonsensical to question the existence of things and tools with which we are immediately coping by being in the world. Heidegger, in the same manner, says; "What is the use of such questions concerning a state of affairs which everybody will in fairness admit immediately, since it is clear as day to all the world that we are standing on the earth and, in our example, face-to-face with a tree?" (Heidegger 1968, 43).

Objectification of "tree" in sciences, and cognizing it in philosophy result in dropping the blooming tree by reducing our relation with the tree to "a pre-scientifically intended relation" (Heidegger 1968, 44).

Heidegger, on the contrary, claims that the concept "stating" was used in the sense of "laying out," "laying before," and "laying to" in ancient Greece (Heidegger 1968, 199). What does he mean?

Heidegger's correlation of "stating" with "laying before" and "laying to" paves the way for an alternative concept of truth. In the next section, I concentrate on Heidegger's alternative to the representational theory of truth.

HEIDEGGER'S ALTERNATIVE

Heidegger's definition of truth diverges from the traditional view that presumes that "truth" is a property of correct propositions, which can be true or false by a correspondence relation with facts. According to Heidegger, the essence of truth is not an empty "generality" of an "abstract" universality, rather it is an inner possibility of the correctness of statements identified as freedom, which in his terminology is directly related to *"letting beings be the beings they are"* (Heidegger 1998, p. 144). "Letting beings be," on the other hand, is not related to indifference or neglect (actually neglect implies the negative meaning of "letting alone"), but is associated with engaging oneself with beings or engaging oneself with open region. Open region, which is identified with "unconcealedness," is associated with truth and freedom. How is it associated with truth and freedom? As discussed earlier, this is related to opening ourselves to things standing in front of us such as a tree to let them disclose, rather than withdraw, their aspects. "Withdrawal" here is in contrast with "a turning to the thing in hand according to its nature, thus letting that nature become manifest by the handling" (Heidegger 1968, 195) (Dreyfus Tue, Nov 20, 2007, Lecture on Truth I). "Nature," on the other hand, should not be understood as the essence of a thing or a tool, which manifests itself in a use, but is related to aspect-seeing. In the process of dealing with tools ready-to-hand we are confronted with aspects of these tools. In case they are used properly they do not withdraw and reveal their aspects (Dreyfus 1991, 65). Here, "properly" is understood as the elimination of automation and alienation, which result in covering up the genuine articulation of tools. Non-transparency is mutual here, in the sense that things, tools and human beings are covered up in their bilateral relations.

Heidegger says "to use" is first "to let a thing be what it is and how it is" (Heidegger 1968, 191). He implies that in case a used thing is cared for in accord with its essential nature, then "the demands which the used thing makes manifest in the given instance" is fulfilled (Heidegger 1968, 191). This is also a key to understanding Heidegger's concept of truth, which requires freedom. His concept of truth is related to freedom in the sense that truth does not manifest itself under the conditions of abuse. In order not to abuse we should let a thing reveal and articulate itself. I correlate it with Habermas's concept of strategic action. Of course, there are significant differences between the view of Habermas and that of Heidegger. However, their views overlap with respect to emphasizing the controlling nature of technical action. Heidegger calls it "technical reason" whereas Habermas calls it technical action and contrasts it with communicative action.

Habermas correlates strategic action with technical action. I draw parallels between what Heidegger calls "logistics" and what Habermas calls strategic action. Our relations with tools, things and human beings lose their transparency in case they are abused strategically so as to cover up their genuineness.

There is a difference between "use" and "abuse." "Use" is associated with "useful." Heidegger quotes one passage from the hymn of Hölderlin's The Ister River;

"It is useful for the rock to have shafts,
And for the earth, furrows,
It would be without welcome, without stay" (Quoted by Heidegger 1968, 192)

He says that "useful" here indicates an essential community of rock and shaft as well as earth and furrow. This community, on the other hand, is determined by welcome and stay.

As mentioned earlier, Heidegger questions the complete separation of object and subject. This separation can be overcome by opening ourselves and our hearts to the world in order to let things and tools display themselves as they are. This is defined as "clearing" or "thankful disposal" (Heidegger 1968, 147 and Dreyfus 1991, 165–166).

Heidegger says that "Considered in regard to the essence of truth, the essence of freedom manifests itself as exposure to the disclosedness of beings:" (Heidegger 1998, 144–145) in other words, "freedom is engagement in the disclosure of beings" (Heidegger 1998, 145). Engagement in the disclosure of beings, on the other hand, is linked with "attunement," attunement, in its turn, is correlated with a bringing into accord. "[A] bringing into accord, prevails throughout and anticipates all the open comportment that flourishes in it. Human comportment is brought into definite accord throughout by the openedness of beings as a whole" (Heidegger 1998, 147).

In what sense this is different from the traditional understanding of truth and how it is connected to knowledge. Since knowledge is an engagement with true propositions, consideration of truth requires correlating it with knowledge. As already said, Heidegger does not define truth as an agreement between the psychical and the physical; he uses the concepts "unconcealed" or "uncovered" in order to explain truth. What do these words signify? Assertion is true in case "it uncovers the entity as it is in itself" (Heidegger 1962, 261 [219]). This is not an agreement between a cognitive process occurring in a knowing subject's mind and a fact. On the contrary, truth or uncovering is possible by Dasein's being-in-the-world. However, this is not an epistemological position, but an existential-ontological position. As Heidegger emphasizes this does not mean that Dasein is introduced to "all the truth," rather it means that *"Dasein is 'in the truth' "* (Heidegger 1962, 263 [221]).

Heidegger is critical of the Cartesian and the cognitivist tradition that separates subject from object. Absorption in the world and absorbed coping with things and tools eliminate the distinction between not only subject and object, but also language and the world as the later Wittgenstein taught us. Therefore, Heidegger correlates truth with freedom. Until now thought and cognition never let the tree stand where it stands, only when this is achieved, can we talk about freedom. Rather than intellectualizing and objectifying the tree we should let it stand where it stands and let ourselves get in touch with an open region, which may free us by affirming life and the free existence of the tree. We may free our minds by means of this affirmation of life, which paves the way for eliminating transcendental illusion and thereby allows us not to question the existence of the tree. It also helps us eliminate an appeal to a private object in our mind to prove the existence of the tree. This Nietzschean affirmation of life will free both our minds and the tree. Heidegger, in a Nietzschean manner, underlines the affirmation of life to question the cognitivist approach and the representational theory of truth based upon logo-centric view of language that presupposes that externalization of "internal outward" is what we expect from language.

Truth is neither below, nor above the world, on the contrary, we encounter truth by being-in-this world. Heidegger says, "Truth is neither somewhere *over* man (as validity in itself), nor is it in man as a psychical subject, but man is *'in' the truth"* (Heidegger 2002, 55). However, truth in this world is covered up. In this world, we are exposed to truth and untruth at an equal strength. In this sense, "everything that lies before us is ambiguous" (Heidegger 1968, 201). Truth is covered up by idle talk, curiosity and ambiguity (Heidegger 1962, 264 [222]). The task of thinking and philosophy is to uncover truth concealed, by questioning. Questioning helps us uncover truth. Therefore, truth does not mean an agreement between a proposition residing in the mind of a subject and a fact taking place in the world, on the contrary, by being-in-the-world Dasein uncovers truth. Of course, this suggests a different methodology.

If our method is not to compare propositions with facts to see whether they agree with one another, then what kind of method or attitude allows us to uncover truth? One method is related to hermeneutics. Heidegger says, "Assertion and its structure (namely, the apophantical 'as') are founded upon interpretation and its structure (viz, the hermeneutical 'as') and also upon understanding-upon Dasein's disclosedness" (Heidegger 1962, 265–266 [223]). He also appeals to phenomeno-logical ontology, which allows us to see connections within referential totality. This point is similar to the later Wittgenstein's assertion that rather than evaluating sentences one by one to see whether they correspond to facts in the external world it is important to see connections. Certainly, Wittgenstein does not appeal to ontology. He refrains from answering questions regarding ontology. Therefore, where Wittgenstein says "This is simply what we do," Heidegger concentrates on the question of how we do it. However, with respect to evaluating individual facts in a whole and in a relational totality, their approaches resemble one another. Their approaches are also similar regarding the critique of the foundationalist presupposition that there is an edifice underneath, supporting and forming a ground for our existence. Hence, it is pointless to attempt to go beyond relational totality, into the context of Being, which provides the support for this totality (Heidegger 1962, 258–259 [216]). Our main starting point is this world; truth can be uncovered in this relational totality and by being in the world. In this sense, truth is not understood by grasping universals, but as Wittgenstein says, when "Light dawns gradually over the whole" (Wittgenstein 1969, §141). In other words, only if we see the connections in relational totality can we uncover truth. As already said, Dasein is in truth, and untruth at the same time. Truth is uncovering in this world. How is it possible to uncover truth?

As already mentioned, according to Heidegger, truth is uncovering by taking entities out of their hiddenness and by letting them be seen in their unhiddenness (Heidegger 1962, 261–262 [219]). Heidegger defines it as a kind of robbery. He says that "Entities get snatched out of their hiddenness. The factical uncoveredness of anything is always, as it were, a kind of *robbery*" (Heidegger 1962, 264–265 [222]). This is similar to taking masks off to reveal a person as it is. As Michael Gelven points out, "we say that our understanding and awareness of someone is true if there are no masks, and we see him as he is" (Gelven 1989, 129).

Dasein is both in truth and in untruth. In order for Dasein to get out of a Platonic Cave and uncover truth, she should not only understand relational totality by means of absorbed coping and circumspective concern, but also take a stand on her life to make a projection. This requires a questioning attitude. Heidegger defines the mission of philosophy as questioning because only a critical or questioning attitude helps us uncover truth. It also requires a way of being in the world. Authentic existence or existential authenticity is a form of life allowing us to uncover truth. "The goddess of Truth who guides Parmenides, puts two pathways before him, one of uncovering, one of hiding.... The way of uncovering is achieved only in...distinguishing between these understandingly, and making one's decision for the one rather than the other" (Heidegger 1962, 265 [223]).

Choosing the way of uncovering requires a genuine existential way of being. Therefore, Heidegger says, "In so far as Dasein *is* its disclosedness essentially, and discloses and uncovers as something disclosed to this extent it is essentially 'true' " (Heidegger 1962, 263 [221]).

Truth, on the other hand, cannot be considered independently of Dasein as independent of any human existence. " '*There is*' *truth only in so far as Dasein is and so long as Dasein is.* Entities are uncovered only *when* Dasein *is*; and only as Dasein *is*, are they disclosed. Newton's laws, the principle of contradiction, any truth whatever-these are true as long as Dasein *is*. Before there was any Dasein, there was no truth; nor will there be any after Dasein is no more" (Heidegger 1962, 269 [227]).

According to the traditional approach, "laws of thought are ...valid independently of the man who performs the individual acts of thinking" (Heidegger 1968, 115). Heidegger is reminiscent of Nietzsche in saying that the rules of logic such as the principle of the excluded middle, the principle of contradiction and the principle of identity are functional devices organizing Dasein's existence in the world, rather than having necessary and causal relations to the structural organization of the thing-in-itself. Hence, it is difficult for us to transcend our phenomenological existence and to be in the position of claiming that they refer to something deeper.

In order to develop his alternative further, Heidegger reinterprets the concept "stating." He correlates "stating" with "laying out," "laying before," and "laying to" (Heidegger 1968, 199). He defines logic as having a relation with "clearing." What does this mean? Heidegger says when we say something about something, in other words, when we state, we make an object visible: for example when we say that "The tree is blooming" we make the blooming tree disclose itself in front of us. So, the essence of *logos* cannot be defined by a correspondence relation between statement and fact, but it is disclosing (Heidegger 1968, 202). What lies in front of us, on the other hand, is actually perceived, and this perceiving is not a passive receiving, rather it is active in the sense that we take what is perceived to heart and keep it at heart (Heidegger 1968, 203). In contrast with the Platonic idea that understanding is grasping, Heidegger asserts that this taking to heart is letting come what lies before us (Heidegger 1968, 211). What is taken in, on the other hand, is safeguarded and kept in memory. In other words, it is kept in our heart first, and then cognized. This taking to heart, on the other hand, is neither grasping, nor apprehending. "For instance, when we let the sea lie before us as it lies, we ...are already engaged in keeping in mind and heart what lies before us. We have already taken to heart what lies before us" (Heidegger 1968, 209). As stated earlier, this is related to Heidegger's critique of the cognitivist approach of trying to bridge the gap between psychical and physical by our intentional attitude. Heidegger reverses the order (if there is any order at all) by trying to get us to see that engaging oneself with an open region is possible by opening our heart to nature and leaving it there as it is, without cognizing or theorizing it. When we do this we leave things where they stand, and have harmonious relations with nature. Only when we allow things to be seen as they are, and only when we let them disclose themselves, can we get a perspicuous representation.

Heidegger's critique of the representational theory of truth is related to his critique of modernity. In the next section I concentrate on this issue.

CRITIQUE OF MODERNITY

Heidegger's critique of the representational theory of truth and the cognitivist approach go hand in hand with his critique of modernity. He sees a connection between an approach regarding logic as the fundamental rules of thinking and logistics "developing into the global system by which all ideas are organized" (Heidegger 1968, 163). Heidegger's critique of modernity is stated in a small article entitled "A Question Concerning Technology" (Heidegger 1977). However, even in *What is called Thinking?* he is critical of the attitude belittling the value of disciplines other than science. This belittling is built upon the idea that science has a representative value, which explains nature and allows us to predict and control it, while other disciplines do not have such a privileged position. However, Heidegger thinks that science does not think in the sense in which thinkers think (Heidegger 1968, 134) and he also says; "The Enlightenment obscures the essential origin of thinking" (Heidegger 1968, 211). What does he mean? Is he an obscurantist philosopher, who does not appreciate the value of scientific knowledge in our time? In what sense, does he say "science does not think"? In order to understand his concern we must briefly concentrate on his analysis of "thinking."

According to Heidegger, thinking cannot be defined, because the mere reflection objectifying "thinking," is fruitless. Just as it is not possible to know what swimming is by reading a treatise on swimming, it is not possible to "remain outside that mere reflection which makes thinking its object" (Heidegger 1968, 21). Heidegger thinks that thinking is "the handicraft *par excellence*" (Heidegger 1968, 23). As mentioned earlier, for Heidegger, the starting point is not intention, but action. As for the later Wittgenstein, Goethe's assertion that "In the beginning was the deed" is valid for Heidegger too. We comport ourselves in this relational totality with our deeds and acts. Hence, thinking is not accomplished by the correspondence relation of a proposition in our minds and a fact in the world, but it is rather the result of our deeds and acts in the process of dealing with tools, things and human beings in this hurly-burly of daily life. Correlatively, understanding is not possible by grasping universals, but it is possible by engaging oneself with an open region so as to take a stand on one's life.

Heidegger diagnoses that the "[m]ost thought provoking in our thought-provoking time is that we are still not thinking" (Heidegger 1968, 6). As mentioned earlier, he also says that science does not think. This is related to Heidegger's critique of modernity and technology. Heidegger depicts a gloomy picture of modernity. He quotes Nietzsche's saying, "The wasteland grows," in order to show the deterioration in literature and the world (Quoted by Heidegger 1968, 29). Just because our age is gloomy, dark, and threatening it is the most thought-provoking age (Heidegger 1968, 29). Heidegger says that "what properly gives food for thought, has long been withdrawing" (Heidegger 1968, 25). He is critical of logistics integrated into technical reason, which results in withdrawal. Withdrawal is the result of not dealing with tools and things in a crafty manner to let them articulate their aspects properly, which in turn provides us with thought. Because of this withdrawal, we find ourselves in a Platonic Cave. Automation which results in alienation eliminates genuine relationship, and leads us to live in untruth.

Heidegger's concern overlaps with those of other philosophers criticizing modernity such as Herbert Marcuse and Jürgen Habermas. Like Habermas, he thinks that technical reason, which is identified as purposive rational action by Habermas, predominates over other forms of knowledge in a way that science and technology are at the top of the pyramid in the hierarchy of knowledge. Besides, because science is in the service of technology, it becomes a mere means in technology's revealing itself; therefore it does not think.

Heidegger not only criticizes science for not evaluating its place in the world properly by doing a projection, he also criticizes "one-track" thinking created in modern societies. He says,

The expression "one-track" has been chosen on purpose. Track has to do with rails, and rails with technology. We would be making matters too easy for ourselves if we simply took the view that the dominion of one-track thinking, which is becoming ever more widespread in various shapes, is one of those unsuspected and inconspicuous forms...in which the essence of technology assumes dominion-because that essence wills and therefore needs absolute univocity (Heidegger 1968, 26).

The essence of modern technology, which is identified as "enframing" (Heidegger 1977, 26–27) by Heidegger is hidden, and it creates "one-track" thinking. As discussed earlier, in the West, thought about thinking resulted in a discipline of logic gathering a special knowledge concerning a special kind of thinking, which is called logistics. Logistics is considered to be the only possible form of strict philosophy because it is integrated with a technological universe, which exercises power over other disciplines in our era. As a result of this, thinking is transformed into a one-track thinking generating an absolute univocity. This dominion of one-track thinking, "reduces everything to a univocity of concepts and specifications the precision of which not only corresponds to, but has the same essential origin as, the precision of technological process" (Heidegger 1968, 34).

Heidegger's critique reminds us of his student Marcuse saying that we become one-dimensional human beings and society in modernity.

CONCLUSION

Heidegger criticizes the representational view of language and truth from several perspectives. He questions the view that there is a correspondence relation between an idea and its object. A complementary assumption of this view is that forming a judgment entails having representational ideas of related objects in our minds. This, assumes that judging and thinking is propositional.

In a Wittgensteinian manner, Heidegger underlines that although every proposition is a sentence, not every sentence is a proposition. Hence, stating is just one form of telling, and there are other types.

The second critique is that when we look at a tree, or a mountain, we stand face to face with the tree and therefore we are out of the realm of science and philosophy. Our position, in this case, is immediate enough not to allow for sophisticated explanation and justification.

Heidegger's alternative is based upon a critique of the Cartesian inclination that assumes a strict separation between subject and object as well as language and the world.

According to Heidegger, the essence of truth is not an empty "generality" of an "abstract" universality; rather it is an inner possibility of the correctness of statements identified as freedom. Freedom is actualized by engaging oneself with open region. In this sense, "stating" is associated with "laying out," or "laying before," which is related to disclosing. Disclosing is accomplished, in case what is perceived is taken to heart.

Heidegger's analysis of truth is related to his critique of technical reason that is implicit in modernity because he thinks that only if we let things disclose themselves as they are, can we uncover truth, which is masked with idle talk, curiosity and ambiguity.

Department of Humanities and Social Sciences, Istanbul Technical University, 34469 Istanbul, Turkey
e-mail: turanliay@itu.edu.tr

REFERENCES

Dreyfus, Hubert. 1991. *Being-in-the-world: A commentary on Heidegger's Being and Time, Division I.* Cambridge, MA: MIT Press.
Dreyfus, Hubert. 2007. Heidegger Podcast. Webcast.berkeley Course Philosophy 185 http://www.learnoutloud.com/Catalog/Philosophy/Modern-Philosophy/Heidegger-Podcast/24272#3
Gelven, Michael. 1989. *A commentary on Heidegger's being and time.* De Kalb: Northern Illinois Press.
Heidegger, Martin. 1962 (2007). *Being and time* (trans: Macquarrie, John, and Edward Robinson). Oxford: Basil Blackwell.
Heidegger, Martin. 1968. *What is called thinking?* (trans: Gray, J.G.). New York: Harper and Row.
Heidegger, Martin. 1977. *The question concerning technology and other essays* (trans: Lovitt, William). New York: Harper Torchbooks.
Heidegger, Martin. 1998 (2006). On the essence of truth. In *Pathmarks*, ed. William McNeill, 136–154. Cambridge: Cambridge University Press.
Heidegger, Martin. 2002. *The essence of truth* (trans: Sadler, Ted). New York: Continuum
Searle, John. 1979 (1999). *Expression and meaning: Studies in the theory of speech acts.* Cambridge: Cambridge University Press.
Wittgenstein, Ludwig. 1967. In *Philosophical investigations*, eds. G.E.M. Anscombe and R. Rhees (trans: Anscombe, G.E.M.), 2nd edn., §§ 23-25-27. Oxford: Basil Blackwell.
Wittgenstein, Ludwig. 1969. In *On certainty*, eds. G.E.M. Anscombe and G.H. von Wright (trans: Paul, D. and G.E.M. Anscombe), §§ 349-350-352. Oxford: Basil Blackwell.

MINA SEHDEV

ORIGIN AND FEATURES OF PSYCHICAL CREATIONS IN AN ONTOPOIETIC PERSPECTIVE

ABSTRACT

In psychology of depth, the unconscious is often opposed to consciousness, as it is the place of what goes beyond rational thought, but at the same time – both in Freudian psychoanalysis and in Jungian analytical psychology – the unconscious is the first cause of consciousness, and also the unlimited memory of culture; in it we can find the numerous symbolic forms through which human thought can express itself. In fact, the productions of collective imagination can be considered an expression of the development of "forming spontaneity" which is rooted in the wide field of phenomenology of life, that is to say "the universe of human existence within the unity-of-everything–there-is-alive"; both the "inward givenness of the life progress common to all living beings" and "cognitive processes of human mind" (in other terms *eidos* and fact, *logos* and *mythos*) simultaneously spring from it. In fact, the logic of self-individualization of life can express itself in human creative actions, by referring to the pre-human; in such an outlook, consciousness and reason appear to be in a close relation to the "world-of-life", as in the archaic periods of human history. We have to bear in mind that both dream phenomena and fantastic creations – associated by the original creativeness and by the transformer energy characterizing them – can be analysed and interpreted only through their stories; this explains the fundamental function of figurative (metaphorical and symbolic) language, which is used in them. According to psychoanalysis, symbols are visual representations of unconscious contents, a sort of phylogenetic heritage referring to ontogenesis; in analytical psychology they become real teleological factors, as the archetypes of the collective unconscious ("a priori" forms common to the whole human kind) can find expression above all through symbols.

INTRODUCTION

In psychology of depth, the unconscious is often contrasted with consciousness, as the place of all that transcends rational thought, but at the same time – both in Freudian psychoanalysis and in Jungian analytical psychology – the unconscious is the first "cause" of consciousness, as well as the unlimited memory of culture, where it is possible to trace the numerous symbolic forms through which human thought expresses itself. In fact, here rational or conceptual thought coexists with

A.-T. Tymieniecka (ed.), Analecta Husserliana CX, 307–314.
DOI 10.1007/978-94-007-1691-9_25, © Springer Science+Business Media B.V. 2011

fantastic or symbolic thought: if the conscious is the detailed memory, for the most part directed to a practical purpose, the unconscious is potential and unbounded memory of culture in which all the symbolic forms are present (...) and all the metaphors with which, in the course of the hominization of history, man has spread his natural cosmos, the narrow riverbed in which nature placed him.[1]

An ontopoietic outlook is particularly congenial to explaining these phenomena, as well as presenting them in ontogenetic and phylogenetic continuity. In fact, from this perspective, the products of collective imagination (including dreams, but above all myths and fairy tales) can be considered expression of the development of the "forming spontaneity" that is unleashed, as the phenomenology of life demonstrates, from the "universe of human existence in the unity-of-every-thing-that-lives" as bearer of the creative function that is capable of sparking the very progress of life. On the other hand, in it are rooted both the putting-into-act-of-life and the most specifically cognitive functions of the human being, both *mythos* and *logos*. It is precisely in the human creative acts that the logic of self-individualization of life is manifested, in which consciousness and modern reason also are placed and thus put into intimate relation with the world-of-life, on the same level as what happened in the most archaic phases of humanity.

CHARACTERISTICS AND STRUCTURES OF THE CREATIONS OF THE PSYCHE

DREAM AND MYTH: SIMILARITIES AND DIFFERENCES

When investigating objects such as dreams, myths, and fairy tales – more in general, the creations of the human psyche –we must always keep in mind their intrinsic nature. For example, in the interpretation of oneiric products, we do not have the opportunity to analyze the original dreams, but only the "stories," the reconstructions provided by the dreamers and formed of their "recollections" (in psychoanalytical terms, the "manifest content" of the dream, which, however, is not the same as its "latent content.") It is evident that the words used to tell the dream have a contingent and in any case instrumental value: they could also not exist, and yet the dream would still remain, at least in the mind of the individual who dreamed it; when there are words, their function in limited to transmitting, to making known to an interlocutor a representative product that is already complete, beyond and independent of the words used to communicate it.[2] The same holds for mythological productions: investigation needs must be conducted on the "narrations of myths," passed from generation to generation orally, which have become the collective patrimony of a people or a society, or have been collected in literary works. Part of the content of oneiric tales and mythological stories, and thus also of their "cognitive value," is inevitably lost. Even so, it can be fruitful to compare them and look for recurring symbols.

Finally, we must not forget that both oneiric products and mythological creations have in common their "original creativity." In fact, both are charged with "poietic," transforming potential, and draw upon the springs of imagination, even if with differences. While the dream is a "spontaneous" product of sleep and an individual creation, mythological construction has a collective value and happens, so to speak, under conscious control: in the myth, every society glimpses the paradigm and the creator crucible of its own culture and, in the final analysis, of the common destiny of all humanity. Perhaps it is not so far-fetched to assert that dreams and myths can be considered associable realities, given that in many myths we find news of dreams (which play a role that is anything but secondary), just as numerous mythic images seem to derive from dreams.

In addition, we should keep in mind – inasmuch as we never study or interpret the myth or dream itself, as it was created, but always a text formed of words – the essential role played by language that is essentially symbolic: the mythic and the oneiric have in common this structure of dual meaning: the dream as nocturnal performance is unknown to us; it is the narration upon awaking that renders it accessible to us and that the analyst interprets by substituting another text that in his way of seeing is the thought of desire, what desire would express freely in a prosopopeia. Since it can be narrated, analysed, and interpreted, it must be admitted that the dream is in and of itself close to language. . . .[3] Both dreams and myths can thus be analyzed exclusively in the form of narrations, often characterized by a spiral structure. In fact, they often present various attempts to rework one theme; it follows that the very personages or forces of the human psyche can take on roles that appear completely different, but that are of the same kind, and comment on, diversify, and clarify each other.[4]

From the perspective of depth psychology, we can compare and associate dreams and myths on two main levels:

– on the level of structure;
– on the level of language.

On the level of structure, both exclude the categories of space, time, and causality. In addition, both psychoanalysis and analytical psychology hold that the mechanisms operating in myths are analogous to those in dreams (condensation, shifting, etc.). In fact, in dreams the first infantile impressions are condensed into atemporal images, enabling the past to unite with the present in a symbolic structure in which diachrony and synchrony fuse; a structure of analogy is also traceable in myths. In addition, from the point of view of content, all the affectively strong symbols (be they of trust or anguish, joy or suffering) have the ability to open to the fundamental themes of existence and express in their singular configuration archetypal contents: the dream of the single individual thus broadens to become the great dream combining the individual and the collective.[5]

Thus it seems possible to hypothesize a reciprocal convertibility between the individual and the collective; to the degree to which the problem of a dream and the attempts at its symbolic reworking are valid for the experience of life of a human group or a people in a certain period or for an entire era, the oneiric symbols of one individual can condense, for example, in the form of poetry, the lived experience of all, and the deeper the representation and the solution offered by a given theme, the more this theme expresses itself radically and substantially in the individual's dreams and poetry, the more universal is its validity and the vaster is the interest it finds in humanity.[6] For that matter, it should be noted that some themes that are central in oneiric experiences as well as in mythological expressions often reflect "conflicts" that belong *a priori* to human existence and that have eternal and universal validity precisely because they can never be resolved definitively: deep down, in both dreams and myths, human beings represent and experience themselves.

On the level of language, both myth and dream have in common the use of symbolic and figurative language, based first of all on the use of archaic images and symbolic portrayals. Symbolism plays an important role both in oneiric phenomena and in mythologic productions, characterized by atemporality and symbolic condensation of humanity's fundamental questions about its existence: seeking to express the forces and conflicts that inform history, staying in the background, the myth remakes *itself into cover memories'*, into certain historic fragments that, owing to their affective, but above all symbolic, density, impress themselves particularly on the collective imagination, and condenses them into atemporal images, representations of the essence of that given human group or in archetypes of human existence.[7]

From the formal point of view, the myth is composed of individual motives that can combine in various ways, and that all tend to the timeless present through cyclic representation of time. This characteristic, which for that matter is typical as much of dreams as it is of myths, is generally explained by depth psychology on the basis of the "repetition compulsion". Just as the dream, returning to the earliest times, seeks to introduce a renewal or a completion of current experience, so the myth, too, does nothing other than revisit events (primordial) of humanity or a people in order to experience, recalling the past to mind, its renewing power and its eternal presence. The instrument for accomplishing this actualisation is, as in the dream, symbolic representation, which, however, now presents itself as actualised, dramatized dream, as rite.[8]

DREAM AND MYTH FROM THE PERSPECTIVE OF DEPTH PSYCHOLOGY

From the perspective of depth psychology, these products of the psyche seem at once to reveal and to hide the unconscious; they are marked not only by a dual nature (expressed in the distinction between manifest content and latent thoughts) but also by overlapping layers. The dream is at once memory of the past, awareness of the present and perspectival harbinger of the future; thus it enables self-representation of the unconscious in all its multiple functions. It should also be borne in mind that though classical psychoanalysis considers it an essentially regressive phenomenon, it can nonetheless also acquire a perspectival value, and in this sense instrumental, for the conscious dimension. For that matter, while certain manifestations of the human psyche such as dreams and myths provide valid instruments for inquiry into the unconscious that help us in the attempt to penetrate its complex nature, the study of the unconscious, in turn, can, if not transform, at least influence our own vision of dream and myth.

It should be specified first of all that the parallel between myth and dream established by depth psychology does not indicate just a relationship of cause and effect; the fact that we can observe dreamed myths, dreams that contain myths, visionary cults or rites with mythic foundations, that is, the fact that myths and dreams mix incessantly, can be explained by the fact that there is a common substratum represented by a world of the soul that makes itself into image.[9] Thus, one cannot simply assert that myths derive from dreams or vice versa, because a bond of reciprocal dependence between them enables us to trace undeniable similarities, explainable on the basis of their common origin. Myths and dreams would thus be experiences that can be associated with each other because of their very nature (the fact that both are classifiable as "creations of the psyche"), even though they present differences that do not allow us to overlap them completely.

Dreams differ from myths first of all because of their purely individual character and because of the regressive tendency that can be found in them (often in the Freudian conception of oneiric phenomena); on the contrary, mythological creations are distinguished by a collective dimension and an essentially progressive tendency, inasmuch as they are oriented more toward the future than the past. Thus, while dreams appear principally suited to represent and interpret individual experiences, myths flow from the projection of certain oneiric images on the life and the lived experience of entire social groups. According to Drewermann, the myth is born when the dream, the vision, the poetry of the individual rise to the rank of great dream, because in this case the symbolic language of oneiric images does not mirror only the sediment of individual experiences, but at the same time condenses the living experiences of a vaster human group, interprets them, or anticipates them.[10] This outlook makes it possible to pass from the individual dream to the collective myth. Thus, in the perspective developed by depth psychology, the dream becomes "model" not only of myth but also of other narrative forms that can be related to it (such as fairy tales, sagas, and legends) and that nonetheless have far from negligible differences. For example, unlike the myth, which, tending to the divine, the religious, is thus by nature non-historic, the fairy tale also expresses atemporal truths, but ones that are "human" and thus not transcendent.

The relationship between dream and myth that both Freudian psychoanalysis and Jungian analytical psychology identify seems definable essentially as a "relation of conjunction," inasmuch as it supports the associability of these two phenomena both on the level of the meaning and function they carry out, and on the level of the structure and language used. This is a "biunivocal relationship," or we could say bidirectional, in the sense that one can find a relationship between myth and dream, as already mentioned, based on reciprocity: in fact, it is possible to identify the presence of mythological motives in dreams, but at the same time many collective mythological creations seem to derive from individual oneiric experiences.

It should be noted that the Freudian method, based as it is on a more objective type of interpretation, perhaps is better suited to analysis of "narrations" from the individual character, such as dreams, while the Jungian method, which tends toward a more subjective type of interpretation, seems more appropriate for analysis of the collective patrimony of peoples, that is, of myths. Even so, the similarities between these two complex realities (dream and myth) and the parallelism identified by depth psychology permit us to assert that "all myths are first of all 'great dreams' of single individuals and, vice versa, all 'little dreams' also have in themselves the power to become the great myth, similarly to the way poetry is truth".[11]

Notwithstanding the difference of approach (essentially causalistic and deterministic as we have seen for Freud and the Freudian school, and in contrast finalistic and perspectival inasmuch as it is oriented toward the search for meaning for Jung and his followers), and in spite of the partially different theory developed for oneiric experience (one can speak of "dream as symptom" for Freud and instead "dream as symbol" for Jung), it should be observed that both acknowledge that the dream has typical characteristics, shared more in general with the processes of the unconscious, the most important of which seem to us to be atemporality and the substitution of external reality with that of psychic reality. The myth is simultaneously connection between the individual and the collectivity, the present with the past and the future, and finally, the human being with external reality and internal reality; it enables a unitary vision of nature and culture, of the divine and the human, the eternal and the temporal.

DREAM AND MYTH FROM AN ONTOPOIETIC PERSPECTIVE

Naturally, we must keep in mind that oneiric phenomena and creations of the imagination, which have in common the original creativity and essentially "poietic", transformative potential that marks them, can be analysed and interpreted only through their narrations, hence, the importance of the role played by the language they use, which is first of all figurative (metaphorical and symbolic). According to psychoanalysis, the symbol is a visual representation of an unconscious idea; thus it is a phylogenetic inheritance referable to ontogenesis, inasmuch as it is an archaic process of thought preceding the development of individual language. Symbols generally represent unconscious ideas subject to removal that would have no other way of emerging to consciousness; Freud thus grasps the essence of the symbol in the constant relationship between the manifest expression of a dream and its latent reference. In analytical psychology, the symbol takes on a more specific dimension, inasmuch as it is thought to derive from the collective unconscious; archetypes (*a priori* forms common to all of humankind) are thought to find expression (in particular, imaginary creations of the human psyche such as myths or fairy tales) in symbols, which can thus be defined as teleological factors that express meanings that are difficult to know and comprehend from the merely rational point of view.

Both psychoanalysis and analytical psychology have contributed to recognizing an "original importance," restoring a deeper value also on the cognitive level, to all the manifestations (individual and collective) of culture, in particular to the products of the unconscious, among which an essential role is played by dream and myth. From analysis of how depth psychology views myth and dream, it emerges that both psychoanalysis and analytical psychology have established a relation of conjunction (analogic for Freud and his school, more clearly dialectic with Jung and his followers) between myth and dream. As we have seen, they have elements in common both on the level of structure and of language, and symbolism plays a fundamental role in both. One of the essential merits of depth psychology is having indicated the essential function of symbolic language found in all creations of the psyche.

On the basis of the association established between dream and myth by depth psychology, it seems possible to us to trace in the latter a value that is hardly negligible on the cognitive level, more accentuated in the Jungian conception than in the Freudian one. In psychoanalysis, myth seems essentially the gratification of removed unconscious desires and the expression of the deepest human impulses that the conscious tries to ignore in order to control them, but that re-emerge in dreams on the individual level and in mythological creations on the more collective one. In contrast, according to analytical psychology, it is not limited to being a kind of substitute satisfaction, but reveals its capacity to express the complex inner reality of a person in all its multiple components, also providing access (inasmuch as it configures as a kind of self revelation) to transcendent truths, and thus enabling the individual to progress in awareness of himself or herself and of the world.

The poietic and transformative nature typical of creations of the psyche find foundation in the phenomenology of life of Anna-Teresa Tymieniecka, which opens the possibility to grasp the *logos* first in its constructive impetus and then in its unfolding in life, which self-individualizes precisely on the measure of the *logos*. The logos of life, in fact, refers to the creativity of the human condition and the creative act, inherent in our condition, taking part in the deepest intense activity of life, and reveals the original "modelling" of the preconscious and reflective functions that characterize human nature in its most intimate essence.

The human condition offers us the key to access being in its living fabric, that is, continually becoming, productive of increasingly more articulated and diversified forms. In fact, human beings not only follow the spontaneous and already traced patterns of universal life, but also incessantly invent and produce new ones, creating devices for life, products of work, works of art, exalting and transfiguring the tremor of existence into the throb of creation. Opening itself to the perspective of the human creative condition, the conscious, which in turn has discovered itself living and vital, thus finds itself witness to the very emergence of life, and at the same time involved in it. When it reaches the level of the human creative condition, therefore, life no longer limits itself to reproduce itself, but in the acts of life of human beings always interprets itself in existence, giving rise to forms of life that not only are new and unimaginable previously, but also are congruent and suitable to the becoming being of life, of which it alone holds the key feature.

Università degli studi di Macerata, Italy
e-mail: m.sehdev@unimc.it

NOTES

[1] M. TREVI, *L'altra lettura di Jung [The Other Reading of Jung]*, Cortina, Milano 1988, p. 90.

[2] G. LAI, *Un sogno di Freud [A Dream of Freud]*, Boringhieri, Torino 1977, p. 62.

[3] RICOEUR, *Dell'interpretazione. Saggio su Freud [On Interpretation. An Essay on Freud]*, Italian translation by A. Renzi, Il Saggiatore, Milano 2002, p. 27.

[4] DREWERMANN, *Psicologia del profondo ed esegesi [Psychology of Depth and Exegesis]*, Italian translation by A. Laldi, Queriniana, Brescia 1996, p. 317.

[5] Ibidem, p. 90.

[6] Ibidem, p. 91.

[7] Ibidem.

[8] Ibidem, p. 93.

[9] E. DREWERMANN, op. cit., p. 90. Cfr. H. ELLENBERGER, *La scoperta dell'inconscio. Storia della psichiatria dinamica [The Discovery of the Unconscious. History of Dynamic Psychiatry]*, Bollati Boringhieri, Torino 1976.

[10] Ibidem, p. 89.

[11] Ibidem, p. 316.

SECTION VI
NATURE, WORLD, CONTINUITY

FRANCESCO TOTARO

NATURE AND ARTIFICE IN
MANIFESTING/PRODUCING THE BEING

ABSTRACT

In the ancient thought, a great importance is given to the notion of nature. Nature is what remains in what becomes and at the same time it is what allows every becoming-being to manifest itself in its proper determination. Nature is thus a principle of identity and a principle for protecting the forms of becoming from contradiction, in so far becoming means the becoming of what is, and the becoming towards what is (in the form of its *telos*). In the modern thought nature is what resists against the transformation promoted by the artifice: therefore nature is a starting point that needs to be overcome to the advantage of the enhancement of what is originally defective. In this framework the way is open towards the criticism of an essentialism, which is stiffened up in the representation of the fixity of nature. However the dynamism of the artifice, which is untied from an orientation to the being of what becomes, can lead to the negation of any eidetic principle, particularly to the negation of the idea of the human being to the advantage of a post-human, that could even contradict the human being in its essential structure, as it is emerged through the historical process. How then can phenomenology and the ontopoietic vision of becoming give value to the dynamism of life and, at the same time, to the exigencies of permanence of what becomes in the process? In order to answer this question, it is necessary to rethink the distinction between generation (as manifestation of the being) and production (as construction of the being), so that the former is not entirely subsumed under the latter.

THE HUMAN BEING AS A PROBLEM

This paper aims at tackling the issue of nature and artifice in relation to the manifestation and production of the being. The being which is investigated here is above all the human being, that is situated in a historical situation characterized by the transition from the human to the post-human. This theme is relevant both for the classical thought and for phenomenology. In the *Crisis of the European Sciences*, as it is well known, Husserl referred to the Greek concept of *human telos* in order to give sense to the elaboration of modern scientific knowledge. In Anna-Teresa Tymieniecka's thought, human finalism is placed at the core of vital dynamism, since the latter cannot be thought without the former. Through the affirmation of artifice the human being becomes a problem. We do not have guarantees that the post-human, which is linked to the artificial technologies, keeps continuity with the human, according to

A.-T. Tymieniecka (ed.), Analecta Husserliana CX, 317–326.
DOI 10.1007/978-94-007-1691-9_26, © Springer Science+Business Media B.V. 2011

the *eidetics* that derives from his history and at the same time consists in some essential components. Wilhelm Dilthey, surely a philosopher of life, has identified these components in the articulation of knowing, feeling and willing. To these connotations are linked the capacities of autonomy and choice. Without these connotations it seems that we can think neither the sphere of individual subjectivity, nor the relationships of individuals to each other. However it is important not to take for granted that this human *eidos* should continue in the future, nor that the continuity between past, present and future (that is the dimension of temporality, which constitutes the *human being*, that hovers between the conscience of finiteness and the tension towards a transcendental horizon) persists. The twofold perspective of the finite and the infinite, that – through either exclusion or inclusion of both terms – has always been the fundamental anthropological tonality, could be cancelled to the advantage of an ontological production entrusted in the power of techniques, which are indifferent to the "conceptual vetero-European apparatus", as Niklas Luhmann defined it. The functional systemic universe would not need the dramatic scenery cultivated by a restless consciences, but rather the docile ability to adapt to already given situations. Probably an ethics of means disconnected from ends would prevail, and the Kantian imperative would be overturned and would sound as follows: treat your post-human essence, whether in your own post-person or in the post-person of any other, always as a mere means and never at the same time as an end. Full reality and full legitimacy would be given to the overturned world represented as an *exemplum vitandum* in Pieter Bruegel the Elder's famous painting of 1559 entitled *Dutch Proverbs*, which is exposed in Berlin *Gemäldegalerie*.

MANIFESTATION   OF   THE   BEING   AND   PRODUCTION

The alternative to the above mentioned catastrophic outcomes is surely an equilibrate relation between nature and artifice, that needs to be carefully calibrated. Thanks to such a measure, what we have always known as *human* and what advances as *post-human* could turn into an ontological and anthropological enhancement and not into an absolute discontinuity, that we could not even indicate through adequate words, being it so extraneous from us. At stake here is the permanence of what is authentically human, its manifestation at higher levels indeed, that do nor negate but complete the having-become-in this way of the human. But this is not an automatic process; it rather depends on an increased ethical awareness, that can both appreciate the contribution that technologies can give to the disclosure of human essence, and at the same time can control technologies in order to avoid the risk that their instrumental role takes the place that is due to the ends.

The ontopoietic dynamics of the human, which has assumed a speed that cannot be compared to the one of previous periods, needs to be consistent with its own possibilities. The latter should be based on the idea that the being that can be produced is measured by the unconditioned being. The unconditioned being is not the object of a production, but is the foundation of any productive effort and provides any construction with the positive direction of meaning. In fact producing, as the historical experience teaches us, can also lead to negative or destructive outcomes. Orientating it towards the positivity of the being depends on us.

In this perspective of ontological positivity, producing can be understood as leading to manifestation the part of the being that is not yet manifest, such as the *hervorbringen* that Martin Heidegger opposed to the *herausfordern* in his work of 1954 entitled *Die Frage nach der Technik*. The manifestation of the being would thus be against a provocation that abuses its power, the construction of the being would lead to light the being itself, that is the part of the being that we are allowed to bring to light, being aware that the being that depends on our conditions is not the unconditioned being.

THE WEAVE OF NATURE AND ARTIFICE

In the background of this precomprehension, let us think in a coherent way the weave of nature and artifice within the human condition. In fact the human nature is both a starting and an arriving point, that is a "vectorial" concept, that connotes a basic equipment, a way to scour and an hoped fulfilment. Therefore nature, in the human, is always *more than* nature. Nature's *ecstatic* character introduces us to the notion of artifice. Artifice belongs to human nature itself: it blends in the intentionality of the *hand* and thus of the entire corporeity. To sum up, the human is at the same time nature and artifice.

This clarification allows us to see, through a brief genealogical reconstruction, how the ideas of nature and artifice have expressed themselves and have developed in the framework of western thought.

NATURE AS A FACT AND NATURE AS ESSENCE

It is important to remark the distinction between nature as a fact and nature as essence. Nature as a fact is the *"what is"* (that which is). Nature as essence is the *"what for"* (that for which) or, better, the *"what for"* in the *"what is"*. This distinction dates back to the Aristotelian vision of the *physis* that presented the concept of nature dynamically, especially in the books of the *Physics* but also in other works. Nature as essence is an end that can be fulfilled, starting from the already given; is does not overlap with an absolutely non-deformable structure.

In our times, the contemporary research in the field of biology has made any firm representation of nature fluid: today we can no longer support a pre-constituted essentialism, that abstracts from the mobility which is attested by a progressing study of phenomena.

The distinction between nature as a (already given) *fact* and nature as essence (an arriving point to strive towards) entails important speculative implications. In fact it highlights an ontological condition which is signed by finiteness and limit. If there were already a full synthesis between the *"what is"* and the *"what for"*, the distinction between these two aspects would disappear. We would thus have the perfect coincidence of existence and essence, such as in the divine condition.

It is the non-adjustment of essence and existence (and it does not matter here to establish whether the essence precedes the existence or vice versa) that allows that the *"what is"* differs from the *"what for"*. This difference makes the *"what is"* restless and puts it in tension with the *"what for"*.

NATURE AND POWER OF THE WILL

The non coincidence, within the being, between the being that already is and the being that is not yet, is maybe the fundamental premise of what we consider as an artifice. If the finite being always needs to be beyond the being that already is, it is always open to the artifice. The artifice, in its constitutive structure, is the intervention on nature as a fact, in view of nature as essence, or in other words it is the intervention on nature *as it is*, in view of nature *as it ought to be*. In the dimension of the artificial, nature as it is disestablished in view of our *idea* of nature.

The artifice, that historically had the function to realize the natural order, has not raised accusations of negation of the human. Problems arose when the intervention of the artifice was no longer lead by the idea of what nature ought to be in continuity to what nature effectively is, but started to be lead by the idea of what we *would* like that nature becomes beyond its objective order, or beyond ends that are not realized yet, but are nonetheless inscribed in the nature itself. Nature, once deprived of its intrinsic form, would be assumed as a *material element* of a form which is dependent from the power of the will, since the latter has knowledge and operative procedure at its disposal, that can allow it to realize its project, or even *any* project.

To sum up, once nature becomes the object of a human free will that finds its law in itself, the will to intervene through the artifice is no longer a bridge between nature as a fact and nature as *eidos*, but is even legitimated to modify the starting and arriving points that are, on the one side, the factuality and, on the other side, the essentiality of nature.

NATURE AND HUMAN CREATIVITY

Let us try to better articulate the character of a position in which nature becomes completely *relative* to the act of will. In this position the initial dimension of nature is recognized only if it is wanted. The begin, if it does not exclusively arise from the will, is not binding with regard to its acceptation. The initial fact or event is not only modified so that it is also wanted, but it can also be refused or annulled if it is not completely convertible with what is wanted. Therefore the will also determinates the ought-to-be of what already is. In this framework, the artifice intervenes as a tool that allows to modify both the initial and the final conditions of the dynamics of development of nature.

These problems cannot be solved through an obtuse reaffirmation of nature as absolutely disconnected from the will. It is important to admit that nature is not inclusive of the entire consistency of the human. Its objective constitution is always mediated through a subjective spontaneity which is in relationship with other subjectivities. We could say that, beyond nature as the set of the already given objectifications, leans out a nature as a dimension of spontaneous and *self-moving* intentionality. Therefore human nature – as we can argue also in the light of the contemporary neuroscientific research when it is not incline to the naturalistic reductionism – is not only an *explanandum* for the human, but always also an *interpretandum*, or better something that *can be interpreted*. It can be approached hermeneutically. The concept of human nature thus entails the role of a threshold-concept: it is something that is given, which refers to a not-given, and the latter is not completely inscribed in a codex *a priori*, and emerges as the fruit of a capacity of an autonomous, or even creative, increase.

PERMANENCE AND MUTATION OF HUMAN NATURE

The previous considerations should make the often apodictic use of the concept of nature less peremptory, and should also favour a greater availability to revise its meaning. It would be necessary to respect both *fires* that are involved in the never-ending interpretative undertaking of human nature. The one fire is the constant reference, within human self-reflection, to something which remains and thus combines the different expressions of the human. The other fire refers to the fact that the statements of the expressions of humanity can change with regard to both time and space: furthermore any single individualisation has an irreducible character and when it gives form to its own self-interpretation, it becomes an autonomous principle of free declination, both in the similarity and the dissimilarity, with respect to nature as a set of already given conditions. The capacity to give an individual form to what is common belongs indeed to the nature proper to the human. The Aristotelian definition of *physis* as «principle of movement and quiet in something» is very apt especially with regard to the human.[1]

GENESIS AND POIESIS

What is natural, in the human, is integrated by the *power* (*Macht*) of the enhancement of the already given conditions. Going on with the analysis of the artifice and tackling it from the above described point of view, we can argue that the artifice discloses itself as the field of human acting in the framework of the enhancement of the already given conditions. In fact the human being can be understood as an indivisible duality of *genesis* and *poiesis*. The *genesis* is a process of manifestation of the being according to internal principles; the *poiesis* is an ontological process that has its principle in an author's *techne*.

The expansion of technique as a systematic and pervading *technological* apparatus has lead to two consequences. The first consequence is that technology has
become the main road for the satisfaction of the normality of life, to the extent that
it makes the natural normality marginal. This shows the paroxystic qualification of
our civilisation as dominated by the essence of technique. The dominance of technique means that the technological normality becomes normative: adapting to the
procedures of technique is not only a *habitus* of living, but also the norm to which
life itself ought to adequate, its *concrete* moral law.

The second consequence is that the technological *poiesis* has widened to the
extent that it entered in circle with the *genesis*. The power of *poiesis* can aspire
to become a *generative* power, breaking the barriers that distinguished the *poiesis*
(which is bound to the use of elements generated by the *physis*) from the *genesis*.
Technique does not come back to nature, once it has fulfilled the task of correcting
its deficiencies or highlighting its performances, but can be itself *naturans*.

The following question arises: is a *naturans poiesis* still governable according to
an idea of permanence of human nature, according to an essential teleology capable
of orienting and binding it, or can it be non responsible with regard to this idea? This
is the core of a match that today is played especially on the mobile field of human
corporeity, and not only in the different ways of intervening (or not intervening) on
the initial and final phases of its manifestation, but also in the management of its
middle states and daily performances.

THE HUMAN BEING AS AN END

The problem, in the technologised human, is to understand which should be the
relationship between the sphere of generating and the sphere of producing, since the
poietic activity can overcome the generative activity and thus lead to an outcome
of non return. The productive activity, even if initially placed in the human, can
disembed itself from its original matrix and can become a *sibi permissa* activity. Can
then nature – understood as the dynamism of generation – maintain its normative
control over production? Or, in the hierarchic overturning of their relation, does
generation reduce itself to a means for productive finalities, that down-grade the
human to a temporary moment of productive operativity? And would it be possible
to contrast such a down-grading?

With regard to these issues, would traditional ethical concepts not become pa-
thetic or illusory? The outcome would then be not the ethical relativism but rather
the disintegration of the ethical codification of the anthropological experience and
the condemnation of our moral vocabulary to insignificance.

The persuasion that has lead humanity, and especially western humanity, till now
is the persuasion of the *insuperability* of the human being in the fulfilment of his
tasks. Considering humanity (as Kant did), whether in one's own person or in the
person of any other, always at the same time as an end and never simply as a means
is a comprehensible prescription only if the human being remains the end of any action and does not become the instrumental means for a being which is different from

the human being himself. Human dignity, human rights and analogous "non negotiable" concepts need to be sustained by such a persuasion. Even in the messages of religious salvation, transcending the human is always in favour of the human. At another level, the Nietzschean figure of the overman, especially in *Also sprach Zarathustra*, does not turn into a cancellation of the human, but represents the *enhancement* of the human in order to avoid the «horrible haphasard» (*grauser Zufall*) and the mutilation of a one-sided development.[2]

THE ANTHROPOLOGICAL PROTECTIONS

However it is not enough to point out unpleasant consequences. The expansion of the technological artifice leads us to ask ourselves how the persuasion of a non transcendible human eidetics (incidentally, a persuasion that has been shared by the current of historicism too) can be justified. Michel Foucault spoke about the human – defined as an empirico-transcendental doublet[3] – as a face drawn on sand, that can be cancelled by the pressing sea wave.[4] How then can we support a will to humanity that is not a nostalgic fancy desire? We cannot escape from the weight to *have to* exhibit the grounds of our capacity to *continue* to be aware or conscious subjects, who are also morally responsible and capable of discerning the good to be done and the bad to be avoid.

In the conviction that the inherited human is also the human to perpetuate, we can refer to the anthropological tradition and defend or protect its fundamental traits: (a) not everything that can be done needs to be done; (b) changes need not to be ends in themselves; (c) we ought to guarantee to our followers at least the same opportunities of choice that we have enjoyed; (d) we ought to contrast the reduction of the human, in any human being, to a mere material for an extrinsic formal principle; (e) we ought not to allow that any individual becomes a mere means for ends that do not belong to him; (f) we ought to allow anyone the expression of each of his faculties or capacities, and of the set of his faculties. The catalogue could go on.

THE LIMITS OF EXPERIENCE AND THE VISUAL OF THE ENTIRE

How is it possible to provide a foundation for these anthropological *protections*, that derive from the traditional self-understanding of the human? We could cling to the religious message, as a message that promises the salvation of the human in God, and the ultimate protection of the aspirations to the human's perpetuation could rest on it. We recognize the legitimacy of the option of entrusting the destiny of anthropological continuity to the hope deriving from a meta-rational announcement; at the same time we think that we need to bracket this option in order to give space to the autonomy of a rational reflection, able to face the threat of the negation of the human being, to which the excess of the artifice could lead. The following question arises: can the defence of human dignity from an instrumental reduction be a task of reason?

This is not an easy matter. At the level of empirical rationality, it is not easy to recognize to the human the power to perpetuate his constitutive *eidos*, or his existential condition. Remaining within the limits of experience, the perspective of the failing of the human does not seem contradictory. Both single experiences and the experience of the human in itself are exposed to a destiny of death.

However the human is the guardian of a radical intentionality, that is the intentionality of the entire. This intentionality goes through experience and thus leans out through a movement that transcends the limits of experience itself. Always situated in the margin of a wait for death, human conscience intentionates the other from death too.

The leaning out on the destiny of death is an eminent case of the function of transcendence that the intentionality of the entire expresses, by opposing the narrow connotations of experience. In fact, any critique of the limits of experience is enabled by the pre-comprehension of experience itself from the visual of the entire. However the critical function of the limits of experience, that pivots on the entire, is an indirect way of searching for a *positivity*, thanks to which the unlimited openness towards the entire can be fully filled. These considerations lead to the following question, that breaks any empirical restriction: is there a *positive* correlate that is on a par with the unlimited openness which is proper to the intentionality of the entire?

THE UNCONDITIONED BEING AS NON PRODUCIBLE

It is the positivity of the unconditioned being that can be put on a par with the openness to the entire.[5] But unconditioned with respect to what? With respect to any possibility to be produced. The unconditioned being is the being that escapes from producibility. It is not the outcome of an instrumental acting by someone or something else. Not only it is not produced: the unconditioned being is not producible too. This does not mean that it must be thought as static and without any dynamism.

If the unconditioned being cannot be produced by the human's productive power or, the other way round, if the human's productive power does not have any power on the unconditioned being, the question on the human condition arises, because the human is a producing subject that is exposed to the risk of being reduced to an object of production. The question is the following: is there an intrinsic relationship between the unconditioned being and the conditioned being? Is there a relationship that can take the human being away from a producibility outcome, to which he is exposed because of the excess of artificiality? In other words, how can the human being enjoy an irreducible unconditionality, being at the same time always conditioned?

THE RELATION BETWEEN CONDITIONED
AND UNCONDITIONED

To the conditioned being can be assigned unconditionality thanks to a relation of participation. The unconditionality of the conditioned being derives from the participation at the unconditionality of the being which is posed as absolute.

In this participation the ontological dignity of the human being can rest on a stable foundation. As well as the unlimited being cannot be subjected to the logics of production by someone or something else, *analogously* the limited being, that is the human being, cannot be reduced to the logics of producibility. Therefore the human being too is a principle in himself and needs to be respected. It cannot become an object of production.

This is the ontological core of the anthropological protections from the absorption into the logics of instrumental production and at the same time it is the foundation of a way of considering production as an authentic ontopoiesis, that is a manifestation of the being that does not exhaust itself in the dominance of human productive power.

MEASURE OF THE ARTIFICE AND CHALLENGES OF THE POST-HUMAN

To sum up: from the idea of an unconditioned *being* that can be participated by the fullness of the human – a fullness which is inscribed in its nature and at the same is time open to a not-yet-given fulfilment – derives the measure of the power of manipulation that the human has, thanks to the disclosure of the artifice. The omnipotence of the artifice and its destructive involutions can be contrasted. The best antidote to the excessiveness of the artifice consists in the capacity to maintain the relation with the non producible being. From here follows a rule of life, which excludes that instrumental production becomes the totality of the experience for a finite being. This rule consists in *taking care* that the artifice does not overcome the limits which are proper to a partial dimension of existence.

The human's ontological dignity then ought to rely on the maturity of consciences that are able to discern between a *manifestative* production, which is open to the fulfilment of the being, and the production of an enslavement to instrumentality, that moves away from that fulfilment. This discernment is at the base of any choice that is done in punctual situations and contingent circumstances, which are never without opacity, uncertainty, and risk.

Thanks to these coordinates we can even face the challenges of the post-human, distinguishing that which is an enhancement of the human (through a coherent use of technologies and a *good* hybridation between the human and his growing artificial equipment) from that which could turn into his negation. It is not a matter of cultivating the "fear of the artificial" on which Emmanuel Mounier poured out his caustic antibourgeois irony.[6] It is rather a matter of rethinking without "reactionary" prejudices the relation between human ontology and artificial dimension, moving towards the new frontiers of a hybridation that is in favour of the process of humanization.

University of Macerata, Macerata, Italy
e-mail: totarofr@unimc.it

NOTES

[1] On this issue the following comment by Robert Spaemann is very precious : «Fin dall'origine, nella filosofia greca, *physis* non significa [...] la pura oggettività di una materia passiva quanto un essere sussistente, pensato in analogia all'esperienza di sé propria dell'uomo: e cioè nel senso di una distinzione di un essere naturale da tutti gli altri, di un sistema vivente, come si direbbe oggi, da un ambiente, inteso come limitazione attiva, come autoaffermazione e autorealizzazione *spontanea*. *Physis*, natura, è secondo Aristotele l'essenza delle cose che hanno il principio, l'inizio del movimento in se stesse. In questo senso *physis* è certamente un concetto che fin dall'origine serve alla distinzione» (R. Spaemann, *'Naturale' e 'innaturale' sono concetti moralmente rilevanti?*, in C. Vigna-S. Zanardo (eds.), *Etica di frontiera. Nuove forme del bene e del male*, Milan: Vita e Pensiero, 2008, p. 88).

[2] A larger analysis in Francesco Totaro, "*Superuomo e senso dell'agire in Nietzsche*", in Totaro (ed.), *Nietzsche e la provocazione del superuomo. Per un'etica della misura*, Rome: Carocci, 2004, pp. 111–133.

[3] Michel Foucault, *Les mots et les choses*, Paris: Gallimard, 1966; engl. transl. *Order of Things*, New York: Vintage Books Edition, 1994, p. 318.

[4] Ibidem, p. 386.

[5] Parmenide's reflection on this issue has not been overcome yet. According to his formulation , «the being cannot not be». From this unconditioned formulation of the being derives the following affirmation too: «thinking and thinking that it is are the same thing», if thinking is above all intentionality of the entire that finds its fulfilment in an unlimited positivity of the being. For this reason Parmenides' intentionality of the entire does not embrace the multiple determinations. A further gain exactly consists in adding that the being can be predicated also with regard to the multiple determinations of the entire, and nonetheless the way in which the determinations are within the entire is not actually manifest, therefore the fact that they concretely belong to the entire does not appear. The being of the determinations, because of their opacity compared to the entire, differs from the being of the entire.

[6] Emmanuel Mounier, *La petite peur du XX siècle*, Paris: Éditions du Seuil, 1949.

MORTEN TØNNESSEN

SEMIOTICS OF BEING AND UEXKÜLLIAN PHENOMENOLOGY [1]

ABSTRACT

German-Baltic biologist Jakob von Uexküll (1864–1944) did not regard himself as a phenomenologist. Neither did he conceive of himself as a semiotician. Nevertheless, his *Umwelt* terminology has of late been utilized and further developed within the framework of semiotics and various other disciplines – and, as I will argue, essential points in his work can fruitfully be taken to represent a distinctive *Uexküllian phenomenology*, characterized not least by an assumption of the (in the realm of life) universal existence of a genuine first person perspective, i.e., of experienced worlds. Uexküllian phenomenology is an example of – a special case of – a semiotics of being, taken to be a study of signs designed so as to emphasize the reality of the phenomena of the living. In the course of this paper, I will relate Uexküllian phenomenology to the eco-existentialism of Peter Wessel Zapffe (1899–1990), eco-phenomenology (including David Abram and Ted Toadvine), and semiotics of nature (biosemiotics, ecosemiotics, zoosemiotics). I will further make a few remarks on the partial resemblance between Uexküllian phenomenology and Tymieniecka's "phenomenology of life", and its difference from the "phaneroscopy" of Peirce.

This paper starts out with the notion of *Uexküllian phenomenology*. The attentive reader will notice throughout this text that the wide-ranging project I am investigating is the relation between phenomenology and semiotics, with the natural world – the world of the living – as a recurring theme. In the first section of this paper, I will make clear why such a phenomenology deserves the name "Uexküllian", and how it differs from the phenomenology proposed by Charles Sanders Peirce (1839–1914). I will then proceed to relate Uexküllian phenomenology to the eco-existentialism of the Norwegian philosopher Peter Wessel Zapffe (1899–1990), to eco-phenomenology, and to various brands of *semiotics of nature*.

Uexküllian phenomenology can be regarded as an example of – a special case of – a semiotics of being. A semiotics of being, in its turn, would be a study of signs (signification, communication, representation) designed so as to emphasize the reality of the countless phenomena of the living (where the latter are acknowledged as the true subjects of the phenomenal world at large). This paper thus presents programmatic statements for both semiotics and phenomenology. The general assumption is that unification of the two fields of inquiry can be mutually enriching.

Before endeavouring to pursue my main objective, however, I will make a few preliminary remarks on the relation between Uexküllian phenomenology and the phenomenology of life proposed by Anna-Teresa Tymieniecka. I am sympathetic

A.-T. Tymieniecka (ed.), Analecta Husserliana CX, 327–340.
DOI 10.1007/978-94-007-1691-9_27, © Springer Science+Business Media B.V. 2011

to Tymieniecka's statement at the 60th International Congress of Phenomenology that what should be fundamentally thematized as primary is life. This contrasts with competing prioritizations of thematizing of being and knowing, respectively (in an anthropocentric sense). That said, I must point out that the notions of "being" and "knowing" applied in the current paper contrasts with Tymieniecka's use of these (and with traditional use of them), and that I have tended to alter the signification of these terms so as to bring them in line. For me, then, "semiotics of *being*" denotes an approach with the whole sphere of life – all that lives in this planet, and possibly beyond – as its area of validity or relevance; and when and if I call any of these creatures "knowing", it is in a sense very different from that of the "knowing human" as it is usually conceived of.

In the interview Torjussen et al. (2009), Tymieniecka was asked about the metaphysical dimension of the ecological crisis, and how she relates to eco-phenomenology. "Actually," replied Tymieniecka, "my account of *ontopoiesis* is an eco-phenomenology." That statement makes sense, given that describing the self-individualization of life, as she calls it, "is the most fundamental ecology that can be done."

Upon pointing out a few commonalities I must disclose my lack of any detailed knowledge of Tymieniecka's phenomenology of life. It is nevertheless clear to me that the two approaches (her being much more developed than mine) share a number of basic convictions. First, I concord that "the order, selfordering, of the course of individualizing life is not a 'neutral,' automatic fitting together of matching elements. To the contrary, this ordering – effected by living intentionality, vis viva – is a sentient selection" (Tymieniecka 2007, p. xxiii). In my context this is related to what I call "semiotic causation". Second, I acknowledge that instead of classifying philosophical problems in separate realms of inquiry we should "approach their common groundwork, which is life itself at its basic onto-metaphysical level (...) wherefrom all scientific and philosophical problems have their common root." (ibid., p. xx). Third, I heartily agree that "the concept of what is 'human' cries out for revision", given that traditionally "the human being has been specified by its 'nature,' that is, identified by the salient features that *distinguish* us from other living beings" (ibid. – my emphasis). Despite these common convictions and aims, I am confident there are a number of points where these two approaches diverge as well.

UEXKÜLLIAN PHENOMENOLOGY

When Jakob von Uexküll extended the reach of the in part phenomenological epistemology and ontology of Immanuel Kant (1724–1804) to the world of biology, i.e., the world of the living, he claimed to represent Kant, rather than to contradict him.[2] In actual fact, he did both. Uexküll's *Umwelt theory* (environmental – or, as we shall see, phenomenological theory) rests not simply on an adoption of certain Kantian terms – such as "phenomenal world" (*Erscheinungswelt*) – but on a radical revision of them.

His explicit, programmatic critique of Kant (cf. Uexküll 1928: 9) points straight to the crucial differences: The phenomenal world springs not from the mind in a rationalistic sense, but rather from the body as a whole. As we can see, Uexküll thus implicitly introduced a notion of *embodied mind* – well before Maurice Merleau-Ponty (1962 [1945]). Uexküll further states that not only humans "have" phenomenal worlds – so do other living creatures. To be a living being implies being someone for which something appears. The concept of *functional cycles/circles* shows how subject (a living being) and object (its relevant surroundings) together forms an organic unit, and, by implication, how any acting creature is actively engaged in its lifeworld. Through this notion in particular, Uexküll demonstrated that "the phenomenal world" is a reality without which no living being can be adequately understood. The life world of an animal is an expression of its ecological situation along with its behavioural capacities, which in turn reflect its physiological constitution. The *subjective biology* Uexküll called for – not to be confused with most modern, objectivist adaptations of ethology – would not least entail theoretical reconstructions of various life worlds.[3]

For Kant, the phenomenal world was a human enterprise, and as such a uniform, singular entity. While a Kantian worldview is in this sense monist (or, if one stresses the category of the thing-in-itself, dualist), the Uexküllian worldview is inescapably pluralist. While for humans there are human-things, as Uexküll would have it, for the cat there are cat-things, for the tick tick-things and so on. One and the same thing can appear as very different phenomenal objects in different Umwelten. This is true not only of different species – and species-specific Umwelten – but furthermore with regard to individual Umwelten.

Admittedly, individuality is an emergent phenomenon which varies greatly in degree in the realm of life. Plants and fungi are diffuse cases. The coordination of their activities is not centralized in the same way as it is for either unicellular organisms or multicellular animals, and their parts may have a higher degree of autonomy with respect to the body (organism) as such. It might not be justifiable to say that plants and fungi *act* – i.e., that they display behaviour – and in consequence, there may not be any Umwelt objects in their lifeworlds (no plant things – no fungi things). Instead, their lifeworlds – their phenomenal worlds – are made up of various *meaning factors*. Some examples are humidity, temperature and light. These meaning factors – which typically fluctuate in strength or concentration, be it regularly or irregularly – may of course be present in the lifeworlds of animals and unicellulars as well. The difference between Umwelt objects and meaning factors is that while the former are identifiable (i.e., stand out, like a figure on a background) and typically require an immediate response, the latter leave traces of influence on the organism over time, and are only in exceptional cases (such as situations of sudden stress) immediately identifiable. Plants and fungi, in other words, respond for the most part to a floating aggregate of influences. Their phenomena are vague, compared with animal phenomena. Nevertheless, plants and fungi, too, constantly interpret and respond to their developing surroundings.

The realm of phenomena, then – and of the semiosis (action of signs, or sign exchange) that go along with them – range from the simpleminded orientation of a nutrition-seeking bacterium, via the plastic improvisation of a plant, to the often incredible gap between conscious identity and actual behavior in the case of the human animal (a sign of great individuality). The realm of life is perfused with appearances and the agendas we apply to categorize them. Generally, animals with a nervous system can be assumed to have phenomenal worlds – Umwelten – of much greater detail and distinction than other creatures. In the human case, language plays a decisive role, further adding to the complexity of the things we are capable of expressing and handling.

As we can see, Uexküll differs from Kant in depicting a world where the spatial and temporal configuration of the phenomenal world, for one thing, varies from life form to life form. In this world of phenomena and phenomenal relations, there is a multitude of perspectives. The human perspective might very well be superior in terms of intelligence and abstraction, but it is not a perspective of total oversight. Instruments of various kinds no doubt add enormously to our knowledge and experience, but our human experience is nevertheless situated, in biological terms, as a limited perspective. There are other creatures that are capable of hearing or seeing above or below our sensory thresholds. And there are senses that we do not have, which other creatures have. These creatures can experience the world in a way we cannot. About these experiences we can acquire data, achieve a certain understanding – but we can never experience the world in that way firsthand.

If Merleau-Ponty (1962: viii–ix) was correct in stating that "we must begin by reawakening the basic experience of the world, of which science is the second-order expression" – and I believe he was – then no doubt firsthand experiences matter. Experiences in the 1st person, and 3rd person descriptions of them – which science offers – are two very different categories. It would be an illusion to think that a scientific worldview can ever achieve total oversight in the sense of "knowing all that there is to know". In this context I adopt Michael Polanyi's (1891–1976) position that the living exhibit "tacit knowing" in and through their actions and doings, cf. Polanyi (2009 [1966]).

A challenge to an Uexküllian worldview is how we can explain the constitution – the formation and structure – of what Polanyi called our "stratified universe". In the world of the living it is clearly all the relations embedded in the life processes – somatic, social, and ecological – that bind it all together. Uexküll's world is a relational world, a world of relations. A related challenge concerns the relation between biological parts and wholes. Like an organ is composed of tissue, which is composed of cells, so is a body composed of organs, and an ecosystem of bodies. The ecosystem constitutes an organic unit of sorts, as can also be claimed of species, populations etc. But there is a fundamental difference between the way an organ is part of a body, and the way an individual is part of an ecosystem.

The difference, which qualifies the level of the individual (organism) as privileged, is that it is the level of the individual which is properly speaking the level of phenomena (experience). I believe it makes sense to talk about aggregate phenomenal worlds, such as the phenomenal world (Umwelt) of human beings, or of bats, or of mammals – but there is a crucial difference between these kinds of phenomenal worlds, and individual phenomenal worlds: In a word, the former are abstractions. No single creature actually experiences the world like that.[4]

Now it is time to make good on my promise to clarify why the phenomenology I am describing is deserving of the brand "Uexküllian". I have previously (Tønnessen 2009a) argued that the Umwelt theory of Uexküll needs to be updated with regard to its neglect of the historical dimension of the life processes. At some other points, as well, his work is too marked by his time and his concrete influences – a case at hand is his relation to Kant. If one looks into the way Uexküll himself tried to generalize his biological findings and make them relevant for politics, the picture gets even gloomier (see Tønnessen 2003 and Harrington 1999). A general disclaimer is in place: Uexküllian phenomenology as I portray it is loyal not to Uexküll's thought in detail but to his essential finding that nature is constituted by the intricate relations of all living creatures, which are all subjects of the phenomenal world at large.

The reason why it makes sense to propagate a phenomenology under the label "Uexküllian" is that Uexküll's fundamental premises about the nature of life are desperately needed in our time – and in the life sciences of our time. While today's life sciences are for the most part reductionist – neglecting the reality of the individual, the primary stakeholder in nature – Uexküll's call for a subjective biology echoes Husserl's call for a return to the things themselves. Whereas biologism is a potential problem in our society of "biological innovation", an Uexküllian worldview is not in my interpretation biologistic, because it portrays society not simply as part of nature but further as an emergent entity within nature, which thus has its unique operational rules on top of the general operational principles of nature.

Uexküllian phenomenology should be rigorously undogmatic. This applies not only to Uexküll's work, but also to semiotics as a scholarly discipline. Its main axiom could here be that the phenomenon is a special case of semiosis. Semiosis, in other words, is the general entity, or process, of which phenomena are part. I will get back to the relation between semiosis and phenomena towards the end. The axiom just mentioned could be taken to imply that phenomenology can be regarded as a subdiscipline of semiotics.

If reading that last sentence provoked you, you should look up how dismissive certain other semioticians can be of phenomenology. A common attitude is that semiotics is more progressed than phenomenology, and many would hold that phenomenology is largely a dated enterprise. In the following I will relate in some detail to the work of Charles Sanders Peirce, the chief source of inspiration for most contemporary semioticians – not least his idea of a field named *Phaneroscopy*. Perhaps my disclaimer on Uexküll should be accompanied at this point with a similar disclaimer concerning Peirce: Uexküllian phenomenology as I conceive of it is loyal not to any specific interpretation of Peirce, nor to his general philosophical outlook, but rather, to the extent that it is of any use, to some basic concepts of his such as those of *symbolicity*, *iconicity* and *indexicality*. It is absolutely crucial that such concepts are not fetishized.[5] As we will see in the following, Uexküllian phenomenology is not necessarily aligned with Peirce's ideas about phenomenology.

In a paper entitled "Is Phaneroscopy as a pre-semiotic science possible?" André de Tienne (2004) treats the prospects of Peirce's phenomenology, which the latter named *Phaneroscopy* in order to distinguish it from Hegel's thought. Peirce's papers on phenomenology and the theory of perception date for the most part from the period 1900–1908 (Luisi 2006). "Phaneroscopy as a research activity", observes de Tienne, "isn't practiced anywhere and hasn't attracted any wide following [. . .] Peirce scholars are divided about what that science is supposed to be and to do, and about how exactly it relates to semiotics."

"Phaneroscopy", said Peirce (1931–1958: 1.284) in his Adirondack lectures in 1905, "is the description of the phaneron; and by the phaneron I mean the collective total of all that is in any way or in any sense present to the mind, quite regardless of whether it corresponds to any real thing or not." Here, the "phaneron" apparently corresponds to the combined phenomena of an individual phenomenal world.

If you ask present when, and to whose mind, I reply that I leave these questions unanswered, never having entertained a doubt that those features of the phaneron that I have found in my mind are present at all times and to all minds. So far as I have developed this science of phaneroscopy, it is occupied with the formal elements of the phaneron.

The common interpretation of Peirce on this point is that his Phaneroscopy was intended to be applicable on the human mind only – a marked difference compared with Uexküllian phenomenology. If that interpretation is correct, Peirce appears to have envisioned a phenomenological world – a world of phenomena – just as limited in its reach as that of Kant. If it is wrong – or, if one were to disagree with Peirce and call for a "Phaneroscopy of the living" – one would have to revise him to the same extent as Uexküll had to rework Kant. We can further observe that Peirce promoted a monistic, not pluralistic, understanding of the phenomenal world – like Kant, but unlike Uexküll, and that his preferred worldview is timeless – like in Kant and Uexküll alike (but unlike an up-to-date Uexküllian phenomenology).

"It will be plain from what has been said", wrote Peirce (1931–1958: 1.286–287) in "Logic viewed as Semeiotics", "that phaneroscopy has nothing at all to do with the question of how far the phanerons it studies correspond to any realities." Uexküllian phenomenology differs from Phaneroscopy (or, it *should* differ) by emphasizing, rather than neglecting, relations between phenomena and the rest of empirical reality. In my take on Uexküll, that implies treating "world history" (human history) as well as "natural history" as organic wholes – the former being part of the latter. The phenomenal world at large has a history, and a reality, and as such it is emergent, historical, and empirical. Distinguishing phenomena that do correspond to something real from those that do not *is* a task for phenomenology – and this task is more crucial than anywhere else in the symbol-heavy human realm. But that enterprise should not be taken lightly. Here, the Peircean notions of symbolic (conventional), indexical (causal) and iconic (similarity-based) relations come in handy, in effect providing the base variants of *semiotic causation* (a dominantly associative logic – cf. Peirce's notion of "abduction" in particular).

ECO-EXISTENTIALISM, ECO-PHENOMENOLOGY, AND SEMIOTICS OF NATURE

In what follows I will treat Uexküllian phenomenology in its relation to the eco-existentialism of Peter Wessel Zapffe, the eco-phenomenology of David Abram and Ted Toadvine, and various brands of semiotics of nature, biosemiotics included.

THE ECO-EXISTENTIALISM OF PETER WESSEL ZAPFFE

On a personal note, it was in Peter Wessel Zapffe's Norwegian language magnus opum *On the Tragic* (Om det tragiske) that I first encountered the Umwelt theory of Jakob von Uexküll (Zapffe 1996 [1941]). Zapffe is one of the three classical eco-philosophers of Norway, along with deep ecologist Arne Næss (1913–2009), who also to some extent referred to the work of Uexküll. For Zapffe, Uexküll was *the* biologist, and thus important for carving out his "biosophy" – philosophy of biological wisdom. From Uexküll, Zapffe learnt that everything alive is fundamentally different from everything not alive (that which is alive is what matters), and that all that lives navigates along the lines of its interests. His infamous pessimism (Zapffe held that humankind should voluntarily stop reproducing) lies in his take on what is characteristic of human interests and abilities. Claiming that we, as a species, are over-equipped in terms of consciousness, his analysis of cultural life amounted to a series of observations of the various ways in which we delude ourselves in order to escape if not our predicament, then at least our awareness of it.

The core contribution from Uexküll in Zapffe's thought was the former's view that in the case of animals, there is a harmonious relationship between ability and need. Zapffe's philosophical innovation is his claim that this is not valid in the case of human beings. He thus establishes man as an exceptional creature in the living world (as have countless others, each in their own way). While the behaviour of most animals is more or less fixed, Zapffe observed, human behaviour is exceptionally unfixed – exceptionally plastic. More precisely, we have become fixed in being unfixed. Instead of having highly specialized limbs or organs, we have acquired an ability to apply tools and technology so as to extend our capabilities. We compensate for our bodily simplicity by innovations and armour. Over time, the specialization of labour and technology has gone so far that the development has long since spun out of control. The technological development is not regulated by any external force, but only by our own choices. Due to our near-global delusion, there is not much hope.

So far Zapffe. Honestly speaking, his portrait of the biological world is very biased, since he everywhere (except for in his humorous prose stories) emphasizes grief and misery and downplays delightful undertakings. He talked of a "brotherhood of suffering", ranging from the amoeba to the dictator or artistic genius. Empathy or sympathy thus has a place in his worldview. But why not a "brotherhood of pure delight" as well?

Zapffe failed to see the true significance of Uexküll's attribution of phenomenological status to other mindful creatures. He was the first major figure in Norwegian
culture to call for conservation measures – but like Uexküll, he did not observe, or
foresee that the apparent harmony between animal ability and need turns out not
to be a timeless fact. In the case of endangered species in volatile ecological situations – such as our current situation – the abilities of any animal can prove to be
insufficient to meet their needs. In short, Zapffe's existentialism did not break with
the tradition of focusing solely on the *human* existence, despite the fact that it –
perhaps for the first time – incorporated the value of nature (though first of all for
recreational purposes) in existentialist thought.

THE ECO-PHENOMENOLOGY OF DAVID ABRAM AND TED TOADVINE

It is a peculiar fact that even proclaimed environmental phenomenologists – eco-
phenomenologists (see Brown and Toadvine (eds.) 2003) – mainly or exclusively
reason from a human point of view. That is not a promising start, as a matter
of methodology, in dealing with issues of ecology. The contribution of Uexküll's
thought, as a possible foundation for eco-phenomenology, is that it carries with
it the theoretically modelled perspective of each and every living being. It offers
elements of a pluralist, ecologically informed worldview in a form which allows
us to come to terms with the manifold diversity of nature. It offers an image of
nature as incredibly much richer than our human perception of nature. If we believe that eco-phenomenology, or environmental philosophy in general, is all about
human perceptions, we commit a categorical mistake, and miss out on the heart of
the matter.

The contemporary eco-phenomenologists David Abram and Ted Toadvine are
highly different in style, method and outlook, yet are both first-rate representatives of this emerging field. I consider *The Spell of the Sensuous: Perception and
Language in a More-Than-Human World* (Abram 1997) to be a modern classic,
a great source of inspiration, and in many ways a work that can help bringing phenomenology forwards. My one reservation – apart from the general point mentioned
above – derives from Abram's defence of animism. I am glad that *someone* is giving
philosophical credibility to the worldview(s) of oral cultures, and Abram is a highly
articulate voice – but in his case the defence of animism gets in the way of an even
richer perspective. To the sensing body, observes Abram, nothing presents itself as
utterly passive or inert. And from that he concludes that nothing *is* utterly passive or
inert. That is animism in a nutshell.

"In the derivation of this word [phenomenology]," wrote Peirce (1931–1958:
2.197) in 1902, " 'phenomenon' is to be understood in the broadest sense conceiv-
able; so that phenomenology might rather be defined as the study of what seems
than as the statement of what appears. It describes the essentially different elements
which seem to present themselves in what seems." Here, both Peirce and Abram are
aligned with a part of the phenomenological tradition which we should break free
from: Namely, the conception that phenomenology should be a study of what seems.
Such a conception is truly deserving of the label "pre-semiotic". In Abram's case, I

consider this a flaw in an otherwise brilliant work. I do, however, sympathize with his project of reengaging with perceptual reality. That is a cornerstone for modern environmentalism, and modern thought in general.

Like Abram, Ted Toadvine is following in the footsteps of Merleau-Ponty. He has investigated to what extent meaning can be attributed to nature (Toadvine 2003: 273), arguing that "the ontological continuity of organic life with the perceived world of nature requires situating sense at a level that is more fundamental than has traditionally been recognized." Much of his project resonates well with biosemiotic and Uexküllian thought. Rather than to the world-subject conjunction, Toadvine theorizes, "sense would be more accurately attributed to the meeting point of world and life. All life carries with it an evaluative projecting into the world. [...] Life values and chooses; it throws a world up before itself and is therefore already intentionally engaged rather than merely causally connected."

SEMIOTICS OF NATURE (BIOSEMIOTICS, ECOSEMIOTICS, ZOOSEMIOTICS)

There are in the main three established brands of semiotics of nature – biosemiotics, ecosemiotics and zoosemiotics – and this is not the place to go into detail about any of them.[6] While the International Society for Biosemiotic Studies is only a few years old, the conference series *Gatherings in Biosemiotics* is now in its tenth year – but biosemiotics as a field dates back to the 1980ies, and zoosemiotics, from which biosemiotics grew, all the way back to the 1960ies. The story can hardly be told without mention of Thomas A. Sebeok (1920–2001), a prominent 20th century semiotician who coined "zoosemiotics" and was a mobilizing figure for biosemiotics.

What is important in the context of this paper is that the Umwelt theory of Jakob von Uexküll has had a renaissance as a work of foundational importance for contemporary semiotics of nature (cf. Kull 2001). For biosemioticians, writes Jesper Hoffmeyer (2004: 89), there is "nothing mysterious about the phenomenal world, for it is deeply embedded in bodily semiotics". That is largely due to the influence in biosemiotics of Jakob von Uexküll. One reason why I find it worthwhile to campaign for an Uexküllian phenomenology – in phenomenological as well as in semiotic circles – is that even in biosemiotics there is a continued need for stringent thought in these matters. Hoffmeyer provides a good example – he is perhaps the one biosemiotician I share most views with; and yet, his thinking around Umwelten (phenomenal worlds) is at times inconsistent.

To the *phenomenological reduction(s)*, at any rate, where the problem with perceptual biases is attempted solved by way of a suspension of judgment – the phenomenological *epoché* – etc., I would add a *biosemiotic reduction*. We could perhaps say that the phenomenological reduction, as it has hitherto been conceived of, aspires only to achieve (or approach) an unbiased perception of *our own*, human Umwelt. We cannot but commence (and continue) our journey into the phenomenal world at large *in*, *by* and *through* the human Umwelt, but current eco-phenomenology is testimony to the fact that a second step, a second reduction, is required in order to reach beyond the domain of human prejudice. The biosemiotic reduction, as I have defined it in Tønnessen (2010), is "the movement in thought whereby we reduce observed material in the life sciences to the meaning-content constituting the lifeworlds (and their constituent parts, down to the level of the cell) of biological organisms. Semantic, syntactic and pragmatic noise is to be done away with."

Earlier in this paper I have established that semiosis (the action of signs, or sign exchange) is the general category of which phenomena are part. In conclusion, we will now consider the relation between semiosis and phenomena in a little more detail, by way of an example. But first, I should delimit the realm of each of these two meaning-constituted notions, however provisionally. For now, I assume that *semiosis* occurs at all levels of biological organization from the cell and up. *Phenomena*, on their hand, occur firsthand (as experiences in the 1st or 2nd person)[7] on the level of the individual (organism) only. In our stratified universe, we can conceive of phenomena as one layer of semiosis, constituted by semiosis at lower levels of biological organization (not least the semiosis of our sense organs, and of our brain).

The example of the tick is classical in Uexküll studies (cf. von Uexküll 1957 [1934]: 7). In a few words, the tick is interesting because it is capable only of recognizing a few elements – such as the butyric acid, hair, and heat. All mammals have butyric acid, so in consequence the tick is able to recognize any mammal – though not to distinguish between them. For the tick, there are no "wolves" and no "sheep", but only "mammals". Uexküll's illustrative point was that the tick is equipped so as to perform exactly the actions it needs to perform in order to get by.

Let us now consider a tick attack on a mammal – say, Larry David. First, receptors of the tick recognize the butyric acid evaporating from Larry David. That is semiosis. At some point – when passing a certain threshold – this semiosis gives rise to a phenomenal experience: The tick senses a (olfactory) sign of a mammal. It responds – acts – accordingly, by letting go of its twig, and fall. After landing somewhere on the surface of Larry David – an event which is reflected in semiosis triggered in tissue surrounding the spot of impact – receptors of the tick may (if the tick is lucky) recognize some hairs (semiosis, converted to a phenomenon). The tick then crawls deeper, until it recognizes the heat radiating from Larry David's skin. That is semiosis – which again gives rise to a phenomenon, as the tick senses yet another sign of the mammal, and responds by penetrating Larry David's skin. Soon thereafter, the tick sucks his blood. At this point Larry David may or may not have become aware of the presence of the tick, or of the pain caused by it. If he has, he has phenomenal experiences (with or without the tick figuring as an Umwelt object). If he has not become aware of the tick or its doings, only the affected tissue is in a state of knowing: That is semiosis.

CLOSING REMARKS

My assertion that semiotics may be conceived of as more comprehensive than phenomenology may strike many as absurd, given that Husserl, for one, held that phenomenology envelops all the phenomena of mind. The difference between

Uexküllian and strictly Husserlian phenomenology on this point is that the former operates with a vastly wider notion of "mind". While a Husserlian phenomenologist may find Uexküllian phenomenology to be absurdly broad, speculative, or conceptually bewildered, an Uexküllian phenomenologist may find Husserlian phenomenology to be unduly narrow.

Within my familiar theoretical framework, the world's "non-reducible presence" is represented by the life worlds – Umwelten etc. – of the organisms of planet Earth, however one chooses to categorize them. Admittedly, it is a paradox that while philosophy has traditionally been devoted to the most general of questions, a pluralistic, ecologically-oriented ontology entails that what *distinguishes* each one kind of the living, and each one individual or cultured population, matters just as much as what we have in common with other living creatures (note that this observation does not, as part of a balanced world view, contradict my former appraisal of Tymieniecka's emphasis on what is common for all that is alive). There can thus be no sharp distinction between philosophy and the life sciences, but rather a gradual transition from the more-or-less philosophical/generic to the more-or-less scientific/specific.

As we have seen, Uexküllian phenomenology differs from the Phaneroscopy of Peirce in that it emphasizes, rather than neglects, relations between phenomena and the rest of empirical reality. In parallel with this point, while Peircean phenomenology is explicitly monistic (as is the "phenomenology" of Kant), Uexküllian phenomenology is as mentioned unequivocally pluralistic. In the line of thought of Uexküllian phenomenology, diversity and differences is to be highlighted, not disregarded. The value of life is perhaps first of all shown in its rich variety. And regardless of the incredible manifold of the living, the human kind remains unique and dignified in its own way – all the while being so deeply intertwined with the situation and existence of other living creatures that to attempt to describe the human species without reference to others would be a truly hopeless task.[8] Such "vacuum-anthropologies" are so remote from life as to be not only philosophically questionable but further ethically *harmful* descriptions of reality. As Francesco Totaro remarked at the Bergen conference, ontology can indeed become a tool for transformation. In other words, how we describe and conceive of the world does indeed influence the way in which the world is turning out to be, in its unfolding process of becoming. In that sense, "the world" – whether qua global ecosystem or qua social system – is at present truly a material extension of our all-too-human thought processes.

I will end this paper with a reply to Ane Faugstad Aarø's critique of my approach in phenomenology at that same congress. She asks whether Uexküllian phenomenology as outlined here is capable of being telling of human reality, given that it tends to present simple, universal models, and pinpoints the absence of a notion of "freedom", which is crucial in human affairs. Part of my response is constituted by biosemiotician Jesper Hoffmeyer's concept of "semiotic freedom" (cf. Hoffmeyer 2008). Semiotic freedom is so to speak our "interpretative freedom", or "perceptual freedom", and it appears wherever there are semiotic agents, i.e. creatures capable of relating to their meaningful surroundings. As I have argued elsewhere (Tønnessen 2009b) in a discussion of the implicit self (embodied in the behaviour of a creature) and the explicit self (manifested in the identity of a creature), while for simple (non-social) creatures the implicit and the explicit self converge, for complex (social) creatures they diverge, to an extent that broadly speaking corresponds to their level of semiotic freedom. With our unprecedented sociality and semiotic freedom, we human beings are apt to experience, at times, an equally unparalleled gap between behaviour and identity. This idea provides us with a biological, or ecological, or evolutionary perspective on alienation etc. Human perception is incredibly more sophisticated – and potentially self-deceiving – than the perception of "lesser" creatures, surrounded as we are in our life worlds by layer upon layer of cultural and sub-cultural filters and amplifiers. This immense freedom in interpretation (and expression), which is usually thoroughly tied up in cultural terms, can easily overwhelm us. We choose who we are to be (some more conventionally than others) – not because we like making choices, but because life forces us to, lest we be lost in eternal qualm. The phenomenology of the human kind is no doubt complex, and any outright telling portrayal of it requires knowledge of both culture and ecology (in Tønnessen 2003 (p. 290) I referred to the *conceptionalized Umwelt experience* of our kind). But at the very foundation it is nonetheless fundamentally similar to the phenomenology of other kinds of life. What distinguishes us from other life forms, I suggest, is not something that is alien to life-as-such, but rather this abovementioned gap between identity and behaviour, which is a product of our immense semiotic freedom. Our human freedom, therefore, is intimately tied to our special stature qua semiotic creature (a creature capable of navigating in a world of meaning) – and a semiotics of being should be able to portray that phenomenon in its proper context.

Department of Semiotics, Institute of Philosophy and Semiotics,
University of Tartu, Tartu, Estonia
e-mail: mortentoennessen@gmail.com
Academic homepage: http://UtopianRealism.blogspot.com

NOTES

[1] The current work has been carried out as part of the research projects. The Cultural Heritage of Environmental Spaces: A Comparative Analysis between Estonia and Norway (EEA–ETF Grant EMP 54), Dynamical Zoosemiotics and Animal Representations (ETF/ESF 7790) and Biosemiotic Models of Semiosis (ETF/ESF 8403), and partaking in the Centre of Excellence in Cultural Theory (CECT).
[2] Kant's treatment of the objects of biology as a scholarly discipline is to be found first of all in *Critique of Judgment* (Kritik der Urteilskraft) (Kant 1987 [1790]), but Uexküll related almost exclusively to *Critique of Pure Reason* (Kritik der reinen Vernunft) [Kant 1996 [1787]].
[3] The other book-length works by Uexküll of foundational importance are Uexküll 1985 [1909, 1921], 1957 [1934] and 1982 [1940].
[4] Naturally, all theoretical reconstructions of lifeworlds are abstractions. My point is that if we model an individual lifeworld, we model something which is itself a model of the world for a particular individual. If, on the other hand, we model an aggregate lifeworld, such as "the Umwelt of 18th century Germans", we model something which does not in itself have a reality in the same sense. Both reconstructions may

very well be telling, but there is a crucial difference between them, and we owe it to ourselves not to get lost in our abstractions.

5   In the cult around Peirce, some followers have built a solipsistic metaphysics around his concepts of Firstness, Secondness and Thirdness. Such concepts may be of value when applied in their right context, but they perform poorly as objects of worship. The same holds true for anything with a triadic structure, which fits in so well with a simplistic Peirce interpretation.

6   Readers with an interest in engaging with these fields are referred to Barbieri (ed.) (2007), Kull et al. (2008) and Hoffmeyer (2008) (biosemiotics), Kull (1998) and Nöth (1998) (ecosemiotics), and Sebeok (1972), Sebeok (1990) and Martinelli and Lehto (Guest Editors) (2009) (zoosemiotics).

7   Whereas a 1st person perspective/experience corresponds to *perception*, i.e. *signification*, a 2nd person perspective corresponds additionally to *communication*, i.e. *social* (or asocial) *behaviour.* *Representations* of Umwelt objects may appear in either domain.

8   This is by and large in line with Tymieniecka's stand that "the human being cannot be defined by its specific nature but by the entire complex of individualizing life, of which complex it is vitally part and parcel" (Tymieniecka 2007, p. xx). In Torjussen et al. (2009) she explains the perspective in the following manner:

(. . .) human being can not be considered in itself as such (. . .) there can be no anthropology that considers human being as such, in the middle of other things almost by chance. On the contrary, human being should be considered as a human condition within the unity of everything there is alive. That means the human being unfolds and generates in a mutual contributive relation to all the other living beings.

REFERENCES

Abram, David. 1997. *The spell of the sensuous: Perception and language in a more-than-human world.* New York: Vintage Books.

Barbieri, Marcello. (ed.). 2007. *Introduction to biosemiotics: The new biological synthesis.* Dordrecht: Springer.

Brown, C.S., and T. Toadvine. (eds.). 2003. *Eco-phenomenology: Back to the earth itself.* Albany, NY: State University of New York Press.

Harrington, Anne. 1999. *Reenchanted science: Holism in German culture from Wilhelm II to Hitler.* Princeton, NJ: Princeton University Press.

Hoffmeyer, Jesper. 2004. Uexküllian Planmässigkeit. *Sign Systems Studies* 31(1/2):73–97.

Hoffmeyer, Jesper. 2008. *Biosemiotics: An examination into the signs of life and the life of signs* (trans: Hoffmeyer, J. and trans: and ed. D. Favareau). Scranton and London: University of Scranton Press.

Kant, Immanuel. 1987 [1790]. *Critique of judgment. Kritik der Urteilskraft* (trans: Pluhar, W.). Indianapolis: Hackett Publishing.

Kant, Immanuel. 1996 [1787]. *Critique of pure reason. Kritik der reinen Vernunft* (second edition) (trans: Pluhar, W.). Indianapolis: Hackett Publishing.

Kull, Kalevi. 1998. Semiotic ecology: Different natures in the semiosphere. *Sign Systems Studies* 26: 344–371.

Kull, Kalevi. 2001. Jakob von Uexküll: An introduction. *Semiotica* 134(1/4):1–59.

Kull, Kalevi, Claus Emmeche, and Donald Favareau. 2008. Biosemiotic questions. *Biosemiotics* 1(1): 41–55.

Luisi, Maria. 2006. Percept and perceptual judgment in Peirce's phenomenology. *Cognito-estudos: Revista Eletrônica de Filosofia* (São Paulo) 3(1):65–70.

Martinelli, Dario, and Otto Lehto (Guest Editors). 2009. Special issue: Zoosemiotics. *Sign Systems Studies* 37(3/4):349–660.

Merleau-Ponty, Maurice. 1962 [1945]. *Phenomenology of perception* (trans: Smith, Colin). London: Routledge & Kegan Paul.

Nöth, Winfried. 1998. Ecosemiotics. *Sign Systems Studies* 26:332–343.

Peirce, Charles Sanders. 1931–1958. In *Collected papers of Charles Sanders Peirce*. 8 volumes. Vols. 1–6, eds. Charles Hartshorne and Paul Weiss, vols. 7–8, ed. Arthur W. Burks. Cambridge, MA: Harvard University Press.

Polanyi, Michael. 2009 [1966]. *The tacit dimension*. Chicago/London: The University of Chicago press.

Sebeok, Thomas A. 1972. *Perspectives in Zoosemiotics* (= *Janua Linguarum. Series Minor* 122). The Hague: Mouton.

Sebeok, Thomas A. 1990. *Essays in Zoosemiotics* (= *Monograph Series of the TSC 5*). Toronto: Toronto Semiotic Circle/Victoria College in the University of Toronto.

de Tienne, André. 2004. *Is phaneroscopy as a pre-semiotic science possible?* Semiotiche no. 2 (Special issue "Fenomeni – Segni – Realtà"). Torino: Ananke.

Toadvine, Ted. 2003. Singing the world in a new key: Merleau-Ponty and the ontology of sense. *Janus Head* 7(2):273–283.

Tønnessen, Morten. 2003. Umwelt ethics. *Sign Systems Studies* 31.1:281–299.

Tønnessen, Morten. 2009a. Umwelt transitions: Uexküll and environmental change. *Biosemiotics* 2(2):47–64.

Tønnessen, Morten. 2009b. Where I end and you begin: The threshold of the self and the intrinsic value of the phenomenal world. In *Communication: understanding/misunderstanding; Proceedings of the 9th congress of the IASS/AIS – Helsinki/Imatra, 11/17 June, 2007* (= Acta Semiotica Fennica XXXIV), vol. III, ed. Eero Tarasti, 1798–1803. Imatra: The International Semiotics Institute.

Tønnessen, Morten. 2010. Steps to a semiotics of being. *Biosemiotics* 3(3):375–392.

Torjussen, Lars Petter, Johannes Servan, and Simen Andersen Øyen. 2009. An interview with Anna-Teresa Tymieniecka. *Phenomenological Inquiry* 32:25–34.

Tymieniecka, Anna-Teresa. 2007. From sentience to consciousness: Sentient intentionality as the thread of living continuity in the ontopoietic unfolding of life. In *Phenomenology of life from the animal soul to the human mind. In search of experience* (= *Analecta Husserliana – The Yearbook of Phenomenological Research* vol. XCIII), ed. Anna-Teresa Tymieniecka, xix–xxv.

von Uexküll, Jakob. 1928. *Theoretische Biologie* [Theoretical biology] (second edition). Berlin: Verlag von Julius Springer. [There is an English translation of this work, but it is of poor standard. A new translation is in process.]

von Uexküll, Jakob. 1957 [1934]. A stroll through the worlds of animal and men: A picture book of invisible worlds. Translation of *Streifzüge durch die Umwelten von Tieren und Menschen* (1934) by Claire H. Schiller, pp. 5–80. In *Instinctive behavior*, ed. C.H. Schiller. New York: International Universities Press. Reprinted in *Semiotica* 89.4 (1992), 319–391.

von Uexküll, Jakob. 1982 [1940]. The theory of meaning. Translation of *Bedeutungslehre* by Barry Stone and Herbert Weiner. *Semiotica* 42.1:25–82.

von Uexküll, Jakob. 1985 [1909, 1921]. Environment [Umwelt] and inner world of animals. Translation (in selection) of *Umwelt und Innenwelt der Tiere*, 222–245. In *Foundations of comparative ethology*, ed. Gordon Burghardt (trans: Mellor, C.J., and D. Gove). New York: Van Nostrand Reinhold.

Zapffe, Peter Wessel. 1996 [1941]. *Om det tragiske* [*On the tragic*]. Oslo: Pax forlag.

SİBEL OKTAR

THE PLACE:WHERE WE SEE THE WORLD AS A LIMITED WHOLE

ABSTRACT

It is known that Wittgenstein read Heidegger and claimed that he could imagine his account of Being and Angst. It is not so surprising that it has been regarded as a scandal that admitting that Wittgenstein understood and even, to an extent, he combines it with his understanding of nonsense in the surprise of the existence of something, which also appears in his description of absolute sense of ethics. In this paper, rather than comparing Heidegger and Wittgenstein, there will be an analysis of our everyday moral acts by considering both Wittgenstein's and Sartre's examples, which will give us an opportunity to understand the phenomenological investigation of a moral dilemma. Later Wittgenstein's "somewhat" phenomenological investigation of moral acts can only be understood by fully comprehending his early works. The focus on questions such as: "How we see the world as a limited whole?" "From where do we observe the world?" will be bound by the concept of "place". By going back to Plato and investigating what "khora" means and whether it has some parallel to the place where we stand in the world in terms of Wittgenstein. Is it the everlasting place where we can see the world as a limited whole?

In 20th century philosophy it is common practice to refer to Wittgenstein and Heidegger. Their names are not only mentioned together because they are the great figures of 20th century philosophy but also because they had somewhat similar pursuits such as "seeking to revolutionize philosophy" by departing from modern rationalization as Stanley Cavell puts it.

It is known that Wittgenstein read Heidegger and claimed that he understood his account of Being and Angst. Friedrich Waismann recorded that, in 1929, at Moritz Schlick's house, Wittgenstein stated that:

To be sure, I can imagine what Heidegger means by being and anxiety. Man feels the urge to run against the limits of language. Think for example of the astonishment that anything at all exists. This astonishment cannot be expressed in the form of a question, and there is no answer whatsoever. Anything we might say is *a priori* bound to be mere nonsense.[1]

It is not so surprising that it has been regarded as a scandal that admitting that Wittgenstein understood and even to an extent combines it with his understanding of nonsense in the surprise of the existence of something, which also appears in his description of absolute sense of ethics. Drawing a parallel with early Wittgenstein and early Heidegger would have been a crime for analytic philosophy, it is running up against the boundaries but not in the sense that Wittgenstein mentioned above. Although, with a totally different insight, Richard Rorty states

341

A.-T. Tymieniecka (ed.), Analecta Husserliana CX, 341–353.
DOI 10.1007/978-94-007-1691-9_28, © Springer Science+Business Media B.V. 2011

that there is no parallelism of their work in this period, but he suggests that Wittgenstein and Heidegger "passed each other in mid-career, going in opposite directions."[2] Meaning that the Wittgenstein of *Tractatus*, or as Rorty puts it, "un-pragmatic", "mystical" and younger Wittgenstein could have more similarities to older Heidegger than the more pragmatic, younger Heidegger or the Heidegger of the *Being and Time*, and vice versa.

Here, rather that comparing Heidegger and Wittgenstein I think, it would be a good opportunity to draw a parallel between Wittgenstein's and Sartre's everyday example of a moral problem to understand the phenomenological investigation of a moral dilemma. Although Wittgenstein discussed this dilemma in his later period, it is necessary to connect his early works and views on ethics. I will concentrate on early Wittgenstein's views on ethics, although limited it is the only topic that Wittgenstein allows himself to talk about on what is unsayable and gives us a chance to draw a parallel with phenomenological tradition.

Wittgenstein's choice of example to explain "taking up an ethical attitude" and Sartre's example of "the state of abandonment" are almost identical. Both examples emphasize a similar ethical dilemma. To understand the nature of that dilemma and to seek a solution will pave the way to understand the way one sees the world and the way one sees oneself in the world, or more specifically for Sartre "in the world" or for Wittgenstein at the "limits of the world."

In *Existentialism Is a Humanism* Sartre describes the condition of his student whose father is believed to be a "collaborator"; his brother died in the German of-fensive of 1940, his mother separated from his father, and lived with her son and depended on him. He, on the other hand, wanted to take revenge for his brother and fight for the independence of his country. He struggled between two choices, either staying with his mother and helping her to live or go to England to join the Free French Forces. Each alternative had both negative and positive consequences. If he chose to join the Free French Forces, he was not even sure whether he would be able to go to England, would be captured and end up in a prison camp or even be killed. He was not even sure about his feelings toward his country and his mother. What would motivate his choice? Sartre defines his situation as: "he was vacillating between two kinds of morality; a morality motivated by sympathy and individual devotion, and another morality with a broader scope, but less likely to be fruitful."[3]

Wittgenstein's case is one of a scientist, who must either leave his wife or aban-don his work on cancer. The man struggles between his two roles, i.e., a husband and a scientist, and if he does not choose one, he will not be able to do either properly; he will be both a bad husband and a bad scientist. The man's attitude would vary according the way he looks at things. He might have the view that he cannot ignore the suffering of humanity so he cannot abandon his research and the wife will get over it. Or he might have a deep love for his wife and if he gives up his work he would not be a good husband anyway. On the other hand, he might think that some-one else could carry on the research and choosing the wife would not be abandoning the suffering of humanity.

So what would help Sartre's student and Wittgenstein's scientist choose between two actions? Upon what would they depend? Christian ethics? Both Wittgenstein and Sartre consider this option. Wittgenstein says that if we consider Christian ethics in this case, we would see that "should he leave his wife or not?" is no problem at all. The answer is clear, Christian doctrine tells him to stay with his wife and be a good husband, there is no other option. Then it alters the problem, now the problem is: "how to make the best of this situation, what he should do in order to be a decent husband in these greatly altered circumstances, and so forth."[4]

For Sartre, Christian doctrine would not serve to clarify, he says that: "The Christian doctrine tells us we must be charitable, love our neighbor, sacrifice ourselves for others, choose the 'narrow way,' et cetera. But what is the narrow way? Whom should we love like a brother – the soldier or the mother?"[5] Looking for certain ethical doctrines, is mainly searching for an "a priori". Searching for the ultimate answer, in Wittgenstein's terminology "an absolute sense" of ethics, that could answer that question. Is there such an absolute sense of good action? Both Wittgenstein's and Sartre's answer is No. Sartre asks: "Who can decide that *a priori*?" and he answers: "No one. No code of ethics on record answers that question" and adds "[n]o general code of ethics can tell you what you ought to do; there are no signs in this world."[6]

To understand the complexity of this situation and the difficulty of the choice we must understand the meaning of such concepts as "abandonment", "anguish" and "despair". All these concepts are also closely related to our assumptions about the existence or non-existence of God.

For Sartre, accepting that God does not exist is a problem. What will happen to fundamental ethical values? Sartre wonders how could it be considered "obligatory *a priori*" to be honest? And he suggests that "if we are to have a morality, a civil society, and a law-abiding world, it is essential that certain values be taken seriously; they must have an *a priori* existence ascribed to them."[7] With a different approach, Wittgenstein also questions such an "a priori existence" and asks: "Can there be any ethics if there is no living being but myself?" and he answers his question with: "If ethics is supposed to be something fundamental, there can."[8] And the absolute sense of value judgements concern ethics as fundamental, independent of our pre-determined standards, regardless of a community's agreement on what good is. For Wittgenstein such an absolute sense of ethics is what cannot be expressed. The distinction between the absolute sense and relative sense of value judgements somewhat helps Wittgenstein to escape the need for such an a priori's existence or better to say to talk about value judgements that need such an a priori existence.

Richard Rorty's distinction of type A and type B entities addresses the same problem. We need type A entities, such as Kantian categories and Platonic Forms, to make type B entities, like Kantian intuitions and Platonic material particulars, knowable or describable. Type B entities are the lower level entities "which stand in need of being related in order to become available ... require contextualization and explanation but cannot themselves contextualize nor explain," on the other hand, type A entities are "their own *rationes cognoscenti*, ... that make themselves available without being related to one another or to anything else."[9] They explain but they cannot be explained. This problem also remains when we try to talk about logical structure, which helps the logical propositions to picture the fact, but it cannot be pictured. Bertrand Russell in his "Introduction" to *Tractatus* explains this in relation to Wittgenstein's doctrine of pure logic. He states that:

[A]ccording to which the logical proposition is a picture (true or false) of the fact, and has in common with the fact a certain structure. It is this common structure which makes it capable of being a picture of fact, but the structure cannot itself be put into words, since it is a structure of the words, as well as of the facts to which they refer.[10]

As Rorty puts it such type A entities "are in the same situation as a transcendent Deity."[11] Thus, if we believe that God does not exist we will not have Rorty's type A entities and we will be left alone with the type B entities that now cannot be related to anything and cannot be explained.

If we believe that God exists we would be able to identify ethical rules and be able to legitimise our acts. Our will and ethical choices will have a standing in God's will. We will have God's guidance in our choices. If God does not exist, then there are no values or commands that legitimise our choices. There is no external source of our moral choices or acts; there is no other justification or "excuse". Thus we are left alone, "without excuse." For Sartre, "man is condemned to be free", which means that "man being condemned to be free carries the weight of the whole world on his shoulders; he is responsible for the world and for himself as a way of being."[12]

The sort of responsibility that "abandonment" puts upon our shoulders limits us to relying upon "that which is within our wills" and there comes the "anguish". Sartre gives the existentialist definition of anguish as follows:

[A] man who commits himself, and who realises that he is not only the individual that he chooses to be, but also a legislator choosing at the same time what humanity as a whole should be, cannot help but be aware of his own full and profound responsibility.[13]

Such a definition associates the sense of dignity with Kant. In the case of obeying a moral law the motive comes from "the idea of the dignity of the rational being, who obeys no law other than that which he himself at the same time gives."[14] According to Kant, the authority of the moral law is duty. In acting in compliance with moral law, because it is a duty, we are obeying because we give the law ourselves. So, the immediate value of compliance with moral law comes from oneself that has the satisfaction of complying with duty, not from outside and even not from our own desires.

Although the sense of freedom has its unique traits in Kant's and Sartre's philosophies, there are still common grounds. There is no doubt that freedom is an important concept in Kant's ethics. Within it, it carries the concept of autonomy with respect to ourselves and respect for moral law. The definition of obligation and duty changes its meaning from the ordinary sense of duty in connection with the idea of freedom. Freedom is defined as "independence from the determining causes of the world of sense (which reason must always ascribe to itself)."[15] Thus, when we talk of a free person we talk of a person whose actions are independent from any external determining sources. This is known as Kant's Copernican Revolution, which changes the centre of laws of reason from an external source to human beings with the capacity of making laws.

Thus, similar to Sartre, Kant's rational being is condemned to be free, who also carries the weight of the whole world on his/her shoulders, even at the cost of his/her happiness because his/her actions not only bind her/himself together but also the world. Kant introduces the law as: "So act that maxim of your will could always hold at the same time as a principle in a giving of universal law."[16] Sartre states how difficult such a situation is as: "So every man ought to be asking himself, 'Am I really a man who is entitled to act in such a way that the entire human race should be measuring itself by my actions?' And if he does not ask himself that, he masks his anguish."[17]

Surely, the main difference is while Kant replaced "a temporal Deity with a temporal subject of experience."[18] Sartre does not replace God with anything else and faces that God does not exist. Sartre also puts it like that "Eighteenth century atheistic philosophers suppressed the idea of God, but not, for all that, the idea that essence precedes existence. We encounter this idea nearly everywhere: in the works of Diderot, Voltaire and even Kant."[19] Thus, when you do not replace God with anything else and try to deal with the fact that God does not exist, there comes the feeling of anguish. Being "thrownly abandoned to the 'world' ",[20] left alone with the feeling of responsibility for all his/her acts, without an excuse. Thus Sartre's student and Wittgenstein's scientist are in such anguish when they are trying to choose the "right" act. So, what is the "right", "good" or "correct" choice/solution in each case?

For Sartre's student, there is no absolute good or bad to guide him, he is left alone and has the burden of his responsibility for his choice. When his student asked Sartre's advice, as for him there is no moral rule that could guide him in what he ought to do, he replied: "You are free, so choose; in other words, invent."[21] Surely, there is no absolute good or bad for Wittgenstein's scientist either. Wittgenstein also states that there are no "higher" values in this world and we cannot talk about an absolute sense of ethical judgements. The situation of the scientist is what Wittgenstein calls "taking up an ethical attitude." Wittgenstein says that "[w]hatever he finally does, the way things then turn out may affect his attitude."[22] This case is related to the attitude of the man towards life.

For Wittgenstein a change in attitude is an important notion in understanding the way ethics manifests itself. Wittgenstein emphasizes the importance of seeing things differently. "Noticing an aspect" is the key to seeing things differently, here noticing the difference is as crucial as noticing the similarity of the things in question. In order to see things differently we must change our "*way of looking at things.*"[23] The notion of seeing things differently was examined to see whether this notion could give us room to have a discourse on ethics. When you change your way of looking at things this change manifests itself in your attitude. Our forms of life somewhat determine the way we look at things. If we accept the role of forms of life as a determinant of our attitude towards the world then we must presuppose the existence of others, the agreement in the language we use and the agreement of our form of life. If we presuppose an agreement on the expression of value judgements in the language we use, then this is the relative sense of ethics. In both examples, it would be easy to choose an alternative if there were the possibility of an absolute sense of ethics. Then what we need to search for is the possibility of a discourse on ethics in the absolute sense, which seems to abandon us to the idea that God does not exist.

Wittgenstein states that when we speak of God, we use a language that "represents him as a human being of great power."[24] In ethical and religious languages we use similes, and in order to legitimately express the value judgements by using "a simile must be the simile for *something*. And if I can describe a fact by means of a simile I must also be able to drop the similie and to describe the facts without it."[25] And Wittgenstein concludes that as we cannot find facts behind the simile, so what seems like a simile turns out to be nonsense. Wittgenstein's description of God as a human being and the notion of a miracle in 1944 seems to resemble "A Lecture on Ethics". Take this remark for example:

A miracle is, as it were, a gesture which God makes. As a man sits quietly & then makes an impressive gesture, God lets the world run on smoothly & then accompanies the words of a Saint by a symbolic occurrence, a gesture of nature. It would be an instance if, when a saint has spoken, the trees around him bowed, as if in reverence. – Now, do I believe that this happens? I don't.[26]

Here, he uses the language of religion and the language he uses represents God as a human being as he says, this is what happens in the language of religion. For Wittgenstein, a miracle "is simply an event the like of which we have never yet seen."[27] Wittgenstein states that he does not believe that such a miracle, that the trees bow to the words of the saint in reference, happens. He says that the reason he does not believe it is that "[t]he only way for me to believe in a miracle in this sense would be to be *impressed* by an occurrence in this particular way."[28] Although he says that he is not impressed he does not say that it is nonsense. But the religious remarks he makes lose their miraculous appearance when he questions them. The method of verification of whether a simile (also a miracle) is nonsense or not, for early Wittgenstein, is to check whether it corresponds to facts or not. For later Wittgenstein, the criterion of verification seems to be the occurrence of a particular example of a language game and believing it. If we look at the following remark by Wittgenstein, we will see how believing effects the meaning of a word:

I am reading: "& no man can say that Jesus is the Lord, but the Holy Ghost." And this is true: I cannot call him *Lord*; because that says absolutely nothing to me. I could call him "the paragon", "God" even or rather: I can understand it when he is so called; but I cannot utter the word "Lord" meaningfully. *Because I do not believe* that he will come to judge me; because *that* says nothing to me. And it could only say something to me if I were to live *quite* differently.[29]

Separating an exemplar (the paragon), a spirit (Holy Ghost) and a supreme being (God) from a Lord seems to be related with the uses of these words. The first three (i.e., the paragon, Holy Ghost and God) are metaphysical uses but the last one, i.e., Lord, is a simile. A simile that makes us believe that the word in use corresponds to actual happenings, there are particular occurrences, practices that we can refer to. If we believe it, it becomes meaningful, but if not, like Wittgenstein, it is not meaningful. If I were to live quite differently then I might have a different attitude that would enable me to believe. This is like the difference between the life (world) of a happy man and an unhappy man.

Considering "the nonsensical use of language", early Wittgenstein's focus was going beyond the boundaries and what cannot be said, whereas later Wittgenstein's focus of attention turned to "the non-rational grounding of religious belief".[30] This is clear when Wittgenstein questions belief in Christ's resurrection. He says: "But if I am to be REALLY redeemed, – I need *certainty* – not wisdom, dreams, speculation – and this certainty is faith. And a faith is faith in what my *heart*, my *soul*, needs, not my speculative intellect."[31] In "A Lecture on Ethics" Wittgenstein says that expressions of ethics and religious belief are not nonsensical because we have not yet found the "correct analysis" of religious and ethical expressions, "but that their nonsensicality was their very essence."[32]

The absolute sense of value could only manifest itself. To be a believer or not makes a difference. Wittgenstein says that:

> If the believer in God looks around & asks "Where does everything I see come from?" "Where does all that come from?" what he hankers after is not a (causal) explanation; and the point of his question is the expression of this hankering. He is expressing, then, a stance towards all explanations. – But how is this manifested in his life?

> It is the attitude of taking a certain matter seriously, but then at a certain point not taking it seriously after all & declaring that something else is still more serious.[33]

The good in the absolute sense manifests itself in our attitudes towards the world. How can we see/notice that the absolute sense of ethics manifests itself? Is it by looking at things in a different way or from a different perspective as Wittgenstein would tell us?

How is it possible to look at things in a different way? Even if we can look at things differently is it possible to see the absolute sense of good in this world? Is it possible to have an absolute sense of good if God does not exist? Or its existence is not relevant at all?

In the "A Lecture on Ethics" Wittgenstein gives the example of an omniscient person, who carries most of God's attributions with just a reporting capacity that knows everything, even "all the states of mind of all human beings that ever lived." And Wittgenstein thinks that if this person writes a book containing "whole description of the world," this book will not contain any ethical judgements because it will only describe the facts and "[t]here are no propositions which, in any absolute sense, are sublime, important, or trivial."[34]

But being omniscient is different than being omnipotent, having unlimited power. An omniscient person does not have any power over what he is reporting. Everything stands on the same level because even Wittgenstein's omniscient person that knows everything will still be an observer that does not interfere with any of the facts s/he describes. Just as resembling the task of philosophy that was described in *Philosophical Investigations*. Louis E. Wolcher, referring this passage states that "[i]n this respect it is not difficult to recognise that the omniscient is a figure for Wittgenstein's own conception of philosophy's task."[35]

Since we are not omniscient observers and obviously not an omnipotent being that could have a "view from nowhere," we are, as Husserl suggested when he claims that perception is perspectival, bound to a spatiotemporal point of view.[36] As we see the object from a certain limited perspective, "the object never appears in its totality"[37] What is the phenomenological insight here? Dan Zahavi states that once we realise that "what appears spatially always appears at a certain distance and from a certain angle, the point should be obvious: There is no pure point of view and there is no view from nowhere, there is only an embodied point of view."[38] Once again without an omnipotent knower we are left alone without a pure point of view, a point of view which might help us to talk about the absolute sense of ethics. With such a limited perspective, we see the world as a limited whole. That is what is mystical for Wittgenstein. In *Tractatus* 6.45 he states that:

> To view the world sub specie aeterni is to view it as a whole – a limited whole.
> Feeling the world as a limited whole- it is this that is mystical.

This is Wittgenstein's fundamental thesis, as Russell emphasizes, "it is impossible to say anything about the world as a whole, and ... whatever can be said has to be about bounded portions of the world."[39] Speaking of the totality of things is speaking of necessity.[40] As what can be viewed is limited by the observer's perspective, what can be said is limited by the propositions of natural sciences. This suggests a kind of awareness of the limits of the world, the limits of language. James C. Edwards suggests that: "To feel the world as a limited whole it is necessary to feel its limit, i.e., to be aware of oneself as that limit of the world"[41]

Then, how could we see the world as a limited whole? How could we view the world sub specie aeterni? How could we change the world without any change in the facts? How does the absolute sense of value manifest itself? At this point, investigating Wittgenstein's understanding of a different sense of seeing would be helpful. To able to look at things in a different way is to be able to see the world *sub specie aeterni*.

We have already mentioned the concept of "noticing an aspect" now recalling it at this stage will give us another insight. Wittgenstein uses the duck-rabbit figure to illustrate the notion of noticing an aspect. The duck-rabbit figure was used by Joseph Jastrow (1863–1944), the American psychologist, to demonstrate that perception is not only a consequence of the stimulus, but also is a product of mental activity. This illustration also clarifies how Wittgenstein makes a distinction between the change of perception and the change of aspect. The duck-rabbit, Figure 1, which can be seen as a duck's or rabbit's head is shown below:

Figure 1. Duck-rabbit[42]

If someone shows you the figure above and asks what it is, you could reply "It is a rabbit", "It is a duck" or "It is a duck-rabbit". For Wittgenstein, these answers are "the report of perception". But on the other hand, if you reply "Now it's a rabbit" your answer is not a report of perception; it is the expression of the change of aspect. While you are looking at the duck-rabbit figure you could see it as a duck and suddenly notice the other aspect and say "Now it is a rabbit". "The expression of a change of aspect is the expression of a new perception and at the same time of the perception's being unchanged."[43] As stated in *Tractatus* 6.43, this is how "the good and the bad exercise of the will" do not alter the facts, but do alter the world.

This means to talk about a new perception, a new perception that suggests a noticing of the change of aspect. Wittgenstein expresses the difference between the usual and different way of looking at things as: "The usual way of looking at things sees objects as it were from the midst of them, the view *sub specie aeternitatis* from outside."[44]

It is now getting more complicated, now we need to position ourselves so that we don't see objects from the midst. Where is this "outside"? Where should I stand to view the objects from outside? From what kind of a place do we need to view the world under the aspect of eternity? From where do we observe the world? Where am I positioned in this world then? Am I placed in the world just at the edge of the limit suggested by Wittgenstein's "eye" analogy? Or is it possible to have a "view from nowhere"?

In the search for a place, a place that could provide us a different view, let us notice the change of aspect, that has a view under the aspect of eternity, we should listen to what Timaeus of Locri in Plato's *Timaeus* when he is explaining why he needs a third kind of discourse which later he named as khôra (χώρα). At a certain point Timaeus realises that the two kinds of discourse he had used to express his account of the universe are not sufficient for the full apprehension of the universe. He says: "The earlier two were sufficient for our previous account: one was proposed as a model, intelligible and always changeless, the second as an imitation of the model, something that possesses were becoming and visible."[45]

This third kind is "a *receptacle* of all becoming," a "wetnurse". It is not another kind of being, being is only used for the first kind, i.e., for the paradigm (the Ideas), whereas the second kind, i.e., copies of these paradigms (the phenomena), is becoming. The third kind is "a kind of kind beyond kind, kind of kind outside of kind."[46]

In a flash of inspiration, the description of "kind of kind outside kind," a need for a third kind of discourse that is outside the kind invokes a reminder of Wittgenstein's positioning himself not in the midst of the objects but outside in order to be able to see them differently. Which suggests a place that is outside that of which lets us see things from the outside. But this is too early a stage to make such connections. We need to let the third kind reveal itself to us.

Timaeus tells us that it is a difficult task to describe the third kind, one of the difficulties of such description is that phenomenal objects are not stable, they are in flux. He mentions the cyclical transformation that can be observed in fire, air, water and earth. "[T]hey transmit their coming to be one to the other in a cycle.

... what we invariably observe becoming different at different times."[47] So the expressions "this" or "that" cannot be used to designate something unless it has a kind of stability. Thus rather than "this" we can only say "what is always such and such."

This is almost the same difficulty as the "ostensive definition" that is described by Wittgenstein. First of all, "an ostensive definition explains the use – the meaning – of a word when the overall role of the word in language is clear."[48] Here, we have Timaeus trying to clarify the language by searching for a way to express the objects in a "reliable" and "stable" account. Even calling, "what is always such and such" might not solve the problem, as Wittgenstein says in *Philosophical Investigations* exegesis 28, "an ostensive definition can be variously interpreted in *every* case." Thus this is not the safe and reliable account.

At this point, Richard D. Mohr introduces the "double aspect" of the phenomena. He says: "The phenomena, then, have a double aspect. On the one hand, they are in flux; on the other hand, they are images of Ideas. Insofar as the phenomena are in flux, nothing whatsoever can be said of them."[49] In relation with this "double aspect" John Sallis states that there are two levels of discourse at hand. At the first level of discourse the word uttered is applied to something "that can in fact only be seen (moving in the cycle of transformations) but *not said*, something that can be, at most, silently pointed out."[50]

That is what Wittgenstein suggests in the opening pages of *Tractatus*: "What can be said at all can be said clearly; and whereof one cannot speak thereof one must be silent."[51] But then later, he adds that although it cannot be said, it transcends the limits of language, it manifests itself. In *Tractatus* 6.522 Wittgenstein combines this with the mystical, he says that "[t]here are, indeed, things that cannot be put into words. They make themselves manifest. They are what is mystical." Timaeus introduced the third kind with the need for a medium in which what cannot be said manifests itself and might even provide a possibility of talking about the third kind. Finally the third kind is described as:

And the third type is space [χώρα], which exists always and cannot be destroyed. It provides a fix state for all things that come to be. It is itself apprehended by a kind of bastard reasoning that does not involve sense perception, and it is hardly even an object of conviction. We look at it as in a dream when we say that everything that exists must of necessity be somewhere, in some place and occupying some space, and that that which doesn't exist somewhere, whether on earth or in heaven, doesn't exist at all.[52]

At this point we need to take a break to our investigation of what the characteristics of χώρα are and clarify what the word χώρα means in the Greek language. Even the translation of the word is disputed. As in the above passage it is, by some scholars be translated as "space" and by others as place, as land and as country. John Sallis in investigating the use of the word in different Platonic dialogues suggests that place rather than space would give a better picture of the word.[53] Following Sallis I too prefer the word "place." Yet we still need to look at its characteristics, what kind of a place χώρα is. The above passage tells us that it exists always and it is stable. Thus it is an "everlasting", "perpetual" place. As it is not apprehended by sense perception it is invisible. Even its invisibility requires a different understanding. It does manifest itself by being the medium of what cannot be said to appear.

Moreover, the third kind, χώρα, is somewhat like the logical structure defined by Wittgenstein, as mentioned before, the logical structure enables logical propositions to picture a fact but it cannot itself be put into words. The third kind is formless, it is "an invisible and characterless sort of thing, one that receives all things and shares in a most perplexing way in what is intelligible."[54]

Finally, being formless, invisible and having no determinations nothing can be said about it and I would agree with Sallis that it makes both the third kind and its name have no meaning. One can only have an illegitimate "bastard" discourse on it. It is not surprising that χώρα is not in the realm of nonsense. What can be said and what cannot be said is the criterion of nonsense in *Tractatus*. Nonsense is in the domain of what cannot be said. And Wittgenstein in Tractatus 6.53 suggests that "whenever someone else wanted to say something metaphysical, to demonstrate to him that he had failed to give a meaning to certain signs in his propositions." This should be a warning for that person that now he is about to transcend the limits of language and go beyond the word and about to have a bastard discourse that will have no meaning.

This timeless, everlasting "place" does not give us an opportunity for a medium that enables us to say what cannot be said. But it is not useless. It provides a medium for what cannot be said to demonstrate/reveal itself. It is a standpoint it is where Wittgenstein could have positioned himself to see the world from outside, to have a new, different than usual, way of looking. It is a place where one could see how what can be said manifests itself, where one realises the double aspect of the phenomena and notice the change of aspect. It is the place where one can view the world as a limited whole. This everlasting place "doesn't exist somewhere, whether on earth or in heaven" and in a dreamlike way it seems like a view from nowhere, that an omnipotent being would have. But as when awakening from the dream we realise that "it doesn't exist at all". Thus we still see the world as a limited whole, but this place provides a new way of looking things that manifest themselves in our attitudes.

Özyeğin University, İstanbul, Turkey, e-mail: sibel.oktar@ozyegin.edu.tr

NOTES

[1] *Ludwig Wittgenstein and the Vienna Circle:* conversations recorded by Friedrich Waismann, ed. B.F. McGuinness (Oxford: Basil Blackwell, 2003), p. 68.
[2] Richard Rorty, "Wittgenstein, Heidegger and the reification of language" in *The Cambridge Companion to Heidegger*, ed. Charles B. Guignon (Cambridge University Press, 1995), p. 339.
[3] Jean-Paul Sartre, *Existentialism Is a Humanism*, ed. John Kulka and trans. Carol Macomber (New Haven: Yale University Press, 2007), p. 31.
[4] Rush Rhees, "Some Developments in Wittgenstein's View of Ethics" *The Philosophical Review*, 74 (Jan., 1965), p. 22.Rush.23.
[5] Jean-Paul Sartre, 2007, p. 31.
[6] Jean-Paul Sartre, 2007, pp. 31–33.
[7] Jean-Paul Sartre, 2007, p. 9.

[8]   Ludwig Wittgenstein, *Notebooks 1914–1916*, trans. and ed. G.E.M. Anscombe (Oxford: Blackwell, 1984), p. 79.

[9]   Richard Rorty, op. cit., p. 342.

[10] Russell's "Introduction" in *Tractatus* in the English translation (edition of 1922) reprinted in 2005 edition. Ludwig Wittgenstein, *Tractatus Logico-Philosophicus*, trans. D.F Pears and B.F. McGuinness (London: Routlege Classics, 2005), p. xxiii.

[11] Richard Rorty, op. cit., p. 342.

[12] Jean-Paul Sartre, *Being and Nothingness,* trans. Hazel E Barnes (London: Routledge Classics, 2009), p. 574.

[13] Jean-Paul Sartre, 2007, p. 25.

[14] Immanuel Kant, *Groundwork of the Metaphysics of Morals*, trans. and ed. Mary Gregor (Cambridge: Cambridge University Press, 2006), p. 42. (AK 4:434). References to Kant *give* the pages in German Academy of Sciences (AK) edition of Kant's collective works.

[15] Immanuel Kant, *Groundwork of the Metaphysics of Morals*, p. 57, (AK 4:452). Kant states that: "With the idea of freedom the concept of autonomy is now inseparably combined and with the concept of autonomy the universal principle of morality" (AK 4:452).

[16] Immanuel Kant, *Critique of Practical Reason*, p. 28 (AK 5:31).

[17] Jean-Paul Sartre, 2007, pp. 26–27.

[18] Richard Rorty, op. cit., p. 340.

[19] Ibid., pp. 21–22.

[20] In *Being and Time*, Heidegger states that: "In its projection it reveals itself as something which has been thrown. It has been thrownly abandoned to the 'world', and falls into it concernfully." Martin Heidegger, *Being and Time*, trans. John Macquarrie and Edward Robinson (New York: Harper and Row, 1962), p. 406.

[21] Jean-Paul Sartre, 2007, p. 33.

[22] Rush Rhees, op. cit., p. 23.

[23] Ludwig Wittgenstein, *Philosophical Investigations,* trans. G.E.M. Anscombe (Oxford: Blackwell, 2005), § 144.

[24] Ludwig Wittgenstein, "A Lecture on Ethics," *The Philosophical Review* 74 (1965), p. 9.

[25] Ludwig Wittgenstein, op. cit., p. 10.

[26] Ludwig Wittgenstein, *Culture and Value*, ed. G.H. von Wright (Oxford: Blackwell, 1998), p. 51.

[27] Ludwig Wittgenstein, op. cit., p. 10.

[28] Ludwig Wittgenstein, *Culture and Value*, p. 51.

[29] Ibid., p. 38.

[30] Cyril Barrett, *Wittgenstein on Ethics and Religious Belief* (Oxford: Blackwell, 1991), p. 193.

[31] Ludwig Wittgenstein, *Culture and Value*, p. 38.

[32] Ludwig Wittgenstein, op. cit., p. 11.

[33] Ludwig Wittgenstein, *Culture and Value*, pp. 96–97.

[34] Ludwig Wittgenstein, op. cit., p. 6.

[35] Cf. Louis E. Wolcher, *Beyond Transcendence in Law and Philosophy* (London: Birkbeck Law Press, 2005), p. 175.

[36] See, Sean D. Kelly, "Edmund Husserl and Phenomenology" in *Continental Philosophy*, eds. Robert C. Solomon and David Sherman (Blackwell, 2003), p. 115.

[37] Dan Zahavi, *Husserl's Phenomenology* (Standford, CA: Standford University Press, 2003), p. 15.

[38] Ibid., p. 98.

[39] Ludwig Wittgenstein, *Tractatus Logico-Philosophicus*, p. xix.

[40] Russell in his "Introduction" to *Tractatus*, states that: "There is no way whatever, according to him, by which we can describe totality of things that can be named. In other words, the totality of what there is in the world. In order to be able to do this we should have to know of some property which must belong to everything by a logical necessity." (*Tractatus*, p. xviii).

[41] James C. Edwards, *Ethics without Philosophy Wittgenstein and the Moral Life* (Florida: University Press of Florida, 1982), p. 46.

[42] Source: The duck-rabbit figure used by Jastrow originally published in *Harper's Weekly* (Nov. 19, 1892, p. 1114. The figure I used is taken from mathworld.wolfram.com viewed 1 February 2008 <http://mathworld.wolfram.com/topics/Illusions.html.>

[43] Ludwig Wittgenstein, *Philosophical Investigations*, p. 167.

[44] Ludwig Wittgenstein, *Notebooks 1914–1916*, p. 83.

[45] Plato, *Complete Works*, ed. John M. Cooper (Indianapolis: Hackett Publishing Company, 1997), p. 1251. (*Timaeus*, [49 a]).

[46] John Sallis, *Chorology* (Indiana: Indiana University Press, 1999), p. 99.

[47] Plato, *Complete Works*, p. 1252. (*Timaeus*, [49 d]).

[48] Wittgenstein, Ludwig, *Philosophical Investigations,* trans. G.E.M. Anscombe (Oxford: Blackwell, 2005), § 30.

[49] Richard D. Mohr, "Image, Flux, and Space in Plato's 'Timaeus'," *Phoenix* Vol. 34, No. 2 (Summer, 1980), p. 142.

[50] Sallis, p. 104.

[51] Ludwig Wittgenstein, *Tractatus Logico-Philosophicus*, trans. D.F Pears and B.F. McGuinness (London: Routlege Classics, 2005), p. 3.

[52] Plato, *Complete Works*, ed. John M. Cooper (Indianapolis: Hackett Publishing Company, 1997), p. 1255. (*Timaeus*, [52 b]). In relation with the 'bastard reasoning' Sallis explains what does bastard means in Athenian usage, that is: "a child of a citizen father and an alien mother." (Sallis, p. 120).

[53] For a sound explanation of this dispute and the related dialogs see Sallis pp. 115–118.

[54] Plato, p. 1254. (*Timaeus*, [51 b]).

CATIA GIACONI

LINES FOR CONTEMPORARY CONSTRUCTIVISM TO REVISIT AND REINTEGRATE THE ANCIENT SENSE OF CONTINUITY BETWEEN MEN AND NATURE

ABSTRACT

This paper is meant to focus the attention on some assumptions of contemporary constructivism which, in line with the groundbreaking thought of Anna-Teresa Tymieniecka's phenomenology of life, allow to revisit and restore the ancient sense of continuity between natural macrocosm and anthropologic microcosm, which, in the scope of the unilaterally objectivist approach of modern epistemology, has fallen out of fashion. To this purpose this paper is essentially comprised of two parts: in the first the author means to outline the complex movement called "constructivism", which finds its place between innatism and empirism and establishes itself as a "third way" where subject and object are no longer the absolute and pre-existing poles of a relation, but the outcomes of a construction taking place in the continuum between natural macrocosm and anthropologic microcosm. In the second part, starting from the above assumptions on contemporary constructivism, the author shall draw some significant *lines of reflection* to restore the continuity between *logos and life phenomenology/ontopoiesis* subject of this International Congress of Phenomenology.

CONSTRUCTIVISM AS A COMPLEX SCENARIO: SUGGESTED READING

The attempt to develop a comprehensive survey of what is currently defined in several areas as "constructivism" outlines as a multi-faceted process which is not always easy to define and most importantly is subject to continuous evolution and dilation in time and space. The same analysis of the semantic spectrum of the term "constructivism" as it manifests itself in different formulas, not yet come to an adequate definition, is extensive and still fruitful, though it may risk to appear as an alluring label, as a suggestive "fashion" or "slogan", rather than the actual acknowledgement of the various meanings that such term, be it from an ontological, epistemological or methodological standpoint, may take in theory and educational and didactic practise.

In general this boils down to a complex epistemological approach revolving around the analysis on the models of knowledge, which admits plural acceptations

355

A.-T. Tymieniecka (ed.), Analecta Husserliana CX, 355–370.
DOI 10.1007/978-94-007-1691-9_29, © Springer Science+Business Media B.V. 2011

and still stands as an *open paradigm*. To explore and provide an explanation to what D. C. Phillips defines as the *nightmarish landscape* (Phillips 2000, p. 7), that is, the intricate constructivist landscape, is a hard and sometimes slippery task, given the intense bundle of disciplines it is laden with and the disciplinary boundaries that in this sense are blurred in nature and not always well defined. The theoretical references are manifold – though we shall attempt at outlining them all – and may be taken from anthropological, ethnological, philosophical, linguistic, mathematical, pedagogical, psychological, sociological, etc. standpoints, though not necessarily connected. Such a trend contributed to coining and spreading several "labels" with reference to different settings and several theoretical branches, often traced back to the thought of several seminal authors.

With the support and reference to recent insight and publications (Giaconi 2008), this article is meant to analyse constructivism from the point of view of the reference scientific literature which from a first look that tends to see constructivism as opposed to previous epistemology according to a dualist logic, moves on to a more complex vision which defines it as a "third way" (Bocchi and Ceruti 1981, p. 256), allowing to re-propose the meaning of ancient issues.

FROM DUAL LOGIC...

Through several publications, scientific literature itself describes and corroborates this topology of theoretical pictures by means of blatantly different conceptual assumptions that alternate in time as the dominating vision of men and knowledge, all the way to the commonly agreed upon structures of the current debate, that is, a combination of old and new generation dualisms:

– Modern and objective vision vs. subjective and romantic vision;
– Endogenous perspective vs. exogenous perspective;
– Empiricism/logical positivism vs. rationalism/idealism;
– Objectivism vs. constructivism;
– Localism vs. globalism;
– Etc.

It follows that the classification logic deployed is "by contrast", where the affinities and convergences between present and past are detected through the analysis of "opposed" movements with regard to the epistemological positions that arose throughout history. To this purpose the work and contributions of several authors take particular importance (Guba, Lincoln, Vattimo, Rovatti, Lyotard, Jameson, Usher, Edwards, Eagleton, Best, Kellner, Ceserani, Terrosi, Chiurazzi, Bauman, Mecacci, Rorty, Bagnall, Goodman, Forman, Pufall, Bernar, Duffy, Jonassen, Steffe, Gale, von Glasersfeld) as they propose a contrastive analysis between the epistemology of the past and the contemporary one, thus highlighting positions markedly identified by "strong" modern, axiomatic, regulatory, nomotetic, logical-formal, universalist, positivist, realist thought as regards to the past , as opposed to "weak", "post-modern", neopragmatist, antidogmatic, logically *fuzzy* or nuanced, relativist, contingent, ideographic and *constructivist* tendencies for the current context. Similarly, R. A. Neymayer, in one of his contributions to the "Journal of Consulting and Clinical Psychology" (Neymayer 1993, pp. 221–234), focuses his attention on the *objectivism-constructivism* duality and the respective positions with regard to nature, validation criteria, cultural traits of knowledge and the very concept of human being and human interaction. Such conceptual binomial is taken in *Epistemologia e psicoterapia* also by M. Ceruti and G. Lo Verso (Ceruti and Lo Verso 1998), who highlight how currently, to a first perspective usually defined as "objectivist" (the world is antecedent to history), there is an opposed one widely defined as "constructivist" (the world is generated through history). In line with the above theories, J. Shotter, in his contribution to a text compiled by L. P. Steffe and E. J. Gale (Steffe and Gale 1995), shows the ontological and epistemological combination at the foundation of the *constructivist* discourse, marked on the one side by a modern and objectivist vision, that is, positivist and post-positivist, and on the other side by a romantic and subjectivist angle (relativist or rationalist). Similarly, in the writings of Kenneth J. Gergen (Gergen 1991) there is a contrastive combination of two perspectives persistently found in the Western philosophical and scientific tradition: on the one side the "exogenous" or "world-centred" one, typical of those theories of knowledge, such as empiricism and logical positivism, that see the outer world as the primary and essential source of the knowledge process; on the other side the "endogenous" or "mind-centred" one, leaning towards those theories of knowledge, such as rationalism/idealism, that give priority to mental process within the knowledge process itself.

Though theories abound that cancel or discredit such "double partitions" in favour of a indistinct condition, most of the history of the Western thought ran along these two major conceptual systems: empiricism or logical positivism and rationalism/idealism. It is within the latter, according to a number of authors, that a new "course" gained way which may be defined as *"constructivism"* and has massive impact in the Nineties thanks to the crisis of the dominating "empiricist/positivist" paradigm and the questioning of the "representationalist" perspective according to which knowledge is nothing but an individual representation of the existing real world per se. To this purpose E. Damiano states that constructivism is «the denomination the new version of idealism gave itself» (Damiano 2006, p. 130), thus opening to a debate which would allow to get rid of the plurality and sometimes dispersion of the phenomena that currently abuse of this label into a more inclusive category such as idealism.

The most recent developments pursue the goal of overcoming both poles and transcending the traditional subject-object duality, as we shall analyse in the following paragraph.

...TO THE "THIRD WAY"

A further attempt to understand the various pictures and directions of contemporary epistemology escape the "contrastive principle", and takes a rather historical criterion that allows to trace back the connatural "paradigmatic and epistemological

transaction" of knowledge that marked our time. The basic theory to new epistemological awareness is hidden in the idea that the paradigms of knowledge are not
replaced, though combine and interweave, and sometimes influence each other and
take "new forms".

In this scope the attempt of the scientific community to represent the position
and sense of that epistemological phenomenon which is more and more frequently
defined as "constructionism/constructivism" has remarkable impact, as it is a multi-
faceted movement going through a time of formidable expansion and at the same
time offers a dynamic configuration which is ceaselessly evolving and articulating
in different theoretical clusters (first and second cognitivism, constructivism, social
cognition, cultural psychology, etc.).

In this second euristic vision, the study on the distinctive traits of the cultural
and scientific context of the present and past is performed within the logic of the
"paradigmatic translations" that allow to grasp evolution and changes at ontological,
epistemological and methodological level in the range of dominating perspectives.
The studies by E. G. Guba (1990), Y. S. Lincoln (1995), T. L. Sexton (in Sexton and
Griffin 1997) and B. B. Bichelmeyer (2000) are good examples of the above. In the
Nineties E. G. Guba (1990) and Y. S. Lincoln (1995) devised and offered the main
trends of the "traditional paradigms" and the "emerging paradigms", that dominated,
in their own view, the modern age and the current post-modern time. The authors
provide a view on such tendencies initially as the expression of a modern and structuralist thought, with regard to positivism and post-positivism, and post-modern and
post-structuralist, with regard to critical theory and constructivism. Finally, in his
2000 work (Guba and Lincoln 2000), Y. S. Lincoln offers a further paradigm, he defined as *participatory*, with regard to the work of J. Heron and P. Reason (Heron and
Reason 1997). As a whole, the comparative analysis is carried out on three levels:
the *ontological* one, that is, of the nature of reality and the knowable; the *epistemological* one, of the nature of knowledge and of the relation between the knower and
the knowable; the *methodological* one, with regard to the systematic approach of the
scientific and educational research (see Guba and Lincoln 2000).

Within the historical analysis on the nature of knowledge by T. L. Sexton (1997)
there is a distinction between the following three phases of human history, each
featuring a different ontological approach: pre-modern, modern and post-modern or
constructivist. In pre-modern age, from the VI century before Christ to the Middle
Ages, the pivotal role is played by faith and religion; in the modern age, from the
Renaissance to the end of the XIX century, the main role is played by empiricism,
logical positivism and the identity between objective truth and validity of scientific
assumptions: scientific knowledge is thus the only source to know the world. Finally,
the third phase, that is the present one, is dominated by the creation, rather than
the discovery, of individual and social realities. The principle of validity (*validity*),
which measures the solidity and reliability of a research, that is true correspondence
between the real world and the conclusions of a research, is replaced by the principle
of viability (*viability*) of assumptions, meant in the Darwinian fashion as "negative
selection", that is, all the elements that are redundant or useless are ruled out, so
that all there is left is "adapt", or viable. These scholars focus on what men think

but also how they think, and underline the importance of human participation in the construction of knowledge: the perspective of the beholder and the observed object are inseparable; the nature of meaning is relative; phenomena are "*context-based*", that is, they may be judged upon the context in which they develop, and the process of knowledge and comprehension is "social, inductive, hermeneutic and qualitative". Reality, thus, may not be considered as objective, independent from the subject that experiences it, because it is the very beholder that gives it sense by actively participating in its construction.

One further analysis perspective is provided by B. B. Bichelmeyer (2000), as she analyses the "educational philosophies" that supported and founded XX century education and didactics. The reference measures of the author are metaphysics – better yet, ontology – epistemology and axiology, through which she tackles the paradigms of behaviourism, constructivism and interactivism. *Behaviourism* considers reality as "objective", permanent, static, unchanging, sees truth as external to individuals and also static and unchanging and finally assesses the actions that receive external awards as beneficial. *Cognitivism*, with reference to "non-ecological first generation" expressed by the HIP model, describes reality as always objective and permanent, though "subjectively experienced" by individuals, sees knowledge as knowable when we compare our internal cognitive patterns with the outer reality and as an instrument to the development of the schematic representations of reality. Constructivism, also considered as "second generation", and "ecological" development of cognitivism as it considers the cognitive processes as they are immersed and integrated with the biological, evolutional, social and technological contexts in which they live and operate, enhances and researches the relation between subject and context, sees the attribution of meanings to things, facts and events and the cognitive act as socially mediated and shared. The focus of the above paradigm is hence directed towards the "subjective reality", that is, on the fact that each individual creates his or her own reality, on truth as a new construct, based on negotiated meanings, on what we deem true and on the agreement on the shared truth as good. The *interactivist* paradigm considers reality as objective but manifold, changing, variable, unforeseeable and "subjectively experienced among individuals"; it sees reality as mutating, changing; it values intentionality (*reflection and action*) through which we master change and unexpected circumstances. The author then researches the position of the three paradigms above and compares them with the emerging "interactivist" one with relation to learning, the role of the teacher, the role of the student and the methods for teaching-learning (see Bichelmeyer 2000).

Beyond single essays, such systematization goes towards the recognition of constructivism and the ensuing epistemology not as a mere alternative to traditional options but rather, to quote Bocchi and Ceruti (1981, p. 256), as a "third way" between the positions of innativism and empiricism, where subject and object are no longer the absolute and pre-existing poles of a relation, but rather the outcomes of a construction that takes place in the continuity between natural macrocosm and anthropological microcosm, a continuity which, in the scope of the unilaterally objectivist approach of modern epistemology, has fallen out of fashion.

In order to further investigate constructivism as the "third way", we shall review the "ways" of constructivism as highlighted by the pedagogical and psychological literature and which allow, in a continuous and synergic vision, to draw some significant *lines of reflection* to retrieve the continuity between *logos and life phenomenology/ontopoiesis.*

THE WAYS OF CONSTRUCTIVISM

THE INTERACTIONIST WAY

A mandatory passage to recognize this way of constructivism goes through the work of J. Piaget, seen by many (von Glasersfeld 1998; Bocchi and Ceruti, 1981; Varela et al. 1993; Varisco 2002) as the "cornerstone" or the father of the XX century constructivist school, with the consequent formation of a constructivist branch inspired directly by Piaget and defined by scholars in different fashions: "cognitive", "interactionist", "operational", constructivism, etc.

Aware that I could not in but a few pages pay due homage to the extensive work of J. Piaget and the intricate knots it raises, which sometimes gave rise to misinterpretations or wrongful translations (Damiano 2006), my dissertation shall focus solely on what G. Bocchi and M. Ceruti (1981) define as the "constructivist itinerary". First of all we start from his epistemological conception which, as again is pointed out by the authors, Piaget himself always defined as "constructivist", construing it more generally as «the search for a "third way", synthesizing and not merely juxtaposing, the positions if innativism and empiricism that long dominated the scientific and epistemological debate, also in our century » (Bocchi and Ceruti 1981, p. 256), that is, we search among those theories that prioritize unilaterally the subjective capacity and those that find the very origin of our knowledge in the environment. It was in this passage, according to N. Filograsso, that the big turn on the Seventies took place: «from an atomistic vision of knowledge, made of aggregates kept together by associative nexuses», to a «systemic, dynamic and constructivist vision where subject and object are interrelated in a continuous transformation process, a standpoint which is not too far from J. Dewey's transactionalism» (Filograsso 1994, p. 55). To this purpose G. Bocchi and M. Ceruti (1981), as they outline the features of Piaget's constructivism, detect a markedly philosophical and general characteristic with regard to its position as a *dialectic constructivism*: «constructivist given the pivotal role played by (. . .) the constituent novelties in the upper development levels and dialectic given the multi-factor and interactionalist nature of the explanation to such development» (Bocchi and Ceruti 1981, p. 260). J. Piaget uses this paradigm to define the general philosophy of knowledge with particular regard to the relations between subject and object and their function in the "growth of knowledge": «the circle of object and subject is taken as primary though not homogeneous, since it is considered from time to time according to specific modalities depending on the levels and fields of knowledge» (*Ibidem*). To this matter Piaget remarks, in *Les courants de l'épistémologie scientifique contemporaine* (Piaget 1967) how the constructivist or dialectic position shelters a concept of knowledge that is «tied to an action that modifies the object and does not reach it but through the transformation triggered by the very action. In this scope (. . .) subject and object are located basically on the same plane, or rather on the same subsequent planes, according to the changes in the spatial scales and the historical and genetic development» (Piaget 1967, p. 124). Nevertheless all these different levels of reality «may be construed unitarily by the reconstruction of the genetic processes in which subject and object are built and defined complementarily» (Piaget 1967, p. 258). To the same extent E. Damiano underlines this concept and maintains that «subject and object are not the outcomes of a construction, they are not the pre-existing poles of a relation» (Damiano 2006, p. 134). Piaget's epistemology refused the empiricism-rationalism dichotomy and the fracture between the innate and the acquired, thus it is «*interactionist* and *constructivist*»: «(. . .) in the relation between subject and environment, it persuasively reports of the action of the subject, the forces withstanding the object and the functional results of such interaction, the *assimilation* – undergone by the subject –, the *accommodation* and finally the *equilibration* – with the subject –» (E. Damiano, in Filograsso, 1994, p. 153). Piaget's "third way" was pursued and achieved, according to G. Bocchi and M. Ceruti, through the key notion of *adaptation* between organisms and environment as the «dynamic equilibration between assimilation and accommodation» (M. Ceruti, in Filograsso 1994, p. 26). E. von Glasersfeld (in Ceruti 1992), too, points out that the value of J. Piaget's speculation lays in the concept of knowledge as a form of "adaptation" resulting from the "necessary interaction between conscious intelligence and environment". According to the author, J. Piaget grounded this instance by maintaining that "the mind arranges the world by arranging itself". This expression should not be wrongfully construed as a philosophical and idealist statement, as it did, because the world the mind arranges does not correspond at all to what idealist philosophers define as reality, but rather as "the world of individual practical experience"; E. von Glasersfeld states that J. Piaget's constructivism and his slightly diverging elaboration, serve the direct purpose of showing how children may ultimately develop knowledge (von Glasersfeld 1989). According to E. von Glasersfeld «Piaget always maintained that cognitive subject experience is moulded by its structures (assimilation) and that these structures are carried forward is they succeed in preserving the subject's inner equilibration, or changed (accommodation) if they do not succeed. Piaget defines it as "adaptation" and I tried to prove that adaptation should not be meant as progress towards better correspondence with the environment but rather in terms of finding viable ways » (E. von Glasersfeld, in Ceruti 1992, p. 200). In a stricter sense, closer to tangible and hard scientific research, Piaget's construtivist position is the result, in the words of G. Bocchi and M. Ceruti of a «local problem-solving strategy with regard to the relations and reductions among different levels of reality» (Bocchi and Ceruti 1981, p. 256). J. Piaget thus tackles the issues of ontogenetic (relative to the stages of intellectual development) and socio-genetic (for instance relative to the development phases of mathematics in different historical times) development, that are problems of relation and reduction among levels. In all the above instances, the author aims at «giving an explanation at once to the existence of factual discontinuity in the development processes and the relevance of the preceding phases to understand the subsequent ones» (*Ibidem*) and is directly drawn to «research the unvarying traits to these solutions and more specifically the general and abstract constructive mechanisms which may operate within genetic development» (*Ibidem*). At a higher level from "constructivist-type local solutions" to generality, the "equilibration theory" may be defined as constructivist, as it «explicitly stands as the unifying moment of all stage-independent issues, hence unvaried with regard to them» (*Ibidem*).

The main trait of Piaget's thought may be detected «in the research for empirical evidence of *knowledge as a form of equilibration*, in evolutional continuity/discontinuity with the forms of equilibration all living forms consist of» (Damiano 2006, p. 115). The interpretation of Piaget's work and the ensuing systemization, such as that carried out by E. Gattico and G. P. Storari (2005), brings out how J. Piaget focuses his entire work, on the epistemological assumption that provides an *isomorphic* relation between biological and cognitive evolution, a comparison tackled by J. Piaget himself in his *Biologie et connaissance* (Piaget 1983) which led to the image of J. Piaget as an "epistemologist" (Damiano 2006, p. 115), as well as between psychogenesis and sociogenesis. To this regard E. Damiano recommends to look at J. Piaget as the «researcher who turned epistemology into an empirical discipline, as he searched some of the unlimited fields one may resort to in order to study it: among them, men in the developmental age, for the construction of structures such as *number, space, time, symbols, object, causality, chance*», that is, cognitive categories that represent some «pivotal notions along the history of science» (*Ibidem*). This is how the Swiss "epistemologist" regards the evolution of the above categories as parallel from a psychogenetic and socio-genetic, individual and collective level, that is, both in the process whereby children become cognitively adult and where the knowledge stored by the scientific communities throughout history has been created. All this is governed by the "functional invariants" that J. Piaget describes as "assimilation, accommodation, and equilibration". Assimilation and accommodation are two different yet connected functions referring respectively to a «process bound from the organism to the environment, from the endogenous to the exogenous (...) assimilation of the external elements» (assimilation) while «a process that goes right in the opposite direction, from the exogenous to the endogenous, from the environment to the organism» (accommodation). As J. Piaget himself explained in *Biologie et connaissance* (Piaget 1983, p. 25): «as well as there is no assimilation without accommodation, there is no accommodation without assimilation: this means that the environment does not simply triggers the recording of prints and the production of copies, but it stimulates active adjustments and as a consequence we speak of accommodation meaning the accommodation of assimilation patterns». To this regard N. Filograsso (1994, p. 65) argues that Piaget's concept of "symbolic representation" is the «result of an important active structuring process» (*Ibidem*) and highlights the role of accommodation which, by determining the adjustment of the assimilation patterns, grows into imitation and gives place to symbolism. The assimilation activity is "internal from the very beginning" as at first it is expressed in the action patterns and then it strips itself of the reference contents and operates regardless of the external model, according to a process J. Piaget defines as an "interiorized imitation", a forerunning behaviour to the mental image and intermediate stage before accomplishing a full-fledged symbolic representation. The series of *echopraxia* phenomena that may be observed on the empirical plane evolves towards forms that are less and less dependant from external references, thus highlighting the «inexpressibility of the assimilation patterns from the modification drives of accommodation» (*Ibidem*). According to N. Filograsso *imitation* itself is construed by J. Piaget from a *constructivist* standpoint; as a matter of fact he writes that «there is no such thing as imitation instinct as well as there is no such thing as the recreational instinct, there is only a schematizing activity which may lean towards the assimilative pole giving place to the recreational phenomenon, or rather towards the accommodative pole, thus producing imitative behaviours», that are later «reintegrated in a constructive balance» (*Ibidem*). The function of the equilibration between assimilation and accommodation is meant as the «arrangement of the subject-system to relate appropriately with the environment through cognitive conflict, partial and discontinuous progression and stable transition to broader and more mature structures» (Damiano 2006, p. 115). This corresponds to *Piaget's stage theory* and the succession of stages as «factors operating at all organic and psychological levels, forming a key continuity factor among them» (Bocchi and Ceruti 1981, p. 274), as well as the evidence of «full-fledged *functional continuity* between the organic and the mental, between life and knowledge» (Bocchi and Ceruti 1981, p. 278) which would explain the presence of countless instances of isomorphism among the organic and cognitive structures. As a matter of fact, as writes E. Damiano (2006, p. 116), J. Piaget «deals with several structures of children and their ways of knowing to seek a confirmation to his theory on continuity between life and knowledge, hence it is not about the entire child nor the entire development». In addition the development of children, that in Piaget's language moves from an "a-dualist" subject to a individual capable of formal thought, stands as «one of the research areas on the forms of equilibration: it is not about observing children's development, but in the progressive growth into a subject capable of mastering knowledge ("*epistemic subject*")» (Damiano 2006, p. 115). Hence a subject as a «cognitive, *epistemic*, that is, *transcendental* entity», who, as Piaget and Beth point out, is common part to all subjects in the same degree of development, whose cognitive structures derive from more general mechanisms of action coordination (see Gattico and Storari 2005). In other words, the characteristics are common to the evolutional phases taking place in every individual, with reference to globally shared general situations. In J. Piaget's genetic structure the *action* is at the foundation of the knowledge process and it is the surfacing awareness of the action that enables the subject to acknowledge itself as such and picture knowledge as the mutual implication of opposites: «Piaget's knower is an agent who evolves necessarily to grow into a theoretical *dualist*, starting from an *a-dualist* condition (...) Piaget, though constructivist, confirms to be (...)'realist'» (Damiano 2006, p. 134). The fundamental concept that tells it apart from traditional approaches lays in the category of "time" given by the construct "genesis" it introduces in the cognitive processes: «(...) at the beginning there is a fleeting *back-and-forth* that progressively takes a direction and arranges itself, resulting in the difference between the subject from the external object through an interactive process – conflictive and from time to time a-symmetrical on the one and/or the other side (*"accommodation"* and/or *"assimilation"*) – to effectively "construct" the two opposed and mutually implied polarities of knowledge (*"equilibration"*)» (Damiano 2006, p. 117). There is an interaction, a constitutive exchange between thought and action: «*action and thought "form" each other*, though following different modalities and by assuring the acknowledgment of their distinction, their relative independence and their intimate correlation» (E. Damiano, in Filograsso 1994, p. 154). According to this perspective the action could not be construed without the thought that regulates it, and if it were not return criterion for thought itself. On the contrary, thought would not be intelligible in its development if it did not produce new actions capable of revealing it, or if it should not be, on its turn, a control criterion for the action. In general the convergence between J. Piaget's constructivist structure of cognitive development and von Foerster's self-arranging theories, Atlans' biological organization, etc. seems to be quite a fruitful one, as they all aim at overcoming the dichotomy of "chance and need", today are taking shape in the research on the dialectic and circular relation between the couples chance/need, continuity/discontinuity and that will refer to the mentioned paradigm of «order from disorder» (Bocchi and Ceruti, 1981, p. 256). Finally, within the scope of constructionism, the epistemological severance with the object, once naively meant as a self-standing presence, may not be considered to suffice: «it is key to escape the temptation of subject» (Morf 1994, p. 40) and for this reason we move on to explain the variants of social and socio-cultural constructivism (Damiano 2006, p. 130).

THE SOCIAL WAY

The contribution of the Russian psychologist Lev Semënovich Vygotskij is pivotal in the current constructivist discourse with regard to the psycho-social and pedagogical fields, most particularly for that branch of constructivism named "socio-constructivism" (Pojaghi 2003) that involves within the knowledge construction process the dimension and mediation of the socio-cultural context and the importance of interpersonal relations. Commonly analysed by scientific literature in parallel with the intellectual dissertations of Piaget e Bruner (Sempio 1998), the Russian thinker focuses his attention on cognitive processes and the essential interaction they produce throughout the development between thought and *language*, a matter tackled with sheer consistency in one of his major works, aptly entitled *Thinking and speech* (Vygotskij 1976) which he develops by resorting to the work of Lurija, and Leont'ev (1975, 1977). From the point of view of the Russian neuropsychologist, between thought and language there are extreme unity and duality. They seem to develop along a path that stretches from the outside to the inside of the object, then, contrary to J. Piaget, they follow a direction that moves from *inter*subjective to *intra*subjective. Such essentiality of speech and social communication in human development is quite evident during infancy, where speech is endowed with a *regulatory* directional function to control behaviour, first as verbal instructions and later as internal self-regulated language, "private" or tacit, which may be defined as self-verbalization. It is thus that thought evolves from an interpersonal, oriented dimension where verbal instructions are external, to an intrapersonal one, interiorized and self-oriented and language displays its regulatory functions on human behaviour, hence, on thought. Any function – writes L. S. Vygotskij (1978, 1980) – appears twice in the cultural development of children, first at a social level and subsequently at individual level, first among individuals and later within the child. All superior functions appear as factual relations among human individuals», following a path going from social to individual, from interpersonal to intrapersonal. L. S. Vygotskij holds into equally important account the interpersonal dialogue and the "dialogic internalization" process, that is, the internal/intra-personal speech, an issue that recent theoretical interpretations value as an analogy to Piaget's theory. Both, according to some authors (Shayer 2003), harbour the individual *internalization* process which, according to L. S. Vygotskij is meant as *dialogic internalization*, subordinate to the social use of speech and which in the mind of J. Piaget is the *interior*/intrapersonal *speech*, primary to social and communicative use, in synergy with the assimilation, accommodation and re-equilibration processes where, in an active, aware and constructive fashion, the mental patterns of subject are transformed and re-arranged for the "conflict" between what is already owned and the new concept.

The theoretical construct of internalization which, stimulated by social interaction pushes the subject to structuring new functions, pushes the thought towards an emergent area and goes back to the basic paradigm of the *proximal development zone*, defined by L. S. Vygotskij (1978, 1980) as the distance between the current level of development as determined by individual problem solving and the level of development as determined through problem solving under the guidance of an adult or in collaboration with more skilled peers or again, in general, on the wake of the support from an adequate cultural and communicative *milieu*, which, besides adults and peers, may include culture, books, communication etcetera. In general the educational and didactic practises that find inspiration in the socio-constructivist branch do not aim at «colonising the knowledge of students by means of the knowledge of scientists» (Damiano 2006, p. 132), but rather to «broaden the scope of possibilities», as it is advocated by H. von Foerster (1990) and to the acknowledgement of the plurality of knowledge games. This principle applies to all, scientists, students and teachers alike. Subject is no longer solitary, unchanging and static in its preordained image, but it is plural, diverse, open to various possibilities, and creates material, technological and procedural constructs, etc.

THE SOCIO-CULTURAL WAY

Today's scientific literature tends to outline constructivism as a broad *socio-cultural* expression and among the most recent contributions we may find the work of L. Moll, J. V. Wertch, D. Newman, P. Griffin, M. Cole, J. Bruner, M. Larochelle, N. Bednarz, J. Garrison, M. B. Varisco, etc. In this paragraph I mean to analyse in detail the contribution of J. Bruner's cultural psychology and M. Cole's approach, that enhance knowledge construction processes with regard to culture. J. Bruner (1997) is a prominent authority for his broad all-encompassing thought, from his juvenile studies on perceptive functionalism to cognitivism in the Sixties and his more recent interest for constructivism. By expressing the position of culturalist psychology, the author, influenced by the theoretical work of the Russian scholars Lurija and Vygotskij, highlights how the construction of knowledge takes place in a hermeneutical process at the backdrop of the meaning taken by the reference culture, which he defines as *perspective*. This is how Bruner's idea that to know means "to do and negotiate meanings" takes shape; here, *to do* refers to a pragmatic, activist meaning, to *agency*, or Piaget's constructivism, while the *"meaning"* refers to the action of attribution of meaning to things which always originates with reference to possible collective and social cultural contexts. The other key parameter for knowledge is *negotiation*, that is, transmitting, mediating and comparing knowledge with culture and the others. Such concept of knowledge, as it is expressed by the author himself in his latest theoretical elaboration, is social in nature as well as intersubjective as it does outline through a personal process, though it is always taken from a cultural perspective or context and in the interaction with the others and culture. Following this approach, to tell, to *narrate*, stands as an action that follows the construction of knowledge, as it gives meaning to men's intersubjective nature. The author pushes this idea further deep and expressly stating the existence of a *narrative thought*, which takes shape just like a *thought mode* and is associated to the other mental work styles. It is mostly through our narration that we build a vision of ourselves and the world, and it is through its narrative that culture provides its members with models of identity and ability to action. In men it recognizes a natural attitude towards composing its experience, the knowledge of facts or things, in a narrative form that does not exclude the individual dynamic components, be it intellective or affective. The main property of narration is found in its intrinsic "sequential" nature, as it is comprised of a sequence of events and the relevant mental states, or "events involving human beings as characters or protagonists". This is where the tangible meaning of things is found, that is, in the context of events and the simultaneous and ceaseless interpretation effort that informs the narrator and its recipient; narration should thus be construed in a scope of verisimilitude rather than realism or certainty and, most importantly, it activates an interpretative mediation between men and the world. With reference to the concept of knowing meant as "to make meaning", the narrative act is therefore enhanced as a process that lead well beyond the mere transmission of information, usually for an entertainment purpose, and it takes a broader cognitive value, a way of feeling that helps children to create a version of the world in which they can envision, at a psychological level, a world of their own, a personal realm. As a consequence it represents a high-level educational instance such as *to form a Weltanschauung*. Nevertheless narration is laden with a much more sizeable cognitive value, as it interprets a way of knowing, a mental strategy oriented towards the interpretation of human events that transcends the pertinence to human things and the historical connection with narrative arts in general that tradition has found in it. Another quite frequent reference within the socio-cultural framework is M. Cole's situationist approach (2004), which further highlights how knowledge takes place most prominently as an act of membership in a community and it is allowed and facilitated by the involvement in its activities. The founder of the *Laboratory of Comparative Human Cognition* (LCHC) at the University of California in San Diego, and developer of Vygotskij's ideas as well as the Russian contributions in the Twenties of the XX Century, Cole gives an essential role to *culture* and its function as a *medium* to the genesis and development of human thought. He maintains that all processes related to the *psychic* realm emerge from culturally mediated practical activities that are susceptible of historical development. The approach proposed by M. Cole revolves around a vision of the "context" as a "system of structured activities" where individuals interact and where individual performance differences must be construed in relation to the array of specific situations where such tasks are required and performed. Cognitive activity is thus meant as an intersubjective, socially organized process that is fulfilled by means of the interaction among individuals in a given context. From the above it follows that learning is a *situated* process, rooted in socially and culturally organized contexts and where the meaning of knowledge is negotiated among those who are involved in a cultural and social practise.

As a whole, the three following conceptual options stand for the most relevant acquisitions in the scope of the overview I am offering and more specifically are:

1. *Mediation through artefacts*, that is, human mental processes that emerge simultaneously with the human ability to modify objects, thus generating *artefacts,* or aspects of the material world that have been modified throughout the history of its incorporation in the human action aimed at a goal, and are at once *ideal* (conceptual) and *material*. They are ideal in that their material shape has been shaped by their partaking to the interactions they were previously part of and now they mediate.

2. *The historical development*, since next to the generation of artefacts, human beings organize into society and are involved in processes of rediscovery of artefacts that have been already generated and existing in the historical memory of every society. As a consequence every single person, in his or her social identity, is the result of what the preceding generation did and left as legacy to the generations to come.

3. *Practical and daily activities* as key for the analysis of the "psychic" and the overcoming of the duality between materialism and idealism, since it is in the activity that individuals experience the material/ideal residue of the activities carried out by the previous generations.

LINES FOR CONTEMPORARY CONSTRUCTIVISM TO REVISIT AND REINTEGRATE THE ANCIENT SENSE OF CONTINUITY BETWEEN MEN AND NATURE

The deep disappointment and theoretical intolerance ensuing the dualist positions of modernity, where the subject of classical rationalism, in its *a priori* shapes and categories, was opposed frontally by determinist trends in their evolutionary history, (since they themselves were "a priori" in an "objective world" preordained with regard to the subject), pushed constructivism beyond the polarity of "subject" and "object" in knowledge, to reach the *"middle way"* that connects them and tells them apart (E. Damiano in Giaconi 2008, p. 11). This is not merely a compromise nor is it a "dialectic synthesis", though an "interaction" that progressively generates, in a natural flow, the acknowledgement of a self, thus favouring a growing individualization of existence. Subject and object are thus no longer the absolute and pre-existing poles of a relation, though they are the outcomes of a construction taking place in the continuity between natural macrocosm and anthropologic microcosm. Knowledge itself establishes itself through complex and non-linear interaction of the subjective factor and the objective one, in a lengthy transaction between countless and composite elements (bodily, emotional, affective, operational, cognitive, symbolic, etc.), capable to shape autopoietic and ontopoietic structures derived though relatively autonomous and self-sustaining. Along this "middle way", there is an evident *continuum* that traditional epistemology (Western, with well-known exceptions) had separated and partitioned (between "body" and "spirit" or "mind" or otherwise designated) or had allowed to proceed deterministically. Within the constructionist perspective, knowledge is *"engraved in the body"* and cognition no longer lives in an isolated condition but is embodied in the physiology of the subject. It is a constitutive integration, according to which *"the subject does not 'have' or does not 'dwell', but it 'is' the body"* (Damiano 2006, p. 12). To state the rearrangement of the knowledge issue in the "middle way", more specifically as *mediation* of the pedagogical jargon, allows to hold into account the respective contribution of the two polarities that do not pre-exist but recognize and complement each other through the exchanges that generate the *co-construction* of knowledge. The focus is thus placed on the interaction process and on the products around which the connections arrange in clusters and give shape and structure on the weave of knowledge. This is how the "return of subject" should be meant, as one of two vectors, jointly necessary and reciprocally implied in activating knowledge. By starting from this assumption, I do believe that what was previously outlined allows to strengthen some meaningful *lines of reflection* to restore the continuity between *logos and life phenomenology/ontopoiesis*, the main focus of this International Congress of Phenomenology. First comes the re-visitation of life ontopoiesis and of human condition as creator, that is, the ability of human beings to activate a "constructive process of individual becoming" which, within its world, is the expression of a "specific type of constructivism", where the cognitive act is the "creative act" and where at once "while being 'generates', it also manifests the logos of its continuous 'letting itself be' ". Secondly the acknowledged mutual pervasion between logos and life, which escapes dualist and static play to position itself in what, with other words, we defined as the "middle way", allowed us to strengthen and restore, from another standpoint, the ancient sense of a synergy between logos e life.

University of Macerata, 62100 Macerata, Italy
e-mail: c.giaconi@unimc.it

REFERENCES

Bichelmeyer, B.B. 2000. *Interactivism: Change, sensory-emotional intelligence, and intentionality.* In Being and learning, paper presented at annual meeting AREA, New Orleans, 24–28 April 2000.

Bocchi, G., and M. Ceruti. 1981. *Disordine e costruzione.Un'interpretazione epistemologica dell'opera di J. Piaget.* Milan: Feltrinelli.

Bruner, J. 1997. *La cultura dell'educazione.* Milan: Feltrinelli.

Ceruti, M. (a cura di). 1992. *Evoluzione e conoscenza. L'epistemologia genetica di Jean Piaget e le prospettive del costruttivismo.* Milan: Lubrina.

Ceruti, M., and G. Lo Verso. 1998. *Epistemologia e psicoterapia.* Milan: Cortina.

Cole, M. 2004. *Psicologia Culturale: una disciplina del passato e del futuro.* Rome: Carlo Amore.

Damiano, E. 2006. *La «nuova alleanza». Temi problemi e prospettive della nuova ricerca didattica.* Milan: La Scuola.

Filograsso, N. (a cura di). 1994. *Mente conoscenza educazione.* Rome: Anicia.

Foerster, H. von. 1990. Etique et cybernétique de second ordre. In *Systèmes, éthique, perspectives en thérapie familiale*, eds. Y. Rey and B. Prieur, 41–54. Paris: ESF.

Gattico, E., and G. P. Storari. 2005. *Costruttivismo e scienze della formazione.* Milan: Unicopli.

Gergen, K.J. 1991. *The saturated self: Dilemmas of identity in contemporary life.* New York: Basic Book.

Giaconi, C. 2008. *Le vie del costruttivismo.* Rome: Armando.

Glasersfeld, E. von. 1989. Introduzione al costruttivismo radicale. In *La realtà inventata. Contributi al costruttivismo*, ed. P. Watzlawick, (a cura di), 17–36. Milan: Feltrinelli.

Glasersfeld, E. von. 1992. Aspetti del costruttivismo: Vico, Berkeley, Piaget. In *Evoluzione e conoscenza. L'epistemologia genetica di J. Piaget e le prospettive del costruttivismo*, ed. M. Ceruti (a cura di). Milan: Lubrina.

Glasersfeld, E. von. 1998. *Il costruttivismo radicale, Una via per apprendere e conoscere*, Quaderni di Metodologia, 6, Società Stampa Sportiva, Rome.

Guba, E.G. 1990. The alternative paradigm dialog. In *The paradigm dialog*, ed. ID., 17–27. Newbury Park, CA: Sage.

Guba, E.G., and Y.S. Lincoln. 2000. Competing paradigmatic in qualitative research. In *Handbook of qualitative research*, eds. N.K. Denzin and Y.S. Lincoln. Thousand Oaks, CA: Sage.

Heron, J., and P. Reason. 1997. A participatory inquiry paradigm. In *Qualitative inquiry*, Vol. 3, 274–294.

Leont'ev, A.N. 1977. *Attività,coscienza e personalità.* Florence: Giunti Barbera (ed. orig. Moskva 1975).

Lincoln, Y.S. 1995. Emerging criteria for quality in qualitative and interpretative research. In *Qualitative inquiry*, Vol. I, 257–289.

Morf, A. 1994. Une épistémologie pour la didactique: spéculations autour d'un aménagement conceptual. In *Revue des Sciences de l'education*, Vol. 1, 29–40.

Neymayer, R.A. 1993. An appraisal of constructivist psychotherapies. *Journal of Consulting and Clinical Psychology* 61: 221–234.

Phillips, D.C. (a cura di). 2000. *Constructivism in education: Opinions and second opinions on controversial issues.* Chicago: NSSE.

Piaget, J. 1967. Nature et méthodes de l'epistémologie. In *Logique et connaissance scientifique*, ed. J. Piaget (a cura di). Paris: Gallimard.

Piaget, J. 1983. *Biologia e conoscenza. Saggio sui rapporti tra le regolazioni organiche e i processi cognitivi.* Turin: Einaudi.

Pojaghi, B. 2003. *Contributi di psicologia sociale in contesti socio-educativi.* Milan: Franco Angeli.

Sempio, O.L. 1998. *Vygotskij, Piaget, Bruner. Concezioni dello sviluppo.* Milan: Cortina.

Sexton, T.L. 1997. Constructivist thinking within the history of ideas: The challenge of a new paradigm. In *Constructivist thinking in counselling practice, research, and training*, eds. T.L. Sexton and B.L. Griffin, 3–18. New York: Teachers College Press.

Shayer, M. 2003. Not just Piaget; not just Vygotskij, and Certainly not Vygotskij as Alternative to Piaget. In *Learning instruction*, Vol. 40, 465–485.

Steffe, L.P., and E.J. Gale. 1995. *Constructivism in education.* Hillsdale, NJ: Erlbaum.

Varela, F.J., E. Thompson, and E. Rosch. 1993. *L'inscription corporel de l'esprit. Sciences cognitives et experiénce humaine*. Paris: Seuil.

Varisco, B.M. 2002. *Costruttivismo socio-culturale. Genesi filosofiche, sviluppi psico-pedagogici, applicazioni didattiche*. Rome: Carocci.

Vygotskij, L.S. 1976. *Pensiero e linguaggio*. Florence: Giunti Barbera.

Vygotskij, L.S. 1980. *Il processo cognitivo*. Torin: Boringhieri (ed. orig. Cambridge, 1978).